EVOLUTIONARY ECOLOGY OF PLANT REPRODUCTIVE STRATEGIES

Evolutionary biologists have produced a solid body of evidence to explain patterns of diversification, both within and among species. Recent textbooks are weighted towards studies of animals, which is surprising given that plants are ideally suited for answering evolutionary questions. Plants do not stand up and walk away and they can easily be cloned, transplanted or potted for experiments. This book aims to set the record straight by placing the wealth of data that have been collected on plants into the unifying framework of game theory. This allows a test of the theory of natural selection in some cases, while in other cases highlighting the need for additional data collection and theoretical development. It is the authors' hope that many students will take on this challenge and help the study of the evolutionary ecology of plants to develop as a mature, predictive science.

TOM DE JONG is a professor in the Institute of Biology, Leiden University. His research focuses on aspects of the ecology and evolution of dioecious plants.

PETER KLINKHAMER is a professor in the Institute of Biology, Leiden University. His interests include the evolution of monocarpy, optimal defence against herbivory and selective embryo abortion.

EVOLUTIONARY ECOLOGY OF PLANT REPRODUCTIVE STRATEGIES

TOM J. DE JONG

and

PETER G. L. KLINKHAMER

Leiden University

CAMBRIDGE
UNIVERSITY PRESS

CAMBRIDGE UNIVERSITY PRESS
Cambridge, New York, Melbourne, Madrid, Cape Town, Singapore, São Paulo

CAMBRIDGE UNIVERSITY PRESS
The Edinburgh Building, Cambridge CB2 2RU, UK
Published in the United States of America by Cambridge University Press, New York

www.cambridge.org
Information on this title: www.cambridge.org/9780521821421

First published 2005

Printed in the United Kingdom at the University Press, Cambridge

A catalogue record for this publication is available from the British Library

Library of Congress Cataloguing in Publication data

ISBN-13 978-0-521-82142-1 hardback
ISBN-10 0-521-82142-8 hardback
ISBN-13 978-0-521-52894-8 paperback
ISBN-10 0-521-52894-1 paperback

Contents

Acknowledgements

Numerous people helped to make this book as it is. David Lloyd pioneered in applying game theory to the evolutionary ecology of plants and his work and his personal encouragement to pick up these matters have been inspirational. Also the approach by Eric Charnov, including his early work on plants with Carl Freeman, had a great influence. A long list of people helped in various ways, by reading chapters, solving problems, providing material, checking maths, bringing up interesting discussion points or just giving encouragement to begin writing a book on such a broad topic in the first place. In alphabetical order they are: Camille Barr, Spencer Barrett, Eke van Batenburg, Frank Berendse, Debby Beugelsdijk, Diane Campbell, Jos van Damme, Tom van Dooren, Henk van Dijk, Michel Durinx, Grit Glawe, Jan van Groenendael, Patsy Haccou, Roger Härdling, Ian Hardy, Piet Heemskerk, Heather Kirk, Hans Koelewijn, Russ Lande, Kirsten Leiss, Evert Meelis, Ed van der Meijden, Hans Metz, Mark Rees, Mike Richardson, Claus Rueffler, Tracey Sanderson, Avi Shmida, Ann Smithson, Irene Till-Bottraud, Jan van der Toorn and Klaas Vrieling. Many students from the course 'Theoretical ecology' at Leiden University gave their comments. Several people from the evoldir group sent reprints. Original pictures were drawn by Martin Brittijn and Frank van der Meer.

Thanks for all your support!

1

Optimization models

About thirty years ago there was much talk that geologists ought only to observe and not theorise; and I well remember someone saying that at this rate a man might as well go into a gravel pit and count the pebbles and describe the colours. How odd is it that anyone should not see that all observation must be for or against some view if it is to be of any service.

Charles Darwin, letter to Henry Fawcett, 18 September 1861

1.1 Introduction

The common garden near my house was initially bare soil, flattened by bulldozers. It was planted with fast-growing plants such as periwinkle (*Vinca minor*) and ivy (*Hedera helix*) with patches of bamboo. After four years, the ground covering plants had won their battle with the dandelions (*Taraxacum officinale*) and the bamboos were man-sized, providing places with shade and privacy. It was delightful. In the fifth year, however, disaster struck. One of the bamboos started to produce flowering shoots and, later in the year, other bamboos did the same. By the autumn all the bamboos were completely dead. In the neighbouring garden the bamboos had also flowered and, looking around, you could see that bamboos all over town had flowered, irrespective of the age of the garden or the date they were planted. It took the local garden centre over a year to catch up with the sudden demand for vegetative bamboos, and by that time our garden was replanted with other species. What is the interesting biological question here? One group of biologists will ask a causal question. What causes the synchronous flowering of bamboos? Do all plants respond to a common environmental cue, such as a dry summer or a very cold winter? Are the plants perhaps communicating with each other? Are they releasing volatiles, such as ethylene, that induce others to flower? These are all relevant questions about the underlying mechanism. Yet another group of biologists asks an evolutionary question. Why do the bamboos have synchronized flowering? Starting from ancient unsynchronized bamboo species, what were the selective factors that favoured the synchronous flowering of all plants? Why isn't flowering synchronized in all plant species and what makes these bamboos so special? The motto of this second group of biologists was stated eloquently by the geneticist Dobzhansky: 'Nothing in biology makes sense, except in the light of evolution.'

Dan Janzen is a bamboo expert with a great interest in evolution. In 1976 he wrote a fascinating article 'Why bamboos wait so long to flower'. From historical records, dating back to the year 919, he concluded that the mainland Chinese bamboo *Phyllostachys bambusoides* flowered once every 120 years, and that flowering of individual plants was synchronized over large areas. Bamboo seeds are at least the size of a rice kernel and have as high a nutritional value. In years of synchronized flowering (so-called mast years) it was not uncommon for bamboo seeds to cover the soil as a thick blanket. Rats and rodents were identified as prime seed-eaters. After masting of bamboo the fertility of rats increased and large litters were produced. However, when the seeding of bamboo had ended and seeds had germinated or disappeared into the soil, a lean period started for the rats and they needed a different food source. Janzen reported that rats destroyed Asian crops after which they died in great numbers, causing famine and infectious diseases such as the plague, which was the reason the mass flowering of bamboos was recorded in the history books in the first place. Janzen hypothesized that mass flowering of bamboos had evolved as a plant strategy to escape from seed predation. In years of flowering, seed production is so abundant that the rats cannot eat everything, even though their population increases rapidly. Some seeds escape and assure survival of the next generation of bamboo plants. Seeds of a bamboo genotype that is out of synchrony and flowers one year too early may all be eaten by the few rats that are around. Seeds of a bamboo genotype that flowers one year too late are certainly eaten by many, hungry rats. Once mast blooming has evolved, there is strong selection for plants to stay in synchrony.

Of course plants do not have a nervous system and certainly do not make conscious decisions about what to do next. Nevertheless, in the case of *Phyllostachys bambusoides* one cannot help thinking that the plant has outsmarted its seed predators. In the past, different bamboo genotypes have played a game against each other. The winner was the genotype with the rather bizarre strategy of flowering once every 120 years, in total synchrony with the others, and devoting all its resources to reproduction. Plant strategies are the subject of this book. Life-history strategies encompass characters such as generation time, the fraction of resources spent on reproduction, and the resources allocated per offspring (seed). Complementary to these are sex allocation strategies which include the allocation to male and female function, attraction of pollinators and so on. The focus of this book will be on general models and data collection that allow testing of hypotheses concerning these strategies. Before going from natural history to the mathematics of these models it is useful to examine the arguments of people who have been sceptical about testing evolutionary hypotheses.

1.2 The spandrels of San Marco or the spaniels of St Marx?

For any student with an interest in the history of science, Gould and Lewontin's (1979) article is a must. The ingenious title is: 'The spandrels of San Marco and the Panglossian paradigm: a critique of the adaptationist programme'. This title certainly needs some explanation. Spandrels are the curved, triangular areas between adjoining arches. Those in St Mark's cathedral in Venice are embellished with mosaics, each depicting one of the four evangelists. When you study these spandrels in isolation, you might come up with the silly idea that they are the reason for the system of arches and domes that surrounds them. Of course this is not true, they are just a logical consequence of the way churches were built in those days. Spandrels were necessary for holding up the dome of the church, they are a by-product of the construction. To give a simpler example; the function of the nose is in smelling and breathing. Someone might come up with the idea that the nose evolved to hold up spectacles, but of course this is nonsense. This function of noses is just an accidental by-product. The Panglossian paradigm is the other element of Gould and Lewontin's (1979) title. A paradigm is an example. Dr Pangloss was a character from Voltaire's book *Candide*, which was written in a satiric style. The doctor's views are perhaps best illustrated by a quote: 'It is demonstrable that things cannot be otherwise than as they are; for as all thing have been created for some end, they must necessarily be created for the best end.' The Gould and Lewontin paper is not about architecture or literature though. It is a critique of evolutionary studies without a sound methodological basis. First, Gould and Lewontin used the spandrels metaphor to argue that characters should not be studied in isolation but as a whole. Second, they used Dr Pangloss to ridicule the biologist who can always dream up an adaptive explanation of some character afterwards, not questioning his basic premise that each character in each organism is fully adapted to its environment. 'Adaptive storytelling' is a term also used in this context. Basically, Gould and Lewontin argued that many evolutionary studies are just-so stories without rigorous hypothesis testing.

Queller was one of the authors who took on Gould and Lewontin's challenge. His 1995 paper was written in the same spirit and was titled: 'The spaniels of St Marx and the Panglossian paradox: a critique of a rhetorical programme' (Queller 1995). A short explanation is again required. One of the sections of Gould and Lewontin's paper was named 'The master's voice re-examined'. While 'the master' referred to Darwin in Gould and Lewontin's paper, Queller intentionally misinterpreted the allusion and substituted Marx. The spaniel refers to the logo of the record label His Master's Voice, which shows a dog listening to the voice of his dead master on a creaky 78. Putting clever metaphors aside, Queller countered the two points made by Gould and Lewontin as follows. With regard to the first point (characters should always be studied in combination), Queller emphasized the success of reductionism in science. He argued that to solve a problem, starting with one factor A and then, once you know its effect, moving on to factor B and the interaction between A and B, is usually better than starting with all possible factors and all interactions at the same time. The second point (no hypothesis testing) is not so easily refuted. Studies with sloppy methodology ending in a long list of possible explanations for an observed phenomenon are found in all times. However, Queller pointed out that many of the hypotheses are, in fact, testable and that many modern evolutionary studies do just that.

Let us look at an example. Janzen's paper on bamboos was a great story. But it was more than that. It had a clear hypothesis that synchronous flowering evolved to satiate predators. Under this hypothesis one would expect that if you induced some bamboo plants to flower outside their mast years, perhaps by supplying plant hormones, their seeds would suffer higher predation than in mast years. An experiment to test this hypothesis could be set up fairly easily. As far as we know Janzen did not do this for *Phyllostachys bambusoides*, nor were systematic data collected for the more than 1200 bamboo species that exist, some with masting over shorter periods and some with ordinary perennial life cycles. But this does not mean we cannot test the hypothesis and other authors have done this in different systems. In New Zealand, Kelly and co-workers (1997, 2001) examined two hypotheses exploring benefits from the highly synchronized flowering in the grass *Chionochloa pallens* (mid-ribbed snow tussock). First, they hypothesized that wind pollination of the grass was better if individuals flowered in synchrony. Second, they hypothesized that plants flowering in synchrony could escape seed predation, as Janzen proposed. By manipulating density they found that the pollination effect was small. Seed predation was high (up to 94%) and masting satiated the insect seed predators but not the birds foraging on seeds.

Kelly and Sork (2002) reviewed the literature and suggested that similar factors played a role in the evolution of mass flowering in unrelated species from different continents. In 16 studies they found support for the cornerstone of Janzen's hypothesis, that there is lower seed predation and therefore higher seed survival in years of high seeding, while three contradictory examples existed. All plants with masting were dominant community species and it is indeed plausible that common species can saturate generalist seed predators more efficiently than rare species. We do not assert that we have answered all questions regarding masting. On the contrary. Kelly and Sork (2002) list many aspects that require additional study. Also we have not answered the question how, starting with an ancient, unsynchronized bamboo species, masting is selected for. Mathematical models that address this question are still quite recent (see Satake and Iwasa 2000). In general, the history of evolution may be difficult to reconstruct. We can, however, go out to the field and collect data that can potentially refute a hypothesis. If seed predation is similar in masting and non-masting years this would refute the idea that seed predation is the driving force behind masting. In this sense evolutionary hypotheses are testable.

1.3 Are evolutionary theories scientific?

The criticism of Gould and Lewontin that hypotheses are not rigorously tested has not disappeared and should be taken seriously. For instance, Murray (2001) wrote a review flavoured with philosophy called: 'Are ecological and evolutionary theories scientific?' In his review he criticized the explanations that biologists employed afterwards, when all data have been collected. One example he mentioned concerned the size differences between the sexes in birds. In hawks and owls, females are bigger than males and apparently no fewer than 14 hypotheses have been proposed for this. Some of these hypotheses were based on a single, specific observation. While all these hypotheses may in principle be rejected or accepted in a study on a new owl or hawk species, chances are that, overwhelmed by the many possible choices, this new study only states which hypothesis best fits the data or, if none of them do, comes up with a fifteenth hypothesis. Murray argued for fewer hypotheses, more general hypotheses and rigorous testing of the predictions of each hypothesis before moving on to the next one.

Although much of traditional biology is descriptive, that does not mean that biology cannot have predictive theories. Towards the end of the article Murray (2001) mentions life-history theory as an area for which this is possible and other authors (Morgan and Schoen 1997) have made the same argument for

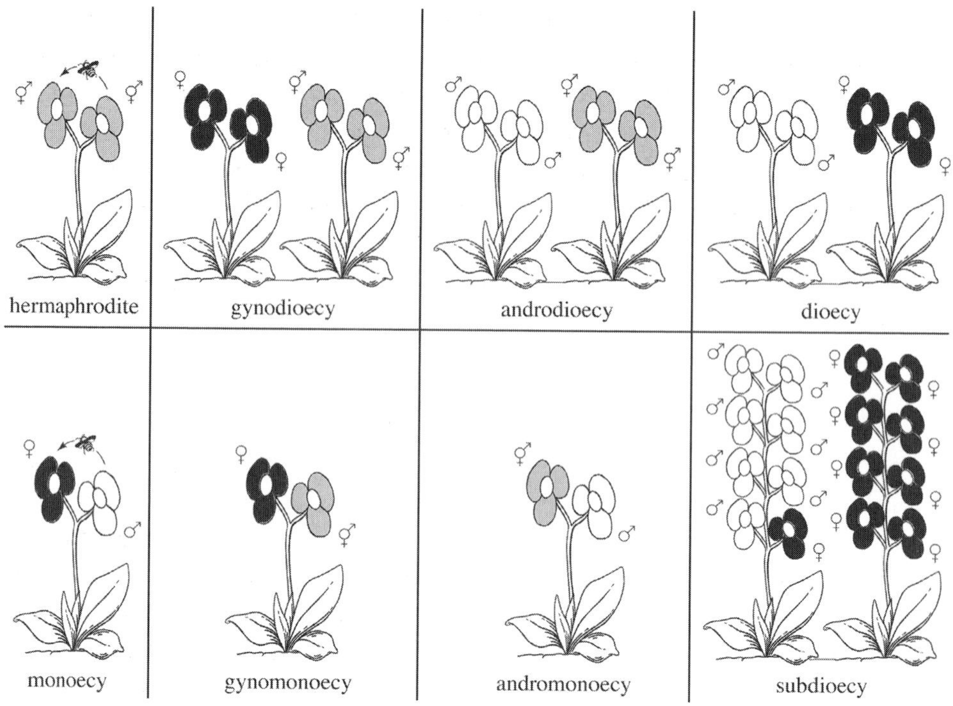

Fig. 1.1 Sex systems in higher plants. Male and female function is either present within the same flower, in different flowers on the same plant or on different individuals. Cosexuality refers to both sexual functions being present on the same individual and therefore includes monoecy and hermaphroditism.

sex allocation theory. Sex systems in plants may be an example, in contrast to animals, which are hermaphrodite or have separate sexes, in the plant kingdom many sex systems are found (Fig. 1.1). Plants with hermaphrodite flowers are most common (about 72% of all species) while dioecious species, i.e. those with separate male and female individuals, are rare (*c*.4%). Gynodioecy, females and hermaphrodites, is relatively common (*c*.7%). On the other hand androdioecy, males and hermaphrodites, is extremely rare and has only been shown for a few plant species (Pannell 2002). The minority sex systems occur at different places in the plant kingdom. Helped by the recent molecular phylogeny of the flowering plants (Soltis *et al.* 1999) we can be confident that the evolution of specific sex systems occurred several times in a number of unrelated events. The theory used to explain the different sex systems is based on the idea of natural selection. This idea implies that there is genetic variation, that such variation leads to differences in success and that this leads to changes in allele frequencies. We will detail this in the coming chapters. As we will see, the theory is not perfect or complete, but it does go a long way in meeting Murray's requirements. The theory is general, it is based on general principles rather than on the available data. The theory applies to all plant species instead of just a few, and it can be tested in a number of specific cases. If our model, for instance, correctly predicts which plant species have become dioecious in one plant family, we could always turn to another plant family with different ecological conditions and a different evolutionary history to see if the predictions are correct for this new situation also. In this sense the theory is testable and scientific.

1.4 100% adaptation?

After reading Gould and Lewontin's critique of the easy acceptance that organisms are in all respects fully adapted to their environment, one might expect that the degree of adaptation is a big issue. Well, it isn't really. No one expects organisms to be 100% adapted. Consider adaptation as climbing a mountain without a map and with limited visual range, say only 100 m. You are determined to make it to the top and the best thing to do would be to go to the highest point you can see and from there look again if you can climb any higher. No doubt, this will get you started up the mountain but you may get stuck at a local peak which is higher than anything within a 100 m radius, but lower than the top. Similarly, organisms may change genetically by small steps. Selection will favour the steps that result in 'better' plants until the point where no small step leads to improvement. We can then say that the organism has reached a local fitness peak, from which it can only move by a large step. Organisms adapt but, as in the case of the mountaineer with limited sight, there is no guarantee an organism will always reach the top of the fitness landscape.

Furthermore, adaptation follows environmental change and plants have the difficulty of having to reach a new area by dispersing their seeds. If long-distance dispersal is rare, this may be a lengthy journey. Reid (see Clark *et al.* 1998) calculated in 1899 that between the glaciation of the UK, some 10 000 years ago, and the Roman occupation, the oak travelled 1000 km north to reach the limit of its northern distribution in Scotland. Later diffusion models calculated that, to travel this distance, daughter oaks would have to grow, on average, 1 km from their maternal parent. This is an unrealistically large distance for an acorn to fall from a tree. This inconsistency between predictions from a simple life-history model and the study of fossil pollen in the soil was called 'Reid's paradox' (Clark *et al.* 1998). It was resolved when it was realized that rare, accidental long-distance dispersal sometimes occurred.

This greatly increases the speed at which tree species travel over the globe. Nevertheless the point remains that plants follow climatic change and we cannot reasonably expect that as soon as a habitat becomes suitable for a certain species, its seeds are immediately present. The climate is now becoming warmer and species ranges are changing accordingly (Parmesan and Yohe 2003). If you transplanted a genotype from the south of Europe to the north and found that it survived or even outcompeted the home genotype, that would be a surprising result. On the other hand, such a result may well be due to limited gene flow so that genetic change is lagging behind environmental change. Such an observation does not alter the principle that under natural selection, the genotype best adapted to the local environment outperforms others.

No organism is probably 100% adapted to its environment, but do we know for certain that plants adapt to their environment? Plants can be transplanted to different environments and are therefore ideal study organisms for tackling such questions. Many early examples of transplant experiments can be found in the studies of Clausen (1951) and co-workers Keck and Hiesey (see also the recent review by Núñez-Farfán and Schlichting 2001). One of their experiments was to collect genotypes of *Potentilla glandulosa* (sticky cinquefoil) at different heights, to clone these genotypes and transplant them to Stanford (just above sea level), Mather (1400 m) and Timberline (3500 m). A single clone looked different when placed at different heights and was therefore plastic in its response to the environment. The home genotype (the squares on the diagonal) was always the largest of the three genotypes (Fig. 1.2), as you would expect if it was best adapted to its own environment. When grown at the Stanford garden differences between genotypes were small, but in the harsh mountain environment at Timberline the lowland genotype was unable to survive. It is quite likely that the differences in performance that Clausen (1951) documented were due to genetics. One could argue, however, that the clones differed in other respects also. Plants from Stanford could, for instance, have lower nitrogen concentrations that made them more vulnerable to frost on the mountain top. To control for this we need to grow at least one generation of seeds in the lab under standardized conditions. This was done by Jan van Groenendael for two genotypes of *Plantago lanceolata* (ribwort), one from a disturbed roadside and the other from an adjacent meadow in the Netherlands (Fig. 1.3) (Kuiper and Bos 1992). After seeds were sown in the greenhouse, plants maintained their own typical growth form. The erect plants in the meadow have to compete with tall grasses while the major problem for the roadside plants is to recover from trampling and mechanical damage. The two growth forms thus seem ideally suited for their own habitat and

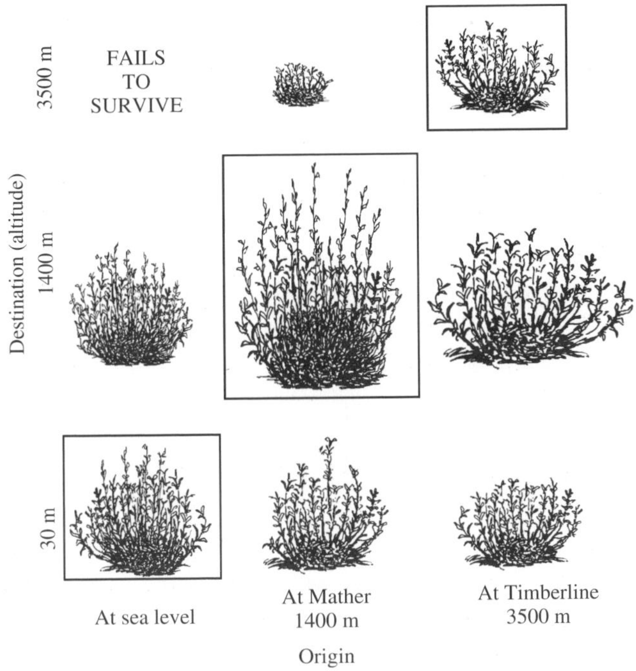

Fig. 1.2 Reciprocal transfer experiment with *Potentilla glandulosa* in three contrasting environments. The home genotype (squares on the diagonal) performs best in its own environment. Redrawn with permission from Clausen (1951), Cornell University Press.

transferring plants between the habitats showed that the home genotype usually outperformed the other genotype (Kuiper and Bos 1992). We can answer the question about genetic adaptation affirmatively for *Plantago lanceolata*. Several other studies have shown that plants show rapid genetic change under strong selection. The clearest examples reviewed by Linhart and Grant (1996) include genetic adaptation of lawn plants to a stringent mowing regime and of plants on mine spill to heavy metals. Adaptation was also found under less extreme conditions. Joshi *et al.* (2001) swapped the common grassland species *Trifolium pratense* (red clover), *Plantago lanceolata* and *Dactylis glomerata* (cocksfoot) between different European countries and found that the home genotype usually won.

1.5 Games plants play . . . against nature

If the critical factor for the plant is to resist an abiotic factor like frost, the optimal strategy is probably independent of the behaviour of other types in the population. This problem is known as straightforward optimization or

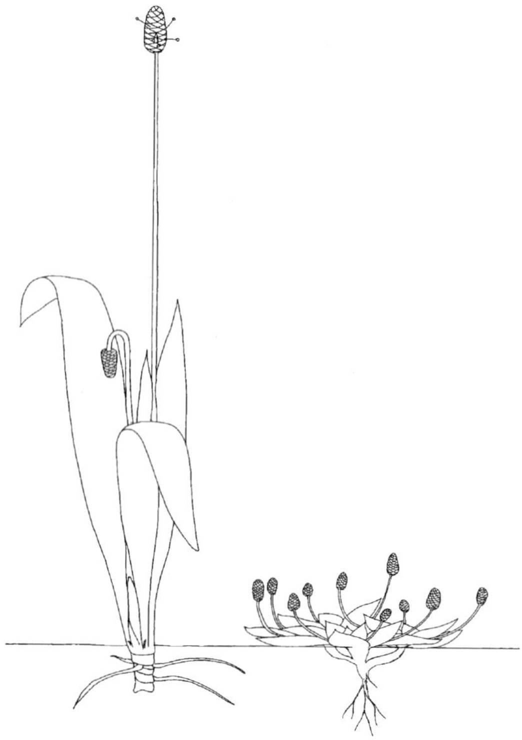

Fig. 1.3 *Plantago lanceolata* genotypes from a hay meadow and a nearby disturbed roadside. The phenotypic differences were maintained when seeds of both types were germinated and grown under identical circumstances. Picture courtesy of Jan van Groenendael.

a 'game against nature'. A nice example is Dan Cohen's (1966) seed bank model. Cohen noticed that in some species less than 100% of the seeds germinate the next spring; a fraction of the seeds germinates in the subsequent years. Cohen considered this a strategy of spreading the risk in an unpredictable environment. Seedling establishment is a risky business. In some years young plants will make it through the vulnerable germination and establishment stage, but in other years a drought may start just after germination, wiping out an entire cohort. If 100% of the seeds germinate at the same time, an entire line could go extinct in a single bad year. Delayed germination of some of the seeds helps to bridge the unfavourable period. Cohen thus predicted low germination fractions in species that live in unpredictable environments with a large variance in seedling establishment. In Cohen's model plants are spreading the risk, just as stockbrokers operate in unpredictable markets by not investing all their money at once.

1.6 ... and against each other

Plants can also play games against each other. A good example of 'plant games against each other' are so-called interference strategies, for which the success depends on the interactions between neighbours (Wilson 1987). Plants may interfere with their neighbours by investing in stem tissue, instead of leaves, shading the neighbours (Falster and Westoby 2003). Trees typically employ this strategy and many plants respond to high density by similarly changing their allocation pattern (Fig. 1.4). Other interference strategies may include the production of toxic litter or the uptake of as much water as possible before the onset of drought, which, in a desert environment, may be an effective way to restrict water uptake of your neighbours and decide competition in your favour. Interference strategies make plants more competitive, but are inefficient when plants grow on their own, because resources are used for purposes other than primary production. Therefore interference strategies reduce the production of the community. You might want to call an interference strategy an adaptation because it helps the plant to win competition, but the term 'adaptation' loses some of its positive meaning in this context. Together, two plants that both invest in interference are worse off, i.e. produce fewer seeds, than two plants that allocate only to primary production.

An example may help to illustrate the usefulness of the game angle for understanding plant strategies. Gersani *et al.* (2001) grew *Glycine max* (soya bean) plants in boxes, with two plants per box (Fig. 1.5). In the first experiment

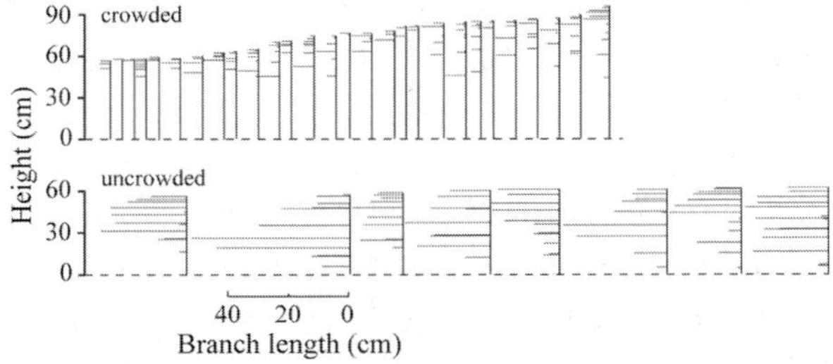

Fig. 1.4 Under crowded conditions plants of *Impatiens pallida* (pale jewelweed) were smaller in diameter with less leaf area, but were nevertheless taller than plants under uncrowded conditions. Vertical lines are main stems, horizontal lines represent primary branch length. Plants have an efficient compact growth strategy when growing alone, but adopt a tall interference strategy in competition, that results in shading the neighbours. Redrawn with permission from Weiner *et al.* (1990), *Journal of Ecology*, Blackwell.

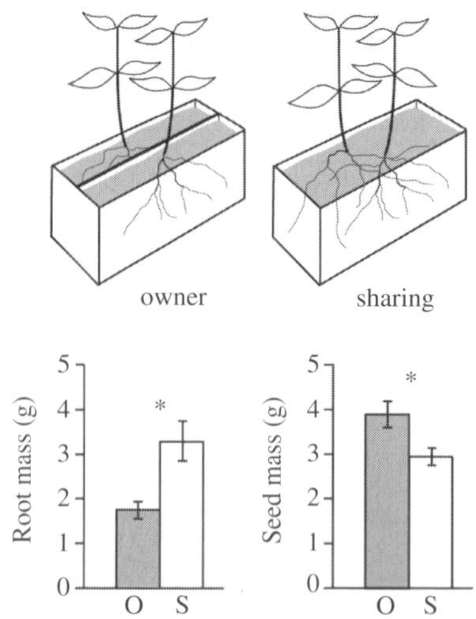

Fig. 1.5 Two soya bean (*Glycine max*) plants either owned their root space or shared it with another plant. Owners (shaded bars) produced few roots and many seeds. Sharing plants (open bars) produced significantly more roots, also in the other compartment, and significantly (*) fewer seeds. Redrawn with permission from Gersani *et al.* (2001), *Journal of Ecology*, Blackwell.

the box was separated by a partition in the middle so that each plant had its own compartment; here the plant is 'owner' of the root compartment. In the second experiment the partition was removed, so that roots of the two plants could enter the other compartment and compete for resources. Sharing resulted in 85% more root production per plant and visual inspection showed that the roots of the two plants completely intertwined. This shift in the root/shoot ratio is inefficient, and Gersani *et al.* found that plants that were owners produced 30% more seeds than plants that shared root space. The behaviour of the plants fits the predictions of game theory. When a plant is an owner it can take up all the nutrients in the compartment with a small root system, and therefore with a small investment, the plant has a high gain. When there is the possibility of interfering with neighbours it becomes profitable to make extra roots in the neighbouring compartment. If the plant makes the same amount of roots in the neighbour's compartment as in its own compartment, it can 'steal' half the nutrients from the neighbour. The same situation holds however for the other plant. If that plant would stick to its own compartment, it would obtain only one-third of the nutrients. When both plants invest more in root, the two plants are both worse off in competition than before when they were owners. Yet when the two soya bean plants are genetically unrelated they each follow the best strategy to pass on relatively more copies of their genes to the next generation than their competitor. In game theory this situation, where two individuals can gain from cooperating but yet each gains more from selfish behaviour, is known as the prisoner's dilemma. The study of Gersani *et al.* (2001) is a nice application of game theory to plants and it would be worth while to investigate if this idea also applies to other plant species.

1.7 Nash equilibrium and the evolutionarily stable strategy

In competitive games, you should play a strategy that cannot be beaten by your opponent. If strategies of players are such that by changing strategy a player can only do worse, this is called a 'Nash equilibrium' after Nobel prizewinner John Nash. In the movie *A Beautiful Mind* the actor Russell Crowe, who plays John Nash, advises his friends that in competition it is not a good strategy for everybody to go after the most desirable object (in the movie a beautiful blonde girl), but to spread attention over the other objects too (her girlfriends in this case). The pigeons he watches in the Princeton courtyard make the point perhaps clearer. When confronted with a variable density of breadcrumbs, a pigeon on its own would do best to forage only in the spot with the highest density of breadcrumbs. However, when a flock of pigeons discovers the bread and all pigeons forage in the best spot, an individual pigeon may do better to go to a less favourable spot where it is less crowded and there is less competition. The best foraging strategy for a pigeon would be to match its time foraging on one place to the food that is available so that, for instance, twice as many pigeons forage in the place where twice as much food becomes available. Matching of foraging time to availability of food is known in biology as the 'ideal free distribution' (Thuijsman *et al.* 1995). In the ideal free distribution, the intake of breadcrumbs per unit of time is equal so that by switching places, no individual pigeon can do better. Reaching the ideal free distribution is probably not so complicated. It is sufficient when pigeons sometimes switch between two sites and then decide to stay a little longer where there is more food.

In theoretical evolutionary biology one is concerned with two questions. First, is there a strategy so that no rare mutant can do better when it plays a different strategy? Second, if so does the population converge to this point with small steps? In total we have four possibilities (Table 1.1). In the first situation the population converges to the stable strategy that cannot be invaded by a rare mutant; this situation is an evolutionarily stable strategy (ESS: Maynard Smith 1982), or continuously stable strategy (CSS). The scenario we have in mind is that in a population all individuals have a certain value for a character,

Table 1.1 *Classification of points with respect to convergence and stability*

		Stable?	
		Yes	No
Convergence?	Yes	ESS	Branching point
	No	Garden of Eden	Repellor

Source: Rueffler *et al.* (2004).

such as height. Next a mutation occurs leading to taller individuals (another mutation could just as well result in shorter individuals). If taller individuals leave more offspring than individuals with the common height, then the mutant allele will completely replace the original allele in the population and individuals in the new population will be taller. Next a new mutation may occur, take over and so on, until the ESS is reached. In the ESS all mutations for shorter or taller individuals are selected against. An ESS is the biological equivalent of the Nash equilibrium from economics, with the difference that biologists always place a greater emphasis on the second condition, that the population needs to converge to the stable point by small mutations (Parker and Hammerstein 1985). An ESS is like an evolutionary trap; once reached, the population stays at that point. ESS is an important concept and we will use it many times in this book. An ESS is not always what would be best for an individual if we did not regard its interactions with others. In the soya bean example it was an ESS to invade the neighbouring compartment, but together the two plants were worse off than when they kept to their own compartment. To give another example, driving on the left side of the road is an ESS in the UK, whereas on the European continent driving on the right side is an ESS. In this case the ESS depends on what most people are doing. Driving on the left is not intrinsically better than driving on the right, and even if a scientific study of traffic showed that fewer accidents occur in countries in which everyone drives on the right, the first person to drive on the right in the UK would have a bleak future.

The ESS has received by far the most attention, but there are also other evolutionary scenarios (Table 1.1). The population may converge with small steps to a so-called branching point which, once it is reached, is unstable. At such points opposite mutants, for instance relatively more male and more female plants, can invade and the population can split up in two types, in two branches. Branching points are clearly of interest in situations when not all plants in the population are the same but two or more types coexist (see for instance Fig. 1.1). Logically there are two more possibilities in Table 1.1. An unstable point is never reached by a population that changes with small steps. Mathematicians call such a point a repellor. Finally, the point may be stable but cannot be reached. Mathematicians have called this a 'Garden of Eden': it can only be reached by creation and the slightest perturbation that brings you out of this state triggers a gradual but inescapable expulsion. Examples of the latter two cases are not included in this book, but the reader can consult Rueffler *et al.* (2004) for biological examples. Geritz *et al.* (1998) clearly explain the maths behind Table 1.1, while Brommer's (2000) review 'The evolution of fitness in life-history theory' sketches the historical perspective of the theory that is now generally referred to as 'adaptive dynamics'. For a mathematically interested reader, the above references are a good entrance to the literature.

So far we have compared plant strategies but were implicit about the underlying genetics. All evolutionary change involves changes in allele frequency and all life-history models are therefore genetic models. In these models, the genes are the replicators, to use the terminology of Dawkins (1976), but organisms are the vehicles on which selection acts. Once we have measured the success of organisms with different strategies we still have to write down the consequences for the different alleles. In some examples the models comparing the different phenotypes correspond closely to genetic models of nuclear genes with several alleles (Charlesworth 1990). Sometimes dominance of different alleles matters somewhat for the outcome of the model but since such details are rarely known, geneticists are usually not worried about using the simpler phenotypic models. For instance, Maynard Smith (1982) wrote:

> it is relevant for the evolution of wing form that the shape which generates a given lift for the minimum induced drag is an elliptical one. If someone were to say 'Maybe, but you do not know that a bird with an elliptical wing is not a heterozygote which cannot breed true?', he would rightly be regarded as unreasonable.

In other cases the relation between phenotype and genotype is not so straightforward and we must be careful in writing down the genetic equations. A clear example is the evolution of altruistic behaviour that we detail in the next paragraph.

1.8 Kin selection

It has long been a puzzle why some animals display altruistic behaviour, i.e. are helping others with a cost of the individual itself. This behaviour begins to make sense when you realize that individuals affect each others' success and when you examine consequences of certain behaviours for the number of allele copies passed on to the next generation.

The geneticist Haldane apparently stated that he would give his life to rescue the life of two drowning brothers. This is puzzling. Wouldn't natural selection support the individual that cares most about its own well being? We can understand the point Haldane had in mind when we examine the genes (alleles) for this behaviour. Suppose the common behaviour is not to do anything when someone is drowning. We focus on the individual that needs to take decisive action. An individual that merely stands by, transfers two copies of the 'non-rescue' allele *a* to the next moment. This individual had two copies of *a* before the event, so per allele a single copy of *a* is transmitted. Individuals with the rare dominant allele *A* jump in the water to save their brothers. When allele *A* is still rare, so all individuals carrying it have the *Aa* genotype, zero copies of *A* are passed on through the focal individual that sacrifices his life. However, each of the brothers has a 50% chance that he also received the *A* allele from one of his parents. On average, if the focal individual gives his life for two brothers one copy of allele *A* is passed on. If you count the number of alleles passed on it does not matter if an individual gives its life or not; in each case, on average, one allele copy is passed on. Allele *A* would disappear from the population if the rescued individuals were randomly chosen from the population, in which case they would mostly have the common allele *a*. However, the rescued individuals were not randomly picked. They were the brothers of the focal individual. If we change the example so that three brothers are rescued, 1.5 copies of *A* will be transferred versus only one copy of *a*: the altruistic allele will increase in frequency and will replace allele *a* in the population. To be able to calculate how a certain type of behaviour is favoured over others, Hamilton introduced the concept of inclusive fitness (Hamilton 1964a; Grafen 1991; Reeve and Keller 1999). Inclusive fitness counts how many allele copies are passed on directly by the focal individual and indirectly by the relatives that are affected by the behaviour.

Plants cannot stand up and walk away, they are stuck with a small set of neighbours for their entire life. It matters for the calculation of inclusive fitness whether these neighbours share alleles with the focal plant, which is the case when neighbours are related to each other or perhaps even genetically identical as in clonal plants. Therefore the ESS may depend on relatedness. If the competing plants in Gersani's experiments (Fig. 1.5) were clones that could recognize each other, most allele copies would be passed on when they avoided competition with their kin and each produced a small root system. Hamilton's method of calculating inclusive fitness applies to such plant examples, perhaps even more directly than to examples of highly social animals in which the link between genes and behaviour is more complex.

1.9 Outline of this book

The first four chapters provide the building blocks for the later ESS models. In Chapters 2 and 3 we first examine the gain curve (the relation between investment and reproductive success). Gain curves are the basis of all optimization models. The shape of gain curves is affected by efficiency factors, independent of other individuals in the population (Chapter 3). The shape can also be affected by the interactions between different plants (Chapter 4). We examine diminishing returns from investment when related males compete for access to females (local mate competition) and related females compete for access to resources (local resource competition). The gain curves are used to illustrate calculation of the ESS for optimal sex allocation in an outcrossing plant species.

The next three chapters are concerned with classical life history theory. Chapter 5 covers the optimal size for flowering. In this chapter fitness is clearly defined. Next we discuss reproductive effort (Chapter 6) and investment per offspring, seed size (Chapter 7).

Chapters 8 to 12 deal with sex allocation theory, extending the example in Chapter 4. The optimal strategy for sex allocation is related to the rate of selfing in Chapter 8. This chapter also compares the different sex systems (Fig. 1.1) found in plants. Chapter 9 examines how plants should plastically change allocation with plant size; this conditional response is called phenotypic plasticity. We discuss under what conditions outcrossing is favoured over self-pollination (Chapter 11) and apply sex allocation theory to the sex ratio in the seeds of dioecious plants (Chapter 10) and to distylous plants (Chapter 12). Selective abortion of fruits of inferior quality (Chapter 13) and the means by which plants interact with pollinators (Chapter 14) complement the chapters on optimal sex allocation.

While the first 14 chapters dealt with the traditional theory, based on individual selection and control by nuclear genes in the mother, the last three chapters discuss modern developments. Chapter 15 shows how the ESS may change if the offspring, rather than the mother, can control the phenotype. This leads to parent–offspring conflict over reproductive investment. This chapter details how Hamilton's method of inclusive fitness can be applied. Cytoplasmic genes are usually transferred through the seeds only, not through the pollen. Some cytoplasms can induce male sterility. If male-sterile plants produce more seeds than plants with viable pollen, more copies of their cytoplasm are passed on to the next generation. Consequently there is intragenomic conflict between cytoplasmic and nuclear genes. Chapter 16 illustrates this conflict in gynodioecious plant species. We end in Chapter 17 by examining how spatial structure and interaction with related neighbours can modify the ESS, as compared to individual selection, again applying Hamilton's method of inclusive fitness.

In all chapters we try both to sketch the theory and to test it. In some chapters there is an extensive empirical literature that allows a test of the theory. In other chapters, especially the chapters towards the end of the book, we are just listing the possible hypotheses, much in the same way that Janzen did, hoping that someone will do the critical experiments.

Questions

(1) What did Gould and Murray think about hypothesis testing in life-history theory?
(2) What do you think that Gould said about the function of the front legs of *Tyrannosaurus rex*?
(3) What is the advantage of phenotypic as compared to explicit genetic models?
(4) For how many half-brothers should Haldane give his life?
(5) What is an ESS?
(6) Suppose the experiment of Gersani *et al.* (Fig. 1.5) was carried out with a clonal plant and two clones of the same plant occupied the pots. What results would you expect?
(7) What is the difference between convergence and stability?

2

Investments, returns and proportionality

Optimization models in biology are strikingly similar to the models that economists use. In economics profits are expressed in dollars or euros, in evolutionary biology they are expressed as the number of allele copies transmitted to the next generation. The basic idea in both economics and evolutionary biology is that the profits are maximized. Evolutionary biologists ask questions like: 'Does this plant achieve the highest fitness by investing in reproduction, or in storage for better survival? Should this plant invest in male or female reproduction and if it reproduces should it produce many small seeds or a few large ones?' It all boils down to the question 'How should a limited amount of resources be divided over different functions?' Such questions are very similar to the worries of a manufacturer who has to decide, given a limited capital, which goods to produce to maximize profits.

2.1 Producing small or large cars

A car manufacturer's production lines allow her to produce a small type of car or a large one. Because she has limited capital, there is a trade-off between the investment in small cars and in large ones. What type of cars should she produce? The answer to the question depends on the costs of producing a small or a large car and the price she can get for them. However, it is not a matter of simply calculating costs and profits. On the one hand, due to increased efficiency the cost per car is likely to decrease with the numbers produced of a given type. Such relations in which costs decrease with quantity are sometimes referred to as 'economy of scale'. On the other hand, when more cars of one type leave the production line, this may saturate the market and it becomes more difficult to sell them at a good price. If the relationships between investment and profit are known, the car manufacturer can calculate the optimal strategy that gives the highest return on investment. Let us, for convenience, assume that the costs of producing a car are independent of the number of cars. Then the relationship between the amount of money invested in producing small or large cars and the total gain will most likely be decelerating for the reasons given above. Mathematically this can be approximated by the equation $G_s = aI_s^b$, in which G_s is the gain for the production of small cars and I_s is the investment in small cars. The gain from the first unit invested equals a and the parameter b determines the shape of the relationship; in our example $0 < b < 1$. For large cars the gain is: $G_l = cI_l^d$. The problem of the car builder can now be solved along two lines of reasoning. Let us think in units of \$1000, so that an investment of \$1000 corresponds with $I = 1$. How should the first \$1000 be invested? For small cars the gain for the first \$1000 invested ($I = 1$) is a, for large cars this gain is c. Therefore production should start with small cars if $a > c$ and with large cars if $c > a$. Then, how should the second thousand dollars be invested? If production started with small cars, one needs to know what the additional gain is when investing an extra \$1000 in small cars and compare this with c, the gain for investing the first \$1000 in large cars. If the gain of producing more small cars is larger than c, our car manufacturer should continue to produce small cars. After repeatedly looking at additional gains, a point will be reached at which additional gains no longer exceed c and at that point large cars should be produced as well. At the point where profits are maximized the gains for the last \$1000 invested in producing small or large cars should be equal. If not, a more profitable investment strategy would have been possible. The additional gain as a function of a small additional investment is called the marginal gain. This is the same as taking the derivative of the investment-gain curve $\mathrm{d}G_s/\mathrm{d}I_s$. The car builder should continue with producing small cars as long as $\mathrm{d}G_s/\mathrm{d}I_s = abI_s^{b-1} > c$. The total number of small and large cars produced that maximizes profit can be found by setting the marginal gains equal:

$$\mathrm{d}G_s/\mathrm{d}I_s = \mathrm{d}G_l/\mathrm{d}I_l \text{ or } abI_s^{b-1} = cdI_l^{d-1} \tag{2.1}$$

This equation can be solved if we know the total investment $I_t = I_s + I_l$ using the trade-off between investment in small and large cars. We can write $I_l = I_t - I_s$ and substitute this in Eq. (2.1), which gives an equation with only one unknown.

$$abI_s^{b-1} - cd(I_t - I_s)^{d-1} = 0 \tag{2.2}$$

Note that this way of approaching the problem only works for particular shapes of the gain curves for which the derivatives of the gain curves are continuously decreasing. For other shapes this approach does not work. For instance, suppose the manufacturer has to invest in a production facility for

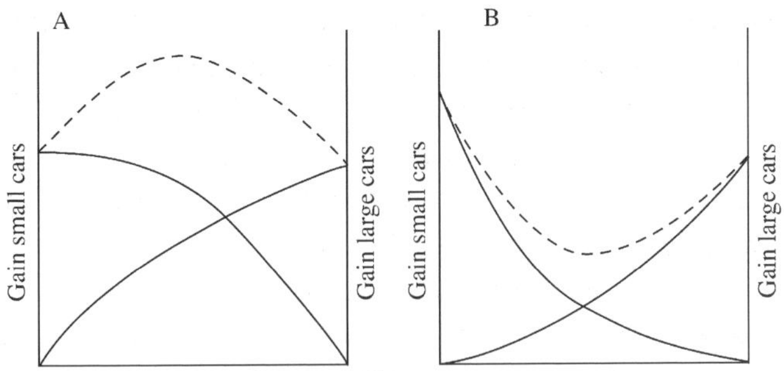

Fig. 2.1 The relationship between gains and the investment in small and large cars, respectively. Total gain (broken line) is found by summing up gains for small and large cars. Total investment is assumed to be constant so that an increase in the investment in small cars is accompanied by an equal decrease in the investment in large cars. (A) Decelerating gain curves ($b < 1$ and $d < 1$) lead to an optimal investment strategy whereby both small and large cars are produced. (B) Accelerating gain curves ($b > 1$ and $d > 1$) lead to an optimal investment whereby only one type of car is produced (in this example small cars).

each separate type of car before the first car can be built, so that there exists a fixed cost before a new type of car can be produced and the gains for the first units invested are zero. The manufacturer will then need to think about how many cars of the new type will be sold before money is made on the investment.

An easier way to solve the problem is to write down the equation for the total gain, given the amount of small and large cars produced: $G = G_s + G_l$, so that $G = aI_s^b + cI_l^d$. Substituting $I_l = I_t - I_s$ gives:

$$G = aI_s^b + c(I_t - I_s)^d \tag{2.3}$$

We find the optimal strategy by three steps: (1) differentiate G with respect to I_s, (2) set $dG/dI_s = 0$, (3) check if this represents an optimum. Setting the derivative equal to zero gives:

$$dG/dI_s = abI_s^{b-1} - cd(I_t - I_s)^{d-1} = 0 \tag{2.4}$$

which exactly equals Eq. (2.2). Under the condition that $b < 1$ and $d < 1$ this gives an optimum because for small values of I_s, $dG/dI_s > 0$ and for large values of I_s, $dG/dI_s < 0$. We assumed that gains were decelerating; $b < 1$ and $d < 1$. In this case the solution to the problem is a mixed strategy (Fig. 2.1): both small and large cars are produced under the condition that the most strongly decelerating curve has the highest gain at low investments (e.g. if $b < d$ then $a > c$). A special case is presented when either b or d is larger than 1. Suppose that instead of overloading the dealer network the results of producing more small cars is that they suddenly become very popular and many people will want one. Our manufacturer can then ask a higher price when more small cars are produced. In that case the total gain will accelerate with investment in small cars ($b > 1$). Under the condition that $b > 1$ and $d > 1$, Eq. (2.4) represents a minimum because for small values of I_s, $dG / dI_s < 0$ and for large values of I_s, $dG / dI_s > 0$. This means that only one type of car should be produced.

We have now answered the question how an individual car manufacturer should divide investment over two types of cars given the relationships between investment and gain for both types. We assumed the manufacturer optimized her strategy on her own, without taking into account what others do; in terms of Chapter 1 this is straightforward optimization or a game against nature. However, the relationships between investment and profit could also be influenced by the total market for small and large cars and thus by the strategies of the other factories. To calculate the optimal strategy for an individual car manufacturer in a market, under the assumption that others continue with their old production strategy, is a useful first step. However, if our factory makes more profit than the other competitors in the market, those others are likely to adapt their strategy. This will change the total market and therefore the profit of our car builder. The market will be stable (that is production strategies will not change any longer) if the strategy of all builders is such that none of the players can come up with a strategy that makes more money. This is what economists called the Nash equilibrium and biologists referred to as the evolutionarily stable strategy (Chapter 1). In Chapter 4 we will show how to calculate an ESS in a concrete example. The essential point to remember now is that the solution to the question how resources should be divided over functions depends critically on the shape of the relationship between costs and gains.

2.2 Testing for proportionality

In the example above, the qualitative answer to the car builder's problem depended on whether the curve describing the relationship between investment and gain was decelerating (the increase in gain is less than proportional to the increase in investment) or accelerating (the increase is more than proportional). When the increase is proportional each additional investment gives the same fitness returns. With a decelerating or accelerating curve each additional investment gives a, respectively, smaller or larger fitness return. In the car manufacturer's example we say that gain G is proportional to investment I if $G = cI$ or $G/I = c$, with c being a constant. A decelerating relationship between G and I means that G/I becomes smaller with increasing I, while an accelerating relationship means that G/I becomes larger with increasing I. A decelerating gain curve most probably leads to a mixed strategy (producing both types of cars); an accelerating curve most probably leads to an all-or-nothing strategy (producing only small or large cars). This problem is very similar to the questions evolutionary biologists are asking. For instance, to answer the question how resources invested in reproduction should be divided over male or female reproduction, we need to know for both modes of reproduction the relationship between resources invested and gains, in this case the number of allele copies passed on to the next generation.

The statistical tool to test for these types of relationships is regression analysis. Tools for regression analysis are available in every standard statistical package, but in many cases the use of regression techniques is not straightforward. It is beyond the scope of this book to give an in-depth overview of the possibilities and limitations of all available regression techniques. Instead we will focus here on some of the problems and possible solutions.

2.2.1 Problems with dependent axes and correlation approaches

The car builder was interested to know how the total sales (in dollars) of small cars increased with increasing investment in small cars. Assuming that the costs of producing a car are independent of the number of cars produced, the car builder needs to know how the total sales depend on the number of cars produced. The car builder can reformulate this problem by asking how the price per car increases with the number of cars produced. Simple algebra shows that with $G_s = aI_s{}^b$, the gain per unit of investment is: $G_s / I_s = aI_s{}^{b-1}$.

With $b > 1$, this gives a positive relation between the gain per unit of investment and the investment, with $b = 1$ there is no relation and with $b < 1$ there is a negative relation. For the car builder this reformulation can be very convenient. There is however a caveat. An evolutionary biologist studying allocation patterns in plants might be interested in the relationship between seed mass produced (G) and vegetative plant mass (V). He or she could hypothesize that, because of economy of scale, G increases more than proportionally with V (i.e. $b > 1$ in the model $G = aV^b$). This problem can be reformulated by asking if seed mass produced per gram vegetative mass (G/V) is positively correlated with V. In literature G/V is referred to as reproductive effort (RE_1) of plants. In some cases $RE_2 = G/(G + V)$ is used. If the biologist finds a positive correlation this seems a confirmation of the original hypothesis of $b > 1$, in the relationship $G = V^b$. However, *the measurement of V is not without error*. V is estimated as $V + \varepsilon_v$ with ε_v being the error term. Errors can include simple mistakes, measurements errors or 'biological' errors, for instance when a large amount of the vegetative part of the plant was eaten by a herbivore just before sampling. In the correlation such errors appear both in the dependent (RE) and the independent (V) variable and these two variables are therefore not independent. This can lead to spurious relationships (i.e. relationships that have no biological basis but result from inappropriate statistics). The problem of spurious relationships will be small if V is measured with only small error. This may be the case if the estimate is an average of a large number of observations. Klinkhamer *et al.* (1990) compared for simulated data sets and nine data sets collected in natural populations the various methods used to test the relationship between RE and plant mass. They concluded that the choice of the statistical test could well influence the probability of rejecting or accepting the hypothesis about proportionality. Especially with Spearman and Pearson correlation between RE and plant mass, the chance of accepting the null hypothesis (H_0, no relationship) while it should have been rejected or the other way around was high (Fig. 2.2). In statistics we call the former a type II error and the latter a type I error. Spreadsheets are convenient tools for generating data sets for which we know the underlying relationship between the variables and the error. Such artificial data can be used as input for statistical programs to test how serious the problem of spurious relationships is.

When testing for proportionality, as a rule, dependent axes should be avoided when both axes share a common error term. When there is no common error term the problem is less serious. For example, a plant physiologist wishes to know if the nitrogen concentration of plants increases with plant mass. On the X-axis he plots plant mass (g) and on the Y-axis he plots nitrogen concentration (mmol)/plant mass (g). At first sight it may seem that there are dependent axes. This is not the case, however. The normal procedure is to measure plant mass by weighing the whole plant, while nitrogen concentration is measured for a carefully taken sub-sample of for instance 10 g plant mass. The errors in both axes are independent and we need not to worry about spurious correlations.

Sometimes the problem of dependent axes can be avoided by choosing the appropriate experimental design. For instance, you want to know if the number of times that a single flower is visited by a pollinator increases with the total number of flowers on the plant. If the plant has several hundreds of flowers of different ages, the number of flowers (F) is an estimate rather than an exact count. If we then count bee visits to all flowers (V) we end up with dependent variables when V/F is plotted against F. The problem can be avoided if a random sample of an exact number of flowers is observed individually for

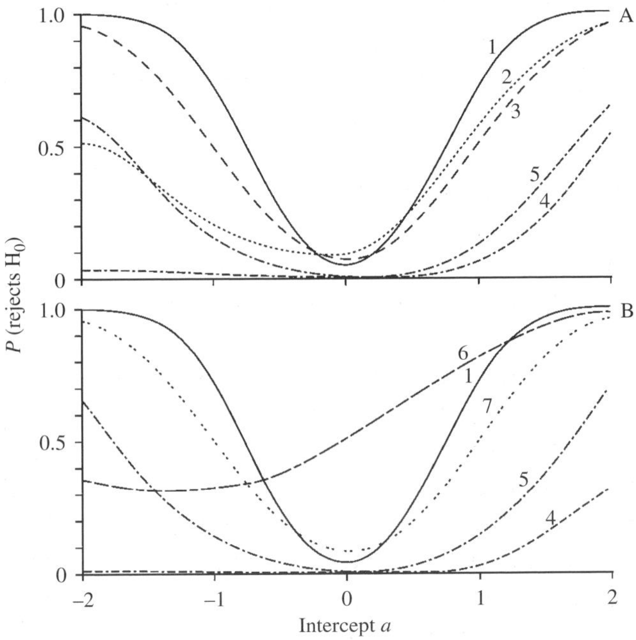

Fig. 2.2 Sometimes statistical tests might not do what you expected them to do. The figure gives a so-called power analysis of different statistical tests applied to generated data sets for which the exact relation between X and Y was known but obscured by adding an error to the independent variable (X). The y-axis gives the probability of rejecting H_0 (i.e. $a=0$, RE is not size-dependent) for several statistical tests. The model used is: $Y_i = a + bX_i + \varepsilon_i$, in which Y_i is the seed mass of plant i, a is the intercept, b is the slope, X_i is the vegetative mass of plant i and ε_i is the error term. If $a=0$, Y increases proportional with X. For each value of a, 500 data sets were generated each with 50 X_i's and ε_i's. (A) The logX_i's are normally distributed with expectation 2 and variance 1. (B) The logX_i's are normally distributed with expectation 1 and variance 1. The ε_i's are normally distributed with expectation 0 and variance 4. $b=1$. (1) F-test (H_0, $a=0$); (2) Spearman rank correlation RE_2 versus X; (3) Spearman rank correlation RE_1 versus X; (4) Pearson correlation RE_2 versus X; (5) Pearson correlation RE_1 versus X; (6) Pearson correlation RE_2 versus log X; (7) Pearson correlation RE_1 versus log X. Note that for $a=0$, the probability of rejecting H_0 should be 0.05; only the F-test comes close to this value. In some cases, for instance line 7 in the bottom graph, the probability of rejecting H_0 is too high for $a=0$ and too low for high values of a. Redrawn with permission from Klinkhamer *et al.* (1990), *American Naturalist*, University of Chicago Press.

bee visits. In that way the (small) measurement error of the dependent and independent variable are not correlated.

Klinkhamer *et al.* (1990) showed that to test for proportionality in the relationship, $G = V^b$, an F-test is superior to the correlation approaches

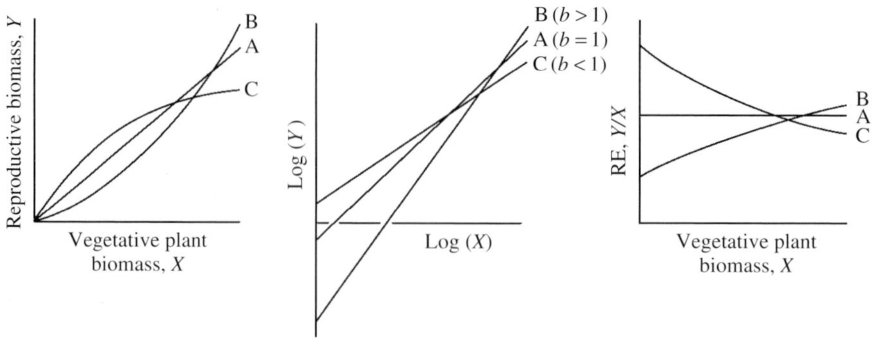

Fig. 2.3 A graphical model of reproductive allocation in plants. Left, reproductive biomass (Y) as a non-linear function $Y = cX^b$ (A: $b = 1$, B: $b > 1$, C: $b < 1$) of vegetative biomass (X). Middle, the linear relationship $\log Y = b\log K + \log c$ for the lines in the graph on the left. The power curve $Y = cX^b$ becomes linear after log-transformation. Right, the proportional-allocation functions (Y/X). Redrawn with permission from Klinkhamer *et al.* (1990), *American Naturalist*, University of Chicago Press.

discussed above (Fig. 2.2). The procedure is as follows. We can write the statistical model as:

$$G_i = aV_i^b e^{\varepsilon_i} \tag{2.5}$$

in which ε_i is the error term. Taking natural logarithms on both sides gives:

$$\log G_i = \log a + b \log V_i + \varepsilon_i \tag{2.6}$$

In this model we now test the null hypothesis that the regression coefficient $b = 1$, no size effect on RE, against an alternative hypothesis either $b > 1$, $b < 1$ or $b \neq 1$ (Fig. 2.3). An F-test for significant deviations from a specified regression coefficient in linear regression is not standard in most statistical packages. However, most statistical packages provide confidence intervals for the slope in model I linear regression, which gives a first estimate to distinguish between the two hypotheses. The procedure for testing on $b = 1$ is given in Box 2.1.

Box 2.1 Testing for disproportionality

Model $y_i = a + bx_i + \varepsilon_i$, $i = 1,2 \ldots, n$; the error term ε_i is normally distributed with variance δ^2 and mean 0 and the ε_i's are independent. H_0: $b = 1$ and H_1: $b \neq 1$. The sums of squares (SS) are calculated over $i = 1$ to n and are as follows: $SS_{xx} = \sum (x_i - \bar{x})^2$, $SS_{yy} = \sum (y_i - \bar{y})^2$, $SS_{xy} = \sum (x_i - \bar{x})(y_i - \bar{y})$ and $SS_{error} = \sum (y_i - a - bx_i)^2$. We write SS_1 and SS_0 for SS_{error} under H_1 and H_0, respectively.

Under H_1: $b = \sum(y_i - \bar{y})(x_i - \bar{x})/\sum(x_i - \bar{x})^2$ and $a = \bar{y} - b\bar{x}$ and $SS_1 = \sum(y_i - a - bx_i)^2$.

Substituting a and b gives:

$SS_1 = \sum(y_i - \bar{y})^2 - [\sum(y_i - \bar{y})(x_i - \bar{x})]^2/\sum(x_i - \bar{x})^2$.

Under H_0: $b = 1$ and $a_0 = \bar{y} - \bar{x}$ and $SS_0 = \sum(y_i - \bar{y} + \bar{x} - x_i)^2$.

The test statistic to accept or reject the null hypothesis $b = 1$ is:

$F = [(SS_0 - SS_1)/SS_1]\,[(n-2)/1] = [(SS_0 - SS_1)/SS_1][n-2]$ with $1, n-2$ degrees of freedom.

The same procedure can be used for comparing with other specified values of b, by substituting these values in the formulas: $a = \bar{y} - b\bar{x}$ and $SS_0 = \sum(y_i - a - bx_i)^2$ to obtain the estimates under H_0 (Sokal and Rohlf 1998). F-tables can be found in most statistical textbooks but are also easily found with standard Internet search engines.

2.2.2 *Choosing the right model when testing for deviations from proportionality*

The statistical null model for proportionality is $y = bx$. However, when testing for proportionality it is not clear beforehand against which alternative model should be tested. Because the test on the value of regression coefficients is not readily available, some authors test for a significant quadratic term or higher-order term in a polynomial regression ($y = a + bx + cx^2 + \ldots nx^n$). In general this is not a good idea because relationships can be disproportional even if a quadratic term does not give a significantly better description of the relationship between the experimental data. Furthermore, the biological interpretation of such a polynomial model is rather unclear.

Preferably the statistical model to be tested should be based on biologically reasonable a priori assumptions. Suppose we want to know if seed number (S) is proportional to plant mass (V). We can use the model (Fig. 2.4):

$$S_i = aV_i^b e^{\varepsilon_i} \tag{2.7}$$

We hypothesize $b > 1$. A biological reason for this hypothesis could be that plants with many flowers attract more pollinators and produce more seeds. The model from Eq. (2.7) is then entirely appropriate. However, with a different biology we might also need a different statistical model. Another biological reason for disproportionality in the relationship between S and V could be that plants need to form a flowering stem first before they can produce the first flower.

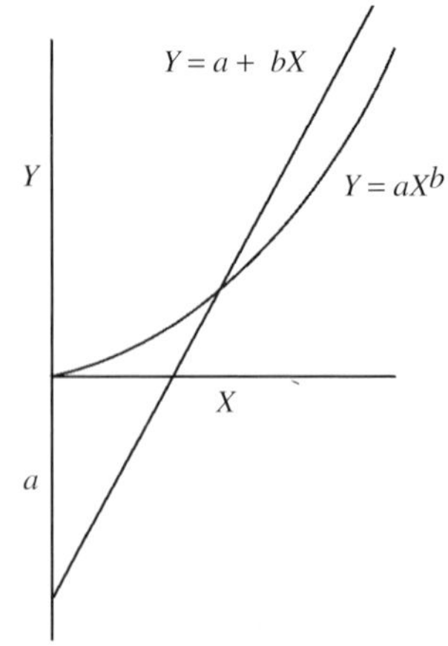

Fig. 2.4 Two alternative models for disproportionality. (1) $Y = a + bX$ ($a < 0$), (2) $Y = aX^b$ ($b > 1$). Although sometimes statistically indistinguishable they have distinct biological interpretations (see text).

This means that an initial investment (fixed cost) would be needed before any returns occur. After the fixed costs have been paid each flower is equally costly to produce (Fig. 2.4). Such an idea can be described as:

$$S_i = a + bV_i + \varepsilon_i \tag{2.8}$$

under the condition $bV_i > -a$ otherwise $S_i < 0$; after all plants cannot have a negative seed production. In this statistical model the parameter a denotes the fixed costs for reproduction. If a is significantly smaller than zero (a positive value is not biologically meaningful) the relationship between S and V is disproportional. Most statistical packages offer tests for $a \neq 0$ and therefore this model has often been used in the study of RE (Samson and Werk 1986). Some authors argued that, in practice, the choice of the test is a rather minor point because for most data sets when $b \neq 1$ in the first model (Eq. 2.7), $a \neq 0$ in the second model (Eq. 2.8) (Rees and Crawley 1989). Our data set for the monocarpic perennial *Cynoglossum officinale* (hound's tongue) showed, however, that this was not always the case (Klinkhamer *et al.* 1990). We advise choosing the model that corresponds most closely to the biology and drawing regression lines that are biologically meaningful.

2.3 Sometimes investment or plant biomass cannot be measured directly

Relationships between investments and returns may depend strongly on how investment is measured. It would often be best to express investment in units of the factor that is limiting reproduction or growth, such as nitrogen or

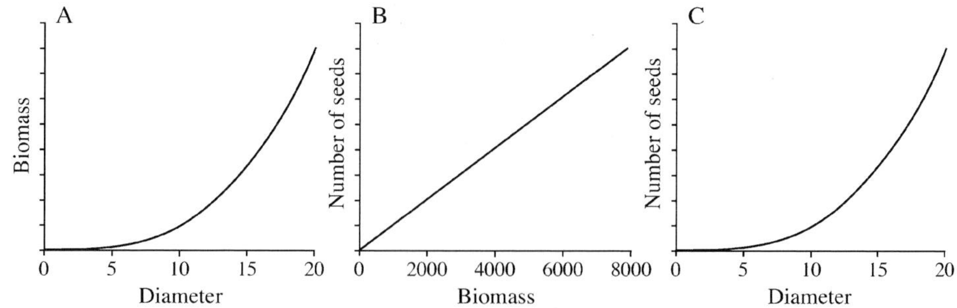

Fig. 2.5 Theoretical relationships between biomass, stem diameter and seed production. (A) Biomass is a non-linear function of stem diameter, (B) seed production is proportional to biomass and, as a result, (C) seed production is an accelerating function of stem diameter. The disproportional relationship in the right-hand graph stems only from the allometric relationship between biomass and stem diameter. When biomass is the correct measure for investment, the gain curve is linear (B). The plot in (C) could lead people to the false conclusion that the gain curve is accelerating.

phosphorus. However, because the limiting factor is often unknown, plant mass is often used under the implicit assumption that the 'true' limiting factor corresponds closely to plant mass. Another problem arises when we can only make an indirect estimate and we have to rely on non-destructive estimates of biomass, like tree stem diameter at breast height or tree height. Suppose that the biomass of the seeds produced is a constant fraction of the biomass of the tree. If someone estimated biomass indirectly by stem diameter, he or she might find an accelerating relationship when seed production is plotted against stem diameter (Fig. 2.5C). The reason is that biomass is not proportional to stem diameter, but rather scales with an exponent between 2 and 3 (Niklas 1994). An accelerating curve for seed production as a function of stem diameter can therefore not be interpreted as evidence for size-dependent RE. The scaling relationship between the indirect estimate of biomass (stem diameter) and biomass needs to be worked out first (Fig. 2.5).

2.4 Estimating regression coefficients

We discussed in Section 2.2.1 how measurement errors in one variable could lead to spurious correlation when that variable figured on both the dependent and the independent axis. Measurement errors in the independent variable cause another problem. Ordinary least squares (LS) regression, which is the most frequently used regression method, assumes that X is measured without error, i.e. that values are specified by the investigator (Wetherill 1986; Niklas 1994).

In field collections this condition will rarely be met. The measurements of X are also subject to natural variation and measurement error. It has been shown that the F-test is fairly robust against violation of this assumption. In other words the F-test can be used to test for disproportionality (Fig. 2.2) (Klinkhamer *et al.* 1990). The slope of the regression coefficient is however systematically biased towards lower values if we apply LS regression. This is clearly stated in most statistical textbooks (for instance, Wetherill 1986). You can easily try for yourself by using a spreadsheet. Apply regression on X vs. Y and then repeat this regression after adding some random error term to X. With error in X, finding the exact value of the regression coefficient becomes a problem. Regression methods accounting for errors in the explanatory variable (X) are referred to as model II regression. Model II regression is a subject of controversy and no definitive recommendations can be given. A useful introduction to the problem is given by Niklas (1994). Two methods are mainly employed in model II regression; major axis regression (MA) and reduced major axis regression

Box 2.2 The relation between ordinary least squares regression (LS or model I) and reduced major axis regression (RMA or model II)

For LS regression the coefficients in $y_i = a + bx_i + \varepsilon_i$ are:

$$b = \frac{\mathrm{SS}_{xy}}{\mathrm{SS}_{xx}} \quad \text{and} \quad a = \bar{y} - b\bar{x} \tag{2.9}$$

The correlation coefficent r is equal to the geometric mean of b for x over y and for y over x:

$$r = \sqrt{\frac{SS_{xy}}{SS_{xx}}\frac{SS_{xy}}{SS_{yy}}} \quad \text{so that} \quad r^2 = \frac{SS_{xy}{}^2}{SS_{xx}SS_{yy}} \tag{2.10}$$

The RMA slope b_{RMA} can be calculated as: $b_{\mathrm{RMA}} = \sqrt{\mathrm{SS}_{yy}/\mathrm{SS}_{xx}}$. As with LS regression $a_{\mathrm{RMA}} = \bar{y} - b_{\mathrm{RMA}}\bar{x}$ and $r = \sqrt{\mathrm{SS}_{xy}{}^2/\mathrm{SS}_{xx}\mathrm{SS}_{yy}}$. Dividing b by r gives b_{RMA}:

$$\frac{b}{r} = \frac{\mathrm{SS}_{xy}}{\mathrm{SS}_{xx}}\frac{\sqrt{\mathrm{SS}_{xx}\mathrm{SS}_{yy}}}{\sqrt{\mathrm{SS}_{xy}\mathrm{SS}_{xy}}} = \sqrt{\frac{\mathrm{SS}_{yy}}{\mathrm{SS}_{xx}}} = b_{\mathrm{RMA}} \tag{2.11}$$

Equation (2.11) shows that you can easily calculate the slope of RMA regression by dividing the slope of LS regression by the correlation coefficient.

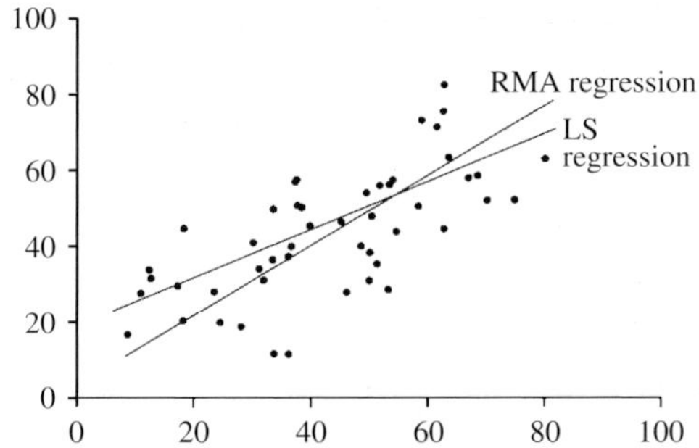

Fig. 2.6 Plot of LS regression and RMA regression for a data set generated by the computer. For each data point $Y=X$ and random error terms were added to both X and Y. The comparison between the two lines shows that RMA is appropriate and gives a close approximation of the 'true' relationship $Y=X$.

(RMA). These methods will not be discussed here in any detail. We do want to point out, however, that the relationship between the regression coefficient (b) estimated by RMA and LS regression is given by a simple formula (Box 2.2): $b_{RMA}=b/r$, with r being the correlation coefficient between Y and X. From this it can be seen that the regression coefficient determined by LS is always smaller than that determined by RMA because $0<r<1$. It can also be seen that the difference will be small if the correlation coefficient is close to one. Equation (2.11) can be used to correct the estimate of the slope of the regression line if the authors used ordinary LS regression, where model II regression was appropriate. Although there are no general rules, it is rather clear what to do in the extreme cases. If X is measured without error or if $\varepsilon_x \ll \varepsilon_y$ LS regression can be used. If the size of the errors in the two variables is within the same range ($\varepsilon_x=\varepsilon_y$) model II regression should be used (Sokal and Rohlf 1998). Their example is plotting the length of the right wing of a fly against length of the left wing. In Fig. 2.6 we worked out an example using similar random error terms added to the 'true' values of the X and Y variable. In this example RMA regression gave a better approximation of the 'true' relationship than LS regression. Another example using biological data is presented in Henry and Aarssen (1999) (Fig. 2.7). In their example LS regression gave a substantially smaller slope of the regression line than RMA regression. In all intermediate cases one should try to reduce ε_x as much as possible, for instance by measuring stem diameter three times instead of once and taking the average. Statistically we can reduce ε_x in a similar way by arranging all data according to increasing X and then calculate average X and Y values for groups of three subsequent data points (Sokal and Rohlf 1998). If a reasonable guess can be made about the distribution of the error in the estimate of X, one can get some idea about the impact of

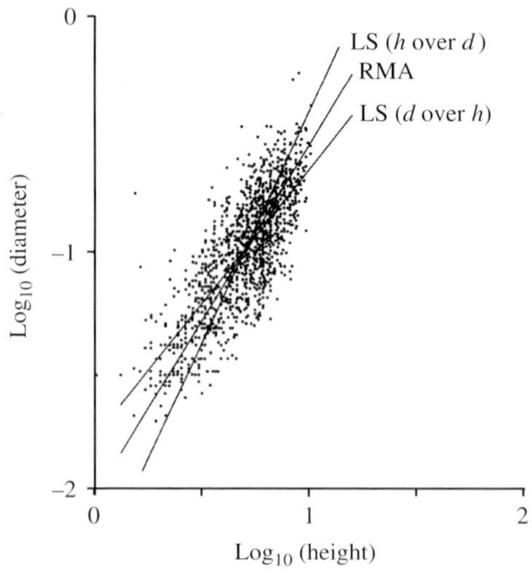

Fig. 2.7 Plot of statistical diameter–height relationships for *Juniperus virginiana* (eastern red cedar) trees. The allometric scaling exponent (b, the slope of the regression line on a log–log scale) is estimated for LS regression with diameter regressed over height ($b = 1.12$), RMA regression ($b = 1.48$) and LS regression with height regressed over diameter ($b = 1.95$). Redrawn with permission from Henry and Aarssen (1999), *Ecology Letters*, Blackwell.

the error on the estimates of the regression coefficient by simulation based on the actual data set.

Henry and Aarssen (1999) also pointed to another feature of LS regression: it matters which of the two variables is taken as the response variable (Fig. 2.7). Or to put it another way: when we swap the dependent and the independent variable, the new slope of the regression is not simply the inverse of the first one. In many cases this will not present a real problem because it is clear which of the two variables changes as a response to the other, but for some allometric relationships this may be less evident.

2.5 Using regression in scaling versus prediction problems

In the above we wanted to know the underlying allometric relationship between two variables. This is often referred to as the scaling problem. A related problem is the prediction problem. For instance, we want to predict plant mass for measured values of stem diameter, because we need a non-destructive measure of plant mass. In contrast to the scaling problem we discussed above, the objective of the prediction problem is not to test a specified relationship and to estimate a scaling parameter but to estimate y for a observed new value $x = X + \varepsilon$. This is exactly the case in which LS regression of y on x works because the independent variable is no longer X but instead x, which is observed without error. In other words, in the prediction problem one is not interested in the underlying relationship between the 'true' values of X and Y, but in the relationship between the measurements x and y including their error, and we can simply apply LS regression. For a more exact summary of the conditions under which this argument holds see Wetherill (1986).

To sum up: when testing for proportionality it is advisable to have independent axes. When the error in X is small, LS regression is appropriate. When errors in X and Y are similar, RMA regression should be applied. When Y needs to be estimated from X and we are not interested in the underlying allometric relation, often LS regression can be used.

Questions

(1) You measure length of the longest leaf of a plant and are interested in estimating dry mass non-destructively for a large sample. For this reason you harvest 20 plants and measure both variables. Which regression method do you use?

(2) You want to know whether estimated dry mass of plants in one year is proportional to dry mass of the same plants in the next year. Which regression method is appropriate?

(3) Why can restricted dispersal of seeds potentially lead to selection for hermaphroditism?

3

Gain curves: efficiency factors

Relations between investment and reproductive success are called fitness gain curves. There are many types of fitness gain curves and we already encountered a non-biological example in the car manufacturer's problem in Chapter 2. Consider, as a biological example, the relationship between investment in a seed and the success of that seed later in life. Since bigger seeds are more likely to survive the seedling stage, this relation is probably an increasing function. Should all plants then make very large seeds, the size of a coconut? No, because there may be diminishing gains from investment, and resources used for one function cannot be used for another. There are trade-offs: the bigger each seed that the plant makes, the fewer seeds it can produce. In this chapter we examine the effect of efficiency factors on the shape of gain curves, assuming that this shape is independent of the behaviour of other plants in the population. The effects of competition on the shape of the gain curve is covered in Chapter 4.

We will return to the problem of optimal seed size in Chapter 7. In this chapter we analyse two different gain curves. First, we examine the relation between the number of seeds produced and female reproductive success. Second, we examine how male reproductive success depends on the number of flowers produced. Because these relationships have been relatively well studied, data are available to illustrate the drawing of the gain curves in practice.

3.1 Definition

We define a gain curve (Charnov 1982a) as the relation between the *absolute* investment I in one function and the pay-off in terms of reproductive success through that function. It is customary to use a power function (Fig. 3.1):

$$\text{reproductive success} = cI^{\alpha} \tag{3.1a}$$

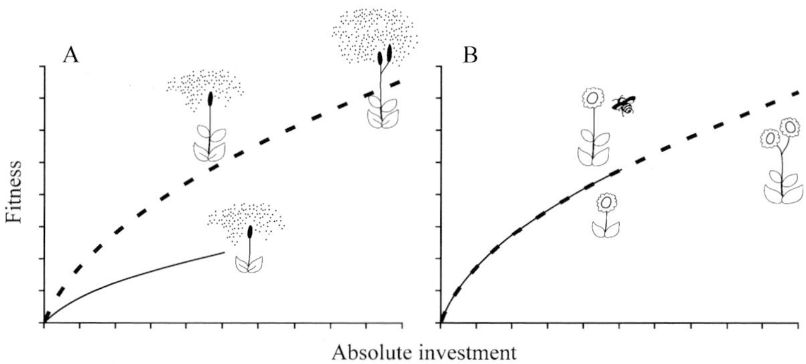

Fig. 3.1 Plant size may affect fitness directly or indirectly. (A) A direct effect means that for a given investment the fitness returns differ for small plants (thin line) and large plants (thick, broken line). (B) If there is no direct effect, the fitness returns for a given absolute investment are equal for small and large plants (Klinkhamer *et al.* 1997a).

The constant c is merely a scaling factor; the exponent α determines the shape of the curve. A doubling of investment I leads to a relative increase in reproductive success of $c(2I)^{\alpha} / cI^{\alpha} = 2^{\alpha}$. When $\alpha = 1$ the function is a straight line through the origin, i.e. reproductive success is proportional to investment. The curve is decelerating when $\alpha < 1$ and accelerating when $\alpha > 1$ (Fig. 2.3). It is convenient to take natural logarithms, so that:

$$\log(\text{reproductive success}) = \log(c) + \alpha\log(I) \quad (3.1\text{b})$$

In this way the exponent α of the power curve can be estimated as the slope on the log–log plot (Fig. 2.3). If a plant has T resources available of which a fraction i is invested in reproduction then $I = iT$, so that:

$$\text{reproductive success} = cI^{\alpha} = cT^{\alpha}i^{\alpha} \quad (3.1\text{c})$$

This is the same as Eq. (3.1a).

3.2 Indirect and direct effects of size on the gain curves

In drawing the female gain curve we ask a question like: 'What would happen to a plant's reproductive success if it doubled its investment in seeds?' To answer this question we would ideally have two genotypes, genotype 2 investing twice as much as genotype 1, the genotypes being grown under identical conditions and producing offspring of a constant quality. In reality such a set-up may be problematic, as when no genetic variation exists in the fraction of resources allocated to seeds. It is more convenient to compare plants in the field that produce different numbers of seeds and compare the reproductive success of these plants. Note however that this assumes that the plants are similar in all other respects. When drawing fitness gain curves based on reproductive success of plants in the population we have to watch out for these confounding effects of size and when they exist we should control for this fact. For instance, if the large plant grew in a habitat with more nutrients then its offspring would also have more nutrients. This is not an aspect of the gain curve and we have to control for it, for instance by transferring plants with different seed productions to a standard habitat after seeds have ripened. This may be impossible in some species, but feasible in others when seeds stay attached to the mother plant for some time after seed ripening. In other cases we wish to check if confounding effects exist. For instance, we want to compare the male reproductive success of a small plant with 10 flowers with that of a large plant with 100 flowers, both pollinated by bees. However, the plant with 100 flowers may also be taller than the plant with only 10 flowers. For estimating the male gain curve height potentially is a confounding factor. By reducing flower number to 10 on both plants we can examine whether the confounding effect of size exists. In the simplest case the pollinators respond to the investment of the plant in flowers, which reflects how much food is available for them. Pollinators may be distracted by the height of the plant, but, in all likelihood, such effects will be small. A small plant investing all its resource in 10 flowers is to pollinators the same as a large plant also with 10 flowers. In this example plants with few and many resources lie on the same gain curve (Fig. 3.1B).

However, size could also have an extra, direct effect on the fitness gain curve apart from large plants simply having more resources to invest. This size effect (Burd and Allen 1988; Klinkhamer *et al.* 1997a) changes the shape of the gain curve so that each size class has its own gain curve. Consider, for instance, reproductive success in wind-pollinated plants. A bigger plant releases more pollen in the wind and is also taller. The pollen from the big plant is released from a greater height so that it may be more easily picked up by the wind and dispersed over a greater area. This would shift the whole gain curve upwards (Fig. 3.1A). This direct effect of size could also shift the female gain curve upwards if, for instance, a wind-dispersed seed is dispersed over a greater distance when released from a tall plant. In this chapter we consider the simplest case (Fig. 3.1B) when only the budget matters and there are no direct effects of size on gain curves. We return to this problem in Chapter 9, when we discuss size-dependent sex allocation in plants.

3.3 Using gain curves

You could write gain curves in absolute (Eq. 3.1a) or relative terms (Eq. 3.1c). The models usually consider the relative allocation. The rationale is that we may try to come up with an evolutionary explanation for the relative allocation to seeds or flowers, the pollen ovule ratio or the relative allocation to reproduction. It seems less likely that we can explain the absolute numbers of seeds or flowers that a plant makes in these terms (Burd and Callahan 2000). At most times, the plants with many flowers and seeds are simply the ones that grow on the best sites. It would be awkward to say that a large plant with more flowers and seeds has a 'better strategy' than a small plant. Modellers like relative allocation best.

However, if we wish to estimate the shape of the gain curve in the field, absolute values on the x-axis are preferred. There are several reasons for this. The first reason is a statistical one; as we have seen in Chapter 2, fractions often blur the interpretation.

The second reason for using absolute investment is a biological one. The gain curves level off due to absolute factors such as more pollen or more seeds being produced and it is most direct to draw the plots in this way.

3.4 Trade-offs

In measuring trade-offs a number of caveats exist (see Stearns (1992) for an extensive review). Most important is the confounding effect of resource level (van Noordwijk and de Jong 1986), as illustrated by the house–car example. Imagine that you can use your salary to buy a house and a car. Both cost money and, with a fixed budget, you would expect a trade-off: buying a bigger car will mean you will have less money left to pay for the house. Nevertheless it would be difficult to find such a trade-off if you just examined the correlation between the sizes of houses that people own and their cars. Instead of a negative trade-off you would almost certainly find a positive correlation. The reason is that people with a greater budget have more money to spend on both their car and their house. To find a trade-off it is necessary to control for the size of the budget.

For plants this means that we have to be very careful that the individuals we compare had exactly the same amount of sunlight and nutrients, the same growth period and so on. Even in the lab this is not easy to achieve. If we are able to control the environment precisely, but one genotype manages to extract more resource from the same environment, then we are faced again with the problem of different budgets.

In the field it is even more difficult to be sure that the plants we compare have the same budget. A method which is frequently used is to take some independent plant measure that estimates the budget and to control for that variable. For instance, plant mass before flowering has been used as such a measure. Pannell (1997) matched male and female plants of the dioecious *Mercurialis annua* (annual mercury) for size and observed that male plants produced four times as many flowers as females. The plants are annual so that the flowers they make are their full reproductive investment. Pannell (1997) concluded that a female flower is four times more expensive to produce than a male flower.

For the hermaphrodite *Cynoglossum officinale* Klinkhamer and de Jong (1987) examined the trade-off between allocation to seeds and allocation to flowers. Conveniently these plants are monocarpic. Plants grow as rosettes for a number of years but when they flower, there is no stopping them. They invest all resources into flowers and seeds and die afterwards. The set of all possible phenotypes contains plants with different numbers of seeds per flower and this may range from zero (no seeds produced, plants are 'male') to four, which is the maximum for the plants in the family Boraginaceae. In the field, plants make about one seed per flower. Klinkhamer and de Jong (1987) used the plant skeleton that remained after senescence as a measure of resource status, and named this straw weight. Plants were selected for similar size and in the lab half the plants were pollinated by hand and the other half were left unpollinated. The plants with pollination set 14.2 seeds per gram of straw weight, while the unpollinated plants set only a few seeds (7.4 per gram straw weight). The latter plants responded by making many extra flowers (44.5 flowers per gram straw weight), compared to the hand-pollinated ones (19.7 flowers per gram straw weight). Assuming that there are no fixed costs and all plants have T resources to spend on either flowers or seeds we have:

$$T = 14.2 \text{ cost per seed} + 19.7 \text{ cost per flower, for the pollinated plants} \tag{3.2a}$$

and

$$T = 7.4 \text{ cost per seed} + 44.5 \text{ cost per flower, for the unpollinated plants} \tag{3.2b}$$

We can solve this by setting Eq. (3.2a) equal to Eq. (3.2b). We then find that the cost of producing one seed is equal to that of 3.64 flowers without seeds. The important point is that matching for resources, in this case by selecting plants of about equal size and by controlling for straw weight, is necessary to be able to calculate the trade-off. If this matching does not occur, one type will have more resources than the other and it is not possible to show the trade-off.

3.5 Accelerating female and male gain curves

With respect to the female and the male gain curve it has been suggested that curves are sometimes accelerating, i.e. that fitness returns *per unit of investment* increase with the investment. With accelerating gain curves a small investment is hardly worthwhile, whereas a big investment yields a disproportional increase in reproductive success.

Givnish (1980) suggested that seed dispersers, such as birds, visit trees with many fruits preferentially, and that this could lead to an accelerating female gain curve. This later became known as the fruit crop size hypothesis. The accelerating gain curve would make it profitable for plants either to invest all energy in fruit production or to invest no energy in fruits at all and allocate it to another function instead. Givnish (1980) argued that an accelerating female gain curve could lead to dioecy (see Chapter 4). He documented for gymnosperms that virtually all (402 out of 447) species with animal-dispersed seeds were dioecious whereas in species with wind-dispersed seeds dioecy was rare (only 18 out of 357 species) and monoecy was the common strategy.

A similar idea with respect to pollinator attraction is known as the floral arms race hypothesis. Bawa (1980) reported that in the tropics, small generalist insect pollinators responded ‘dramatically’ to changes in floral resources, suggesting an accelerating gain curve, which could select for dioecy. Schaffer and Schaffer (1979) suggested that pollinators prefer large inflorescences of *Agave* or *Yucca* over small ones. They argued that such relations could lead to an accelerating gain curve and the evolution of monocarpy. More pollinator visitation may improve fertilization and seed set, but even if this is not the case (for instance, Udovic 1981) attracting more pollinators could potentially increase pollen export and in this way enhance male reproductive success.

We will now list, in a general way, the factors that contribute to the relation between fruit number and female reproductive success and between flower number and male reproductive success. As noted at the beginning of this chapter, the assumption is that the gain curves we draw are independent of the behaviour of other plants in the population.

3.6 Seed number and female reproductive success

The fruit crop size hypothesis (Givnish 1980) was received favourably. Several authors reported that more birds arrived at large fruit crops and that these

Table 3.1 *Total fruits removed by birds versus fruit crop size*

	β	r^2	n	P	Reference
Caeseria corymbosa	1.09	0.90	17	0.31	Howe and vande Kerckhove (1979)
Crataegus monogyna	1.09–1.17	0.71–0.77	51–56	?	Sallabanks (1992)
Ilex opaca	0.90	0.75	63	?	Carr (1992)
Phyllerea latifolia	1.81	0.97	4	0.07	Herrera and Jordano (1994)
Pistacia terebinthus	1.03	0.15	13	0.96	Traveset (1994)
Prunus virginiana	1.21	0.78	32	0.08	Parciak (2002)
Sapium sebiferum	0.95–1.08	0.87–0.95	15–17	0.23–0.91	Renne *et al.* (2003)

Notes: Data were tested for proportionality after log–log regression. β is the exponent of the female gain curve; it was estimated as the slope β of the least squares regression on the log-transformed values. P denotes significance towards $\beta = 1$ in an F-test (Chapter 2, Box 2.1).

crops dispersed more seeds. However, it was soon realized that, in order for the gain curve to be accelerating , the increase in seed dispersal with increased fruit crop needs to be more than proportional: the fraction of all fruits dispersed should go up with the size of the fruit crop (Ortiz-Pulido and Rico-Gray 2000). We can write the gain curve for fruits dispersed ($N_{\text{dispersed}}$) versus investment ($N =$ number of fruits produced) as:

$$\text{reproductive success} = N_{\text{dispersed}} = cN^{\beta} \tag{3.3a}$$

so that

$$\log N_{\text{dispersed}} = \log c + \beta \log N \tag{3.3b}$$

Testing the hypothesis that large crops disperse more seeds requires that $\beta > 0$. Testing the hypothesis that the gain curve is accelerating so that large crops disperse a greater fraction of the seeds requires that $\beta > 1$. Data in the literature that allowed the second statistical test are listed in Table 3.1. Studies were included that reported on fruit removal over the whole period that fruits were available to birds. We excluded studies that only reported on, for instance, the speed at which fruits were removed in a one-week period. The trend in Table 3.1 is clear, the relation is proportional ($\beta = 1$) or slightly accelerating ($\beta > 1$). Model I regression was used here but if number of fruits is estimated with considerable error, RMA regression will be more appropriate. This will lead to higher estimates of exponent β, but does not change the *P*-values given in Table 3.1 (Chapter 2). Other studies do not allow our proportionality test from Chapter 2, but are still relevant for our purposes. For instance, a significantly positive rank correlation between a fruit abundance score and the fraction of fruits removed was found by Sallabanks (2003); this shows that $\beta > 1$ but its exact value remains unclear. In another example, Davidar and Morton (1986) calculated the Pearson correlation coefficient of fruit crop size versus the percentage fruits removed for six plant species. In all cases the correlation coefficient was not significantly different from zero, suggesting that the relation between total fruits removed and crop size was proportional ($\beta = 1$). Only two species in Table 3.1 are dioecious (*Pistacia terebinthus*, terebinth, and *Ilex opaca*, American holly) so there is no obvious trend that exponents of the female gain curve are systematically higher in dioecious species. Carr's (1992) detailed study of *Ilex opaca* deserves special mention. He found that the number of fruits removed was a decelerating function of crop size in one year (1986) in which birds were uncommon and many plants, small or large, dispersed zero seeds. In 1986 the exponent of the gain curve was only 0.21 with an r^2 of only 19%. In 1987 more birds foraged on the holly plants and the relation between fruit crop size and fruits removed was proportional with a much higher explained variance. Carr (1992) also compiled some older literature data (with only one study (Howe and vande Kerckhove 1979) overlap with Table 3.1). In 17 out of 20 cases he analysed, fruits removed increased in proportion to fruit crop size, while in three cases the relation was slightly accelerating. Carr (1992) suggested that fleshy fruits may facilitate evolution towards dioecy, by relieving plants from competition between seeds from the same plant (see Chapter 4). Carr (1992) also concluded that it is highly unlikely that fleshy fruit on its own drives the evolution of dioecy. Table 3.1 supports Carr's views on the evolution of dioecy. We return to the problem of the evolution of dioecy in Chapter 8.

Of course, plants with many fruits may not only attract more seed dispersers, these plants may also attract more seed predators. Results on studies on moths and butterflies that lay their eggs in the flowerheads of thistles suggested a disproportionate preference for large plants (Bullock *et al.* 1994; Ohashi and Yahara 1998). In this case the fraction of the heads escaping visitation should be equated to reproductive success; the data on *Cirsium vulgare* (spear thistle) (Bullock *et al.* 1994) showed a significantly decelerating gain curve with $\beta = 0.66$.

3.7 Flower number and male reproductive success in animal-pollinated plants

While in wind-pollinated plants the number of pollen grains landing on a stigma of a conspecific may simply be proportional to how much pollen is released in the air, this relation is more complex and involves a number of distinct steps in animal-pollinated plants. First, flower number is important for the number of pollinators arriving at a plant and for how many flowers they visit during their stay. Visits per flower is then a first estimate of reproductive success. Second, we could focus on the number of pollen grains on the pollinator once it leaves the plant. For self-incompatible plants, this pollen exported to other plants seems the most important component of male reproductive success. Third, we could estimate the pollen from a focal plant that is deposited on stigmas on neighbouring plants. Finally, we might use methods from molecular genetics to determine paternity and detail the relation between flower number and the number of seeds sired on neighbouring plants. This last step is detailed in the next chapter; plant success at pollinating neighbouring plants clearly depends on strategies of other plants, which also place competing pollen on the stigmas. While in the last step we can determine the 'real' fitness of the plant, studying the other steps is still useful as this gives information about the different mechanisms underlying the shape of the gain curve.

3.7.1 Flower number and pollinator visitation

In order to have an accelerating gain curve, each additional flower produced should have a higher probability of being visited than the previous flower. This requires that total flowers visited increases more than proportionally with flower number. It is insufficient when plants with many flowers simply receive more visits as some of the early papers (for instance, Augspurger 1980) reported. For instance, compare plant A with plant B. Plant B has twice the number of open flowers as A. Plant B receives 50% more bee approaches and after arrival each bee visits 20% more flowers on B. The number of flowers that a bee visits in sequence after arriving at a plant is referred to as the length of its foraging bout or simply bout length. Now the total number of flower visits to B relative to A is $1.5 \times 1.2 = 1.8$. This may seem to be an advantage for B, but remember that B had twice as many flowers. Per flower it received only 0.9 times the visits that A received. If the number of bee visits the flower receives is any indication of its reproductive success, then flowers on B are less well pollinated than on A! It is convenient to write power curves for visitation versus flower number F:

$$\text{number of approaches} = c_1 F^{\alpha_1} \tag{3.4}$$

$$\text{bout length} = c_2 F^{\alpha_2} \tag{3.5}$$

$$\text{total flowers visited} = \text{approaches} \times \text{bout length} = c_1 c_2 F^{(\alpha_1 + \alpha_2)} \tag{3.6}$$

$$\text{visits per flower} = \text{total flowers visited}/F = c_1 c_2 F^{(\alpha_1 + \alpha_2 - 1)} \tag{3.7}$$

Here c_1 and c_2 are just scaling factors, the shape being determined by the exponent α. From these equations you see why it is convenient to plot the logarithm of flower number against the logarithm of either approaches, bout length or total flowers visited as we did in Fig. 3.2. The slope on the log–log plot corresponds with the exponent α. The slope $\alpha_1 + \alpha_2$ estimates how the total number of flowers visited increases with flower number. If $\alpha_1 + \alpha_2 = 1$, the number of visits per flower is constant and the gain curve is linear. If $\alpha_1 + \alpha_2 > 1$, as in Fig. 3.2, large inflorescences receive more visits per flower. In Fig. 3.2 the gain curve is accelerating and it may indeed be advantageous to invest heavily in only one function as Bawa (1980) and Schaffer and Schaffer (1979) hypothesized. If $\alpha_1 + \alpha_2 < 1$, then large inflorescences receive fewer visits per flower and the gain curve is decelerating. Another advantage of the log–log plot is that the values of α_1 and α_2 quantify how the pollinator responds to more flowers by either approaching the plant more frequently, by increasing its bout length or both.

Ohashi and Yahara (2001) analysed literature data on the relation between flower number, approaches and bout length. In 23 out of 26 cases plants with many flowers received more approaches, and in 26 out of 27 cases bout length was longer on plants with more flowers. In itself each factor increased less than proportionally to flower number. Total number of flowers visited was proportional to flower number F in 15 cases ($\alpha_1 + \alpha_2 = 1$). A decelerating gain curve was found in five cases ($\alpha_1 + \alpha_2 < 1$), and an accelerating gain curve (as in Fig. 3.2) in five cases ($\alpha_1 + \alpha_2 > 1$).

The mechanism behind the behavioural response of the pollinator to flower number is not at all clear. Ohashi and Yahara (2001) and Goulson (2000) argued that when flowers become too numerous, bees cannot remember all flowers visited. In that case flowers may be revisited. Once flowers are revisited the bee might do better to leave, rather than waste time in trying to find the last unvisited flowers. This argument may explain why the relation between flower number and bout length is decelerating, but it is only half the story, the other half involving the approach rate of bees to plants.

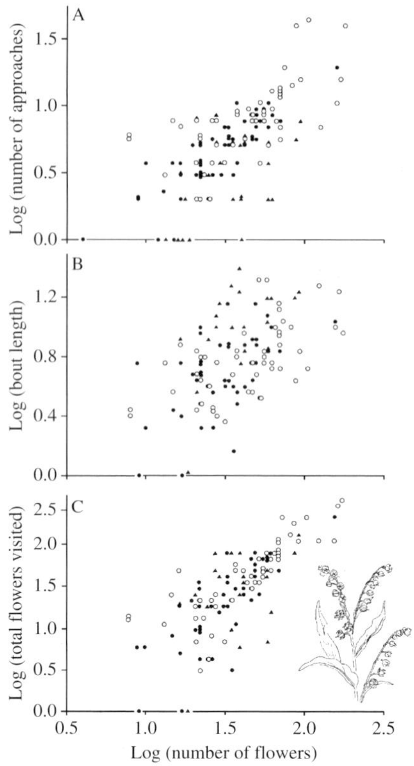

Fig. 3.2 Scatterplots of ^{10}log (number of flowers) versus (A) ^{10}log (number of approaches), (B) ^{10}log (number of flowers visited after an approach) and (C) ^{10}log (total flowers visited) by bumble-bees visiting *Cynoglossum officinale* during 30 minutes. Three situations are indicated: unshaded habitats (open circles), shaded habitats (closed circles) and isolated plants (triangles). The slopes of LS regression on the log–log plot indicate deviations from proportionality (slope = 1); α_1 is the slope in the top graph, α_2 in the middle graph and α ($= \alpha_1 + \alpha_2$) in the lower graph. Unshaded habitats: $\alpha_1 = 0.63$, $\alpha_2 = 0.48$, $\alpha = 1.11$; shaded habitats $\alpha_1 = 0.84$, $\alpha_2 = 0.59$, $\alpha = 1.43$; isolated plants $\alpha_1 = 0.62$, $\alpha_2 = 0.82$, $\alpha = 1.44$. From the figure we may conclude that bees approach inflorescences with many flowers more often. After a bee arrives it visits more flowers in succession (longer bout length) on a plant with many flowers. However, both the increase in approaches and in bout length is less than proportional to flower number. The combined effect of more approaches and longer bout length is that a flower on a plant with many open flowers tends to be visited more often ($\alpha > 1$), a potential advantage of presenting many flowers at the same time (Klinkhamer *et al.* 1989). The slopes were fitted by LS (model I) regression, which assumes that the variable on the x-axis is estimated much more precisely than the variable on the y-axis. Since flowers can be counted accurately on this species and the number of bees arriving per 30 minutes is quite variable this is a reasonable assumption. Redrawn with permission from Klinkhamer *et al.* (1989), *Oikos*.

Instead of examining pollinator visitation, one could also check directly how much pollen is deposited on the stigmas of flowers on small and large plants, or how much pollen is removed per flower. When pollinators are rare this method saves time. Willson and Price (1977) found in *Asclepias syriaca* (common milkweed) that pollen packages, called pollinia, were removed in proportion to the number of flowers ($\alpha = 1$).

In conclusion, when we consider visits per flower as a measure of reproductive success, in most cases flower number is proportional to success. Only in a few cases were flowers on plants with many flowers visited more often, which could lead to an accelerating gain curve.

3.7.2 Flower number and pollen exported

Now instead of flowers visited we examine how pollen export to other plants changes in relation to flower number. In self-incompatible plants, pollen export seems a good estimate of the male reproductive success of a plant. In self-compatible plants, we also have to take into account the fitness returns from selfing.

Ohashi and Yahara (2001) reviewed papers showing that bout length nearly always increased with the number of open flowers. As a result, plants with many flowers will have more transfer of pollen between flowers on the same plant (geitonogamy). Because the bee constantly loses pollen while foraging on the plant, pollen export per bee will be a saturating function of the number of flowers visited. This effect is unique to animal-pollinated plants; it does not occur in plants with abiotic pollination. Is the deceleration of the male gain quantitatively important in plants with animal pollination?

Pollen loss between flowers is typically measured by having a pollinator visit a donor flower with a marker (Fig. 3.3). The marker could be pollen or fluorescent dye particles, much smaller than the pollen grains. The use of dye grains is convenient because by using different colours, a researcher can distinguish between different plants. Next the pollinator is induced to visit a sequence of flowers. From these flowers the style is collected and the marker is counted on the stigma. Pollen (or dye) deposited on the stigma should reflect the amount on the body of the pollinator. This amount deposited usually follows an exponential decline (Fig. 3.3). A fraction k of the pollen on the pollinator is lost between two flowers, $(1 - k)$ is the fraction of pollen remaining on the pollinator after it has moved between two flowers and this is called 'pollen carry-over'. For bumble-bees, k ranged between 2% and 47% (listed in de Jong *et al.* (1992) and Robertson (1992)). The estimates of k for hummingbirds seem to be lower, in the order of 3% per flower (Fig. 3.3). For bumble-bees pollen losses are due to grooming, pollen falling

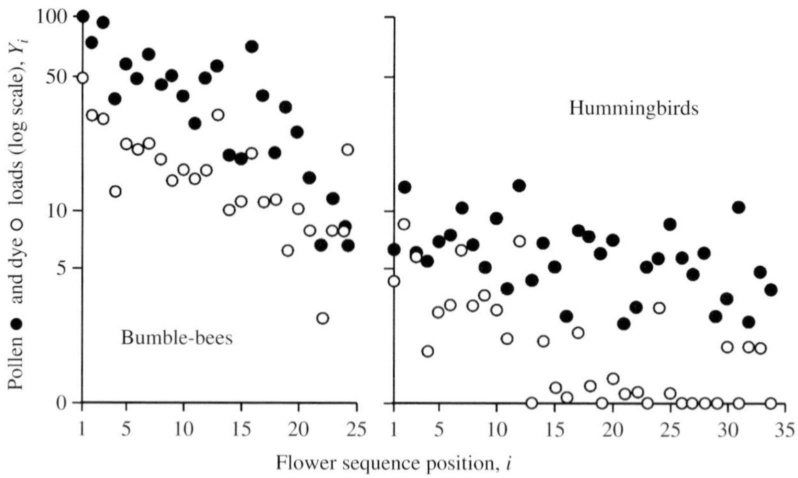

Fig 3.3 Experiment to estimate losses of pollen from the pollinator between flower visits. Data from Waser (1988) for *Delphinium nelsonii* (Nelson's larkspur). In a target flower the pollinator picks up pollen or dye. It then visits an array of emasculated flowers; in these flowers the pollinator cannot pick up any new pollen or dye during its foraging bout. Deposition on the stigma (Y_i) of all subsequently visited flowers is then estimated. The deposition on the stigma reflects the pollen load on the body of the pollinator. The curve can be fitted as a continuous exponential decay, $\log(Y_i) = \text{constant} - \gamma i$, in which γ is the slope of the regression line. It can be calculated that the fraction pollen lost between flower, k, equals $1 - e^{-\gamma}$ (de Jong *et al.* 1992). In *Delphinium nelsonii* the decay rate was similar for dye and pollen (see Dafni 1992 and Kearns and Inouye 1993 for more information on this technique). Redrawn with permission from Waser (1988), *Functional Ecology*, Blackwell.

to the ground or adhering to other flower parts and, probably in the last place, due to deposition on the stigma. For instance, in *Echium vulgare* (viper's bugloss) it was estimated that 6% of the pollen on the bee was lost between flowers ($k = 0.06$) and of this loss only 2.5% was on the stigma and 97.5% occurred otherwise (Rademaker *et al.* 1997).

With these parameters we can estimate pollen export. The pollinator has a bout length of n flowers. After the first flower visited, a fraction k of the pollen that adhered to the pollinator in the first flower is lost during each of the $n-1$ subsequent flower visits. In the end the pollinator leaves the plant with only a fraction $(1-k)^{n-1}$ of the pollen picked up in the first flower. One unit of pollen is the number of pollen grains that adheres to the bee in a single flower. Only for the last flower visited does the bee fly off with all the pollen picked up in that flower. To calculate the total amount of pollen on the bee when it leaves the plant we must sum over all n flowers visited in a single foraging bout. By

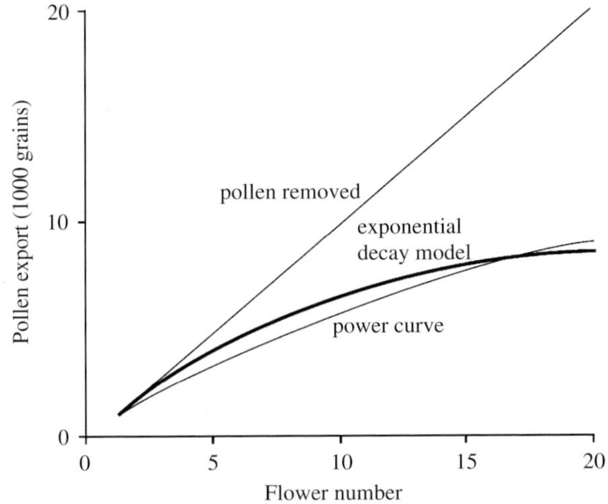

Fig. 3.4 Pollen removed is proportional to bout length n, the number of flowers visited in sequence by a pollinator, whereas pollen export, the amount on the bee when it leaves the plants is a decelerating function of n. Consequently the efficiency of pollen export always decreases with n. Pollen export was calculated using the mechanistic Eq. 3.8 with $k = 0.1$ (thick line), but can also be fitted by a power curve (thin line, $y = 1.09x^{0.70}$).

applying the formula for the sum of a geometric progression (de Jong *et al.* 1992) we calculate that the pollinator leaves the plant with:

$$E_n = [1 - (1 - k)^n]/k \text{ units of pollen grains} \qquad (3.8)$$

which is an increasing but decelerating function of bout length n. Equation (3.8) can also be approximated as a power curve:

$$E_n = cn^{\alpha} \text{ or } \log E_n = \log c + \alpha \log n \qquad (3.9)$$

We can relate the statistical parameter α to the fraction of pollen lost between flowers k by first computing E_n (Eq. 3.8) and plotting this against flower number n (Fig. 3.4). Next we compute the parameters of the power curve that best fits the data. The easiest way to do this is to estimate α as the slope of the plot of the logarithms of computed E_n versus n. The fit of the regression is excellent ($r^2 = 0.99$ for n ranging between 1 and 20) so that a power curve may well be used to approximate the more complicated Eq. (3.8) (Fig. 3.4). In Table 3.2 you can then look up how values for k correspond to exponents of the gain curves. For the k-values given in the literature (between 0.02 and 0.50), the fraction of pollen lost between flowers is significant and for self-incompatible plants this would lead to a decelerating curve for bout length versus pollen sent away to other plants in the population, with an exponent

Table 3.2 *Correspondence between the fraction of pollen lost per flower by the pollinator* (k) *and the exponent* α *of the curve of flowers visited versus pollen exported to other plants (see text)*

k	α
0.02	0.93
0.05	0.84
0.10	0.70
0.20	0.49
0.30	0.34
0.40	0.24
0.50	0.17

between 0.93 and 0.17. When bout length increases in proportion to flower number, α (from Eq. 3.9) is also the exponent of the gain curve. In most cases the increase will be less than proportional (for instance, Fig. 3.2B) and in such cases the calculation of the exponent of the male gain curve involves some more steps.

The efficiency of pollen export if one pollinator approaches the plant equals the amount of pollen exported divided by the amount that is removed from the anthers. If from all n visited flowers one unit of pollen is removed:

$$\text{efficiency of pollen export} = E_n/n = cn^{\alpha}/n = cn^{\alpha-1} \qquad (3.10)$$

Thus efficiency of pollen export depends on bout length only; because $\alpha < 1$, $\alpha - 1 < 0$ so that the efficiency always decreases with bout length n. You can also see this from Eq. (3.8) directly. If a bee visits a very large number of flowers on a plant $(1-k)^n$ approaches zero and E_n approaches $1/k$. Consequently, the efficiency of pollen export becomes $1/kn$, which is a declining function of n (Fig. 3.4).

When large inflorescences receive more approaches, the calculation of pollen export becomes slightly more complicated (see Iwasa *et al.* 1995). With very few pollinators, the extra visits to inflorescences with many flowers may contribute substantially to pollen export. With many pollinators most pollen is removed from the anthers anyway and it is only the efficiency of pollen export that matters: from the point of view of the plant the pollinator should leave as soon as possible.

Some authors have assumed that a pollinator can be 'saturated' with pollen so that, at some point, it is not physically possible to place more pollen on it. This is a misconception with respect to honeybees and bumble-bees and perhaps also for other pollinators. Bees are constantly grooming themselves during flight, brushing pollen away to the baskets on the legs. Bees typically carry a few thousand pollen grains on their body (Rademaker *et al.* 1999) while they could easily hold millions of grains. Pollen export as a function of flower number is levelling off but this is due to more pollen loss during foraging on plants with many flowers, not to 'saturation' of the bee.

3.7.3 Flower number and pollen deposited on stigmas of other plants

We calculated with how much pollen the pollinator leaves the plant. Now how much of this pollen will be deposited on stigmas of plants that the bee visits? This is not necessarily a constant fraction of the load on the bee when it flies off to another plant.

Harder and Thomson (1989) did an interesting experiment in this context with *Erythronium* lilies. They estimated how much pollen was removed per flower, and plotted this against how many grains were deposited on stigmas of the next 20 flowers visited. Harder and Thomson (1989) found that the exponent of this relation was 0.329, i.e. pollen deposited was a strongly decelerating function of pollen removed.

For *Ipomopsis aggregata* (scarlet gilia) Campbell (1989) found that dye delivery to stigmas of neighbouring plants was proportional to flower number, i.e. that the exponent of the gain curve was 1. Using dyes to estimate pollen delivery on stigmas Rademaker and de Jong (2000) found that the exponent of the power curve was 0.89 in *Echium vulgare* and 0.96 in *Cynoglossum officinale*. For *Cynoglossum officinale*, apparently the higher number of visits per flower in large plants (Fig. 3.2) was cancelled out by the fact that these extra visits contributed little to pollen export (bees remove most of the pollen anyway), and geitonogamy is more frequent in large plants. Rademaker and de Jong (1998) found an exponent of 0.24 with a low explained variance ($r^2 = 0.24$) for the relation between flowers visited and pollen exported to other plants for *Oenothera erythrosepala* (evening primrose). In this species bumble-bees stay in the large flowers (petals are 5–6 cm) for as long as 30 seconds or longer to drink nectar. The pollen clogs together and is typically a nuisance for the bees, so that they try to get rid of the pollen as soon as possible by cleaning themselves. As a result of the cleaning, pollen export to stigmas of other plants is only slightly higher for bees visiting many flowers than for bees visiting only a single *Oenothera* flower.

3.7.4 Flower number and geitonogamous selfing

If bees stay longer on the plant, they constantly lose outcross pollen from their bodies and self pollen builds up. The first flower visited will thus be outcrossed, but the later flowers in the sequence are more and more likely to be selfed. It is convenient to assume that all pollen flows are a fraction of the amount of pollen in the donor compartment; this is called the exponential decay model of pollen transfer (Fig. 3.5) (see Morris *et al.* (1994) for more complex models with more parameters). If the pollinator removes a constant amount of pollen in each flower, and loses a fraction k of the pollen it carries between flowers, the amount on its body will reach an equilibrium. In that case we can then derive that the average fraction of self pollen (S) deposited on stigmas of all flowers visited on the plants depends only on bout length n and the rate k at which bees lose pollen when going from one flower to another:

$$S(n) = 1 - \frac{[1 - (1 - k)^n]}{kn} \tag{3.11}$$

(For the derivation of Eq. (3.11) see de Jong *et al.* (1992).) When self and outcross pollen are equally competitive, Eq. (3.11) describes the rate of geitonogamous selfing of the plant. An analogous equation for wind-pollinated species was given by Gregorius *et al.* (1987). Note, however, that if self pollen is outcompeted by outcross pollen, the fraction selfing among the seeds may be lower than suggested by Eq. (3.11) (Rademaker *et al.* 1999).

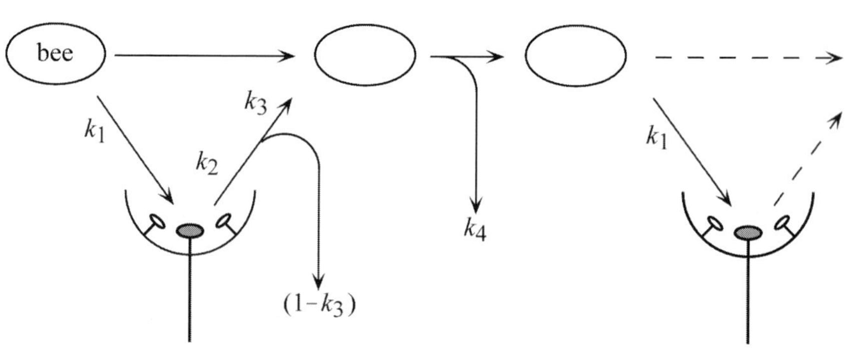

Fig. 3.5 The exponential decay model of pollen transfer. All flows are fractions from the donor compartment. Arrow labelled k_1 denotes the fraction of pollen on the body of the pollinator that is deposited on the stigma, k_2 is the fraction of pollen removed from the anthers, k_3 is the fraction of pollen removed that adheres to the pollinator and k_4 is the fraction of pollen that is lost from the pollinator between flowers (grooming, passive loss, smeared on flower). Redrawn with permission from Rademaker *et al.* (1997), *Functional Ecology*, Blackwell.

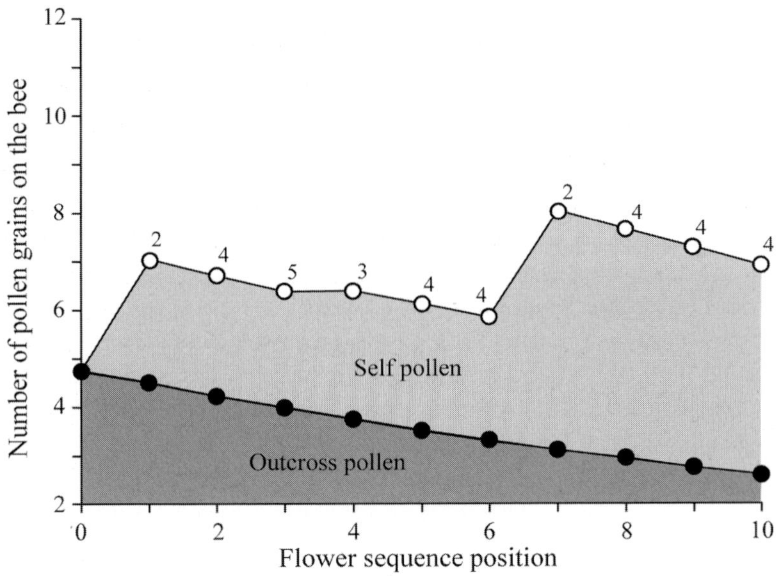

Fig. 3.6 Pollen on a bee when visiting ten flowers of different ages of *Echium vulgare* in sequence. The flowers were divided into five stages, from young to old. Stages are denoted in the figure. After a bee visits a young flower (stage 2; stage 1 was not encountered in this simulation run), the amount of self pollen on the bee increases steeply. Calculations were made using the model in Fig. 3.5 and exact parameters are given in de Jong (2000). Redrawn with permission from de Jong (2000), *Plant Species Biology*, Blackwell.

When flowers vary in how much pollen they contain, for instance because some are old and have been visited before and others are still fresh, the situation is slightly different. The amount of pollen on a bee will then fluctuate around an average value (Fig. 3.6). The amount of pollen on the bee increases if it visits a fresh flower full of pollen and it decreases when old 'empty' flowers are visited. Nevertheless calculations showed that Eq. (3.11) then still gives a good approximation of the average selfing rate (de Jong 2000).

To sum up: fitness gain curves were drawn for number of fruits produced versus fruits removed and for number of flowers versus male reproductive success. For fruits eaten by birds, we found the relationship to be either proportional or slightly accelerating; there seem to be advantages of large fruit crops but these are small. In very few cases, visits per flower increases with total number of open flowers in animal-pollinated plants. Moreover, there is a distinct disadvantage of having many flowers; when pollinators stay longer, more geitonogamy occurs, leading to less efficient pollen export and more selfing. Therefore, in contrast to plants with abiotic pollination, the number of pollen grains delivered on stigmas of conspecifics is likely to be a decelerating function of flower number in animal-pollinated plants. The fruit crop size hypothesis and the floral arms race hypothesis, both based on accelerating gain curves, lack empirical support. It seems that other hypotheses must be considered for the evolution of dioecy, monocarpy and giant inflorescences.

Questions

(1) What does the fruit crop size hypothesis state?

(2) By supplying only self pollen to stigmas, a plant species could be easily selfed in the lab. In the field bees visit many flowers in succession and geitonogamy clearly occurs. Yet you find with molecular genetic methods that in the field very few of the seeds are selfed. What could be the cause?

(3) When comparing wind pollination with insect pollination, how do you expect the curves to compare for pollen export as a function of flower number?

(4) What could be the disadvantage of using dye as a pollen marker? In what situation can the dyes still be useful?

4

Evolutionarily stable strategies: sex allocation strategies as an example

In the imaginary town Surf City there is abundant food for the human population. However, the major hobby is surfing, which is so risky that each couple can raise just two children until maturity, on average one boy and one girl, so that the population numbers remain constant. Due to climatic changes the waves grow higher and higher, each year taking a greater toll on male and female surfers. Each couple can now only raise 1.8 children, so that on average each mother with children leaves only 0.9 daughters. The population is heading for extinction and the government needs to take measures. Everybody is having too much fun with the surfing, so the government cannot forbid that. A biologist in the government then suggests changing the sex ratio to two girls for every boy. Newly opened clinics use a variety of techniques to change the sex ratio at birth to two girls to one boy. Now provided each guy is happy entertaining two girls on the beach, each mother still has 1.8 children of which two-thirds is a female who will also have children: the net result is that the population grows with a factor 1.2! The problem of population decline is solved. The example confirms what some already suspected, males are almost entirely useless. Males need to fertilize females but for this purpose there is no need for a balanced sex ratio. A strongly female-biased ratio is best for the group, for population growth.

If a female bias is better for the group then why is the sex ratio often so close to 50% males, as it is in humans? Darwin wrote about this problem that it was better left to future generations, although he came close to solving it (Seger and Stubblefield 2002). To understand the equal sex ratio we need to consider how many copies of genes (alleles) for a certain sex ratio are passed on to the next generation. We have to consider the sex ratio problem as a game in which one strategy tries to do better than all the others. In game theory this is known as a 'game against the field' (Sigmund 1993). The winning strategy is often not the best one for the group. In most plants individuals are both male and female (cosexual), so their problem is how much to allocate to pollen versus seeds. This allocation problem in cosexuals is similar to the sex ratio problem in organisms with separate sexes like humans or dioecious plants (Chapter 10), only the mathematics is more straightforward, as we will see below.

In this chapter we cover quite some ground. We show how Shaw and Mohler (1953) solved the sex allocation problem, and at the same time, illustrate methods of finding an evolutionarily stable strategy (ESS). In Chapter 3 we examined how efficiency factors could change the shape of the gain curve. In this chapter we will show how more competition between related pollen grains or seeds can also lead to decelerating gain curves. Competition between related pollen grains is known as local mate competition (LMC), and competition between related seeds is known as local resource competition (LRC). Next we show how gain curves can be used to predict the sex system and to compute the ESS for allocation in fully outcrossing cosexual plants. We end by listing data on sex allocation in outcrossers, which show large variation around 50% male allocation and a tendency for animal-pollinated species to have more female-biased allocation than wind-pollinated species.

4.1 Evolutionarily stable strategies

A strategy is considered an ESS (Maynard Smith 1982) if a rare mutant with a different trait value cannot do better than the common type and if the population converges to such a stable point (Chapter 1). Finding an ESS that is locally stable (Fig. 4.1A, B) can be done in a step-by-step way; later we discuss an analytical short cut that can often be used.

(1) Define which phenotypes are possible (the 'phenotype set') and be specific about which evolutionary variable can change. The evolutionary variable could be any variable for which there is variation, like generation time, seed size or reproductive effort. In most cases a higher investment in one function will be at the cost of investing in other functions, so that we need to assign trade-offs.
(2) Write down a fitness equation for the common type and for a rare mutant and be specific about how genes (alleles) are transferred to the next generation.
(3) Examine for each common phenotype whether mutants can invade or not. By trial and error you can find a candidate ESS at the point where fitness does not change with a small change in the evolutionary variable. Check whether starting from different places in the parameter space affects the point to which the population converges.
(4) Examine whether the candidate ESS is a fitness maximum or minimum. Only in the first case is the point a real ESS; once the population has arrived at this point, it will remain trapped there and evolutionary change with small steps is no longer possible.

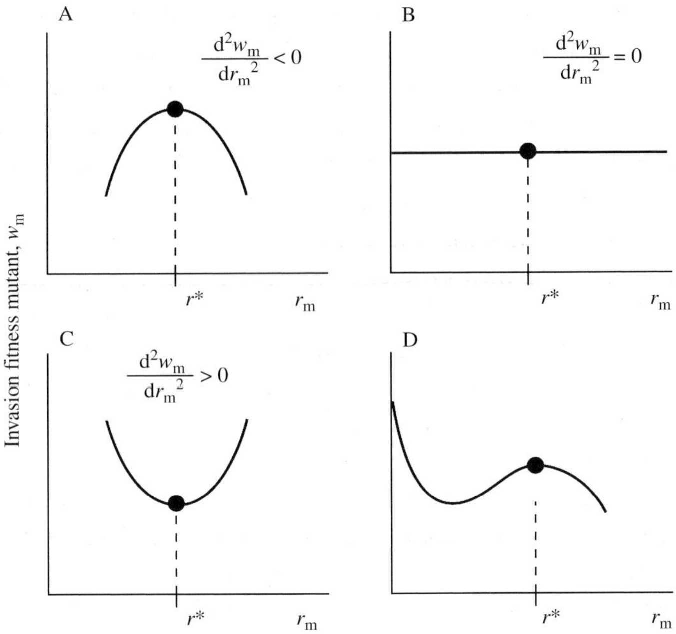

Fig. 4.1 Is cosexuality stable? Invasion fitness is plotted for mutants with different allocation strategies r_m in a population at the candidate ESS r^*. (A) Fitness of all mutants is lower than that of the common type; there exists a fitness maximum and r^* is a strong ESS; in this case the second derivative of fitness of the mutant with respect to r_m is negative. (B) Fitness of all mutants is equal to that of the common type. The strategy r^* is a weak ESS and the second derivative is zero. (C) The fitness of all mutants is larger than that of the common type, there exists a fitness minimum, i.e. strategy r^* is not stable. The second derivative is positive, at r^* branching may occur. (D) Strategy r^* is locally stable as in case A, but not globally stable. Small mutations cannot invade in a population with strategy r^*, but a large mutation might.

This is all rather abstract, but the four points can be illustrated in the problem of optimal allocation of resources to either male or female function in cosexual plants.

4.2 Optimal sex allocation in an outcrossing plant

4.2.1 Phenotype set

Consider a plant that can allocate resources to male function or female function. Hermaphrodite plants are most common but to avoid any confusion with defining the phenotype set let us start with a monoecious species. Monoecious

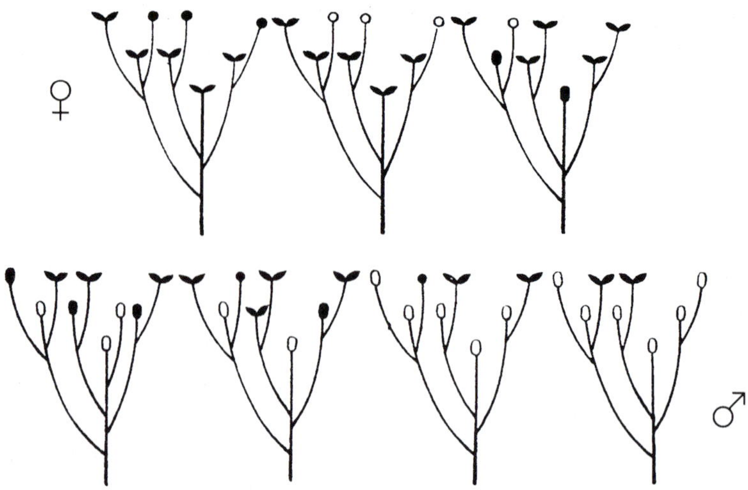

Fig. 4.2 *Acer platanoides* (Norway maple) is a monoecious plant. Inflorescences range from female to male (female flowers dark, male flowers open), with many intermediates. After Correns (1928).

plants have separate male and female flowers (Fig. 4.2). Familiar examples of monoecious plants are *Zea mays* (maize), *Quercus* spp. (oak) and *Cucumis sativus* (cucumber). These species apparently have a potent hormone system that can turn on male or female function in a flowering bud, one function excluding the other. If genotypes differ in hormone production one can easily imagine the whole spectrum from only male to only female flowers being produced. Assume a male flowers costs $C_{♂}$ units of resource and a female flower costs $C_{♀}$. In general, female flowers, seeds and fruits included, are heavier and more costly to produce than male flowers. We assume that pollination is assured and that one plant produces so much pollen that it can be the father of seeds on many plants, which is an entirely reasonable assumption for plants. The seed output of the population as a whole (and of each plant individually) would then be highest if just enough male flowers were produced to fertilize all female flowers, i.e. if the sex allocation is extremely female-biased. However, we must consider the strategy of plants in relation to each other. We also assume that each plant has T resources it can divide over male and female flowers. The evolutionary variable is the fraction of resource allocated to male flowers, r. If a plant allocates a fraction r to male, $1 - r$ remains for investment in the female function. This is where the trade-off kicks in. The plant then makes $(1 - r)T/C_{♀}$ female flowers and $rT/C_{♂}$ male flowers. A fraction f of the resource invested per female flower goes into seeds, a fraction g of the resource per male flower goes into pollen. A single seed costs s units of resource, a pollen grain costs t. Plants then make $(1 - r)Tf/C_{♀}s$ seeds and

$rTg/C_{♂}t$ pollen grains. We abbreviate this to $(1-r)T\nu$ seeds and $rT\omega$ pollen, in which the multipliers ν and ω convert the resource invested in actual numbers of seeds and pollen grains produced, respectively. The phenotype set contains types with r anywhere between 0 (only female flowers) and 1 (only male flowers).

4.2.2 Fitness

Absolute fitness (W) is how many copies of the allele that corresponds with a certain allocation strategy are passed on to the next generation, through seed production or through siring seeds on other plants. Calculating fitness in this way, we count the number of successful gametes through male and female function. The full genetic model that produces the same end result is detailed in Box 4.3 at the end of this chapter. Now assume the population consists of a large number (say 10^6) of plants, all of type A except one mutant B, and is fully outcrossing. The pollen goes to an outcross pollen pool, which is assumed to be well mixed, so that each pollen grain is equally likely to sire a seed on each of the million plants. Animal ecologists often refer to panmixis, when making this assumption. The common type has strategy r and its average fitness does not depend on the presence of the rare mutant. The common type produces $(1-r)T\nu$ seeds. Each seed has one mother and one father, each contributing one allele copy. Therefore, on average, each individual of the common type also sires $(1-r)T\nu$ seeds. Hence the absolute fitness of the common type (resident) is:

$$W = 2(1-r)T\nu \tag{4.1}$$

If we optimize W on its own, the fitness is highest if r is almost zero, i.e. if just enough pollen is produced to fertilize all seeds. Is this then the solution? No, we have to consider a mutant that can invade the population with a certain value of r. For example, in a strongly female-biased population mutants with more male allocation, a higher value of r would do very well in siring seeds and would pass on many allele copies to the next generation.

The mutant type has a different strategy r_m. It produces $(1-r_m)T\nu$ seeds. This already gives the female component of its fitness. What is the male component? We first calculate how many seeds are formed in the population and then calculate the share of all these seeds that the mutant will sire. There were 10^6 plants in the population, all of the common type with strategy r, allocating $1-r$ to seeds. If we neglect the single mutant, $10^6(1-r)T\nu$ seeds are produced in the population. Each plant of the common type contributes $rT\omega$ pollen grains to the outcross pollen pool. The outcross pollen pool then contains pollen from the million plants of the common type and of the one mutant. This sums up to $10^6 rT\omega + r_m T\omega$ pollen, which is approximately $r10^6 T\omega$ if we again neglect the mutant. The mutant's chance of success in siring a seed is determined by how much pollen it makes ($r_m T\omega$), relative to the total amount of pollen in the outcross pollen pool ($r10^6 T\omega$), with which it will be competing. The chance that the mutant sires a randomly picked seed is then $r_m T\omega / 10^6 rT\omega = 10^{-6}(r_m / r)$. Lloyd (1984) referred to r_m/r as the 'competitive share' of the mutant. If r_m equals r the mutant sires as many seeds as the common type, if $r_m = 2r$ the mutant sires twice as many seeds. If we now multiply the chance that the mutant sires a randomly picked seed with the total number of seeds in the population ($10^6(1-r)T\nu$), we obtain the expected number of seeds sired by the mutant which is $10^{-6}(r_m/r) \times 10^6(1-r)T\nu = (r_m/r)(1-r)T\nu$. This is the male component of fitness of the mutant. Summing up female and male fitness components for the rare mutant in a population with strategy r gives the absolute fitness of the mutant:

$$W_m = \left[(1 - r_m) + \frac{r_m}{r}(1 - r)\right]T\nu \qquad (4.2)$$

This is generally known as the Shaw–Mohler equation (Shaw and Mohler 1953; Charnov 1982a). The relative fitness or invasion fitness w_m of the mutant is defined as the per capita growth rate of a rare mutant (allele) in a stable resident population (Geritz *et al.* 1998):

$$w_m = W_m / W \qquad (4.3)$$

By dividing Eq. (4.2) by Eq. (4.1) we obtain:

$$w_m = 0.5\frac{1 - r_m}{1 - r} + 0.5\frac{r_m}{r} \qquad (4.4)$$

If $w_m < 1$ the mutant decreases in frequency, if $w_m > 1$ the mutant increases in frequency and can invade. Remember that the invasion fitness of the mutant depends on its own strategy (r_m in this case) and that of the common type (r). To emphasize this, different notations for invasion fitness have been used; $w(r_m)$ or $w(r_m, r)$. Our notation saves paper but keep in mind that w_m always refers to a rare mutant in a large population of individuals with another strategy. The scaling factors ν and ω have disappeared from the fitness equation; they are unimportant for the calculation of the ESS. For this reason, and to be as brief as possible, we will further leave them out of the fitness equations. Often in the literature Eq. (4.4) is used as a starting-point (for instance, Charnov 1982a), but writing absolute fitness of the mutant (Eq. 4.2) has it merits also. It brings out most clearly that pollen competes for access to ovules, but seeds do not compete among each other. In the Shaw–Mohler equation each extra seed produced by the mutant contributes $1 - r_m$ to fitness. Female investment is a sure thing and male investment is a gamble; the fitness returns from male investment depend on the ratio of females to males in the population $((1 - r)/r)$.

4.2.3 Finding the candidate ESS

With the two equations (4.1) and (4.2) (or with 4.4. only) we are in business! We can take any value of r, say 0.4, calculate fitness W of the common type, and then try out values of r_m for the rare mutant of 0.39, 0.41 etc. to see if one or more of these mutants can invade ($W_m > W$). A spreadsheet is convenient for this calculation. The population starting at 0.4 can be invaded and taken over by a mutant with $r_m = 0.41$. When 0.41 has become common, $r_m = 0.42$ can invade and so on. The spreadsheet tells us that the population creeps to $r^* = 0.5$; this point is a so-called attractor.

4.2.4 Evolutionary stability

Now what happens when the population has arrived at the candidate ESS $r^* = 0.5$? After a while small mutations might occur that generate mutants with, for instance, $r_m = 0.49$ or $r_m = 0.51$. The spreadsheet tells us that as long as these mutants are rare, their fitness is equal to that of the common type. The situation is apparently as sketched in Fig. 4.1 case B, in which $r^* = 0.5$ is an ESS; it does not lose from other strategies, but it is a so-called weak ESS. The ESS is to invest exactly the same amount of resources in pollen and seed. If one sex is cheaper to produce than the other, it should be produced in greater numbers so that resource invested is equal (Fisher 1930).

PARKER'S METHOD

Box 4.1 Differentiation of the fitness function for finding the ESS

In this simple case there is also an analytical way to calculate the ESS. The method for finding it is called 'the differentiation of the fitness function' (Parker 1985). This is a standard tool that we refer to in this book as Parker's method. Parker (1985) was not the first author who used the method; there are several earlier biological (for instance Maynard Smith 1982) and non-biological applications, but Parker's accessible and enthusiastic account of the ESS concept is well worth consulting. First we find the candidate ESS, second we test for invasion stability and third we test for convergence.

The candidate ESS The steps are as follows. Differentiate invasion fitness w_m with respect to the evolutionary variable; in this case r_m. Set the derivative equal to zero. Because at the ESS the best mutant should be similar to the common type we can set $r^* = r_m = r$. Rewrite the equation so that it is of the form $r^* = \ldots$; this point is a candidate ESS. Mathematicians refer to r^* as a singular point; in this point fitness does not change with a small chance in the evolutionary variable r_m. When we apply this method to Eq. (4.4) we obtain:

$$\frac{dw_m}{dr_m} = 0.5\left(\frac{-1}{1-r} + \frac{1}{r}\right) = 0 \qquad (4.5)$$

This yields the candidate ESS $r^* = 0.5$. The absolute fitness W of the common type does not depend on the mutant strategy; Eq. (4.1) does not contain r_m. Therefore it does not matter if we differentiate invasion fitness ($w_m = W_m / W$) or differentiate W_m with respect to r_m to calculate the candidate ESS. Both methods give identical results and since differentiating absolute fitness of the mutant involves fewer terms we adopt the simpler method at many places in the book.

Invasion stability To calculate when the candidate ESS is stable or unstable we need to compute the second derivative of invasion fitness of the mutant with respect to r_m. If $d^2w_m/dr_m{}^2 < 0$ then strategy r^* is a fitness maximum, corresponding to a strong ESS (Fig. 4.1 case A or D). If the second derivative is equal to zero, there exists a weak ESS (Fig. 4.1 case B). In our example of sex allocation in an outcrossing monoecious plant species, the first derivative (Eq. 4.5) contains no r_m. The second derivative of w_m with respect to r_m is therefore equal to zero: r^* is a weak ESS. Again it is of no consequence if we differentiate W_m or w_m with respect to r_m.

Convergence To show that $r^* = 0.5$ is an ESS we also need to show that the population converges to this point with small steps. Geritz *et al.* (1998) derived the condition for convergence:

$$\frac{d^2w_m}{dr^2} > \frac{d^2w_m}{dr_m{}^2} \qquad (4.6)$$

In words: the second derivative of invasion fitness of the mutant with respect to r needs to be larger than the second derivative with respect to r_m. We already calculated that the second derivative of invasion fitness with respect to r_m was zero in our example, so we need to show that $d^2w_m/dr^2 > 0$. Taking the second derivative of fitness with respect to r and then setting $r_m = r$ yields $d^2w_m/dr = 1/(1-r)^2 + 1/r^2$, which is always positive. The condition from Eq. (4.6) is always met. This confirms what we already noticed in our spreadsheet; mutants closer to the ESS can invade and, with small steps, the population converges towards the attractor $r^* = 0.5$. Note that for convergence we should always differentiate invasion fitness with respect to r. Appendix 1 provides some help with the elementary rules of differentiation and links to computer programs like Maple that can help you with the maths.

4.3 Optimal sex allocation with decelerating or accelerating gain curves

In Chapters 2 and 3 we introduced gain curves; success through male or female function could level off, for instance, because pollen export from plants that produced much pollen was less efficient. It is customary to introduce power functions into the equations. How does this affect the ESS for sex allocation? Instead of writing r_{m} for fitness returns through male function, we now write the power function $r_{\mathrm{m}}{}^{\alpha}$. Similarly, the power function $(1-r_{\mathrm{m}})^{\beta}$ replaces $1-r_{\mathrm{m}}$ for female fitness returns. Exponents α and β of the power functions are assumed not to depend on the behaviour of other plants in the population. Gain curves are levelling off when α or β are between 0 and 1. It is also possible that the gain curve is a straight line through the origin ($\alpha=\beta=1$) or is accelerating (α or $\beta>1$).

We now return to the problem of optimal allocation in a monoecious plant, for which we calculated in Box 4.1 that, with $\alpha=\beta=1$, the ESS was to allocate 50% to male and 50% to female flowers. The absolute fitness of the mutant is now (compare Eq. (4.2); we left out the conversion factors ν and T):

$$W_{\mathrm{m}} = (1-r_{\mathrm{m}})^{\beta} + \frac{r_{\mathrm{m}}{}^{\alpha}}{r^{\alpha}}(1-r)^{\beta} \tag{4.7}$$

The absolute fitness of the common type is $W=2(1-r)^{\beta}$. We use Parker's method (Box 4.1) of differentiating the absolute fitness (or invasion fitness) of the mutant (Eq. 4.7) to calculate the candidate ESS. Differentiating Eq. (4.7) with respect to r_{m} yields $\mathrm{d}W_{\mathrm{m}}/\mathrm{d}r_{\mathrm{m}} = -\beta(1-r_{\mathrm{m}})^{\beta-1} + \alpha(r_{\mathrm{m}}{}^{\alpha-1}/r^{\alpha})(1-r)^{\beta} = 0$. By setting $r_{\mathrm{m}}=r=r^*$ and dividing by $(1-r)^{\beta}$, we obtain:

$$r^*/(1-r^*) = \alpha/\beta \tag{4.8}$$

At the candidate ESS, the ratio of allocation to male and female function is in proportion to the ratio of the exponents of the gain curves. If the male fitness returns are levelling off more quickly than the female returns ($\alpha<\beta$), then the plant should allocate relatively more to female function, a result that makes intuitive sense. Rewriting Eq. (4.8) gives:

$$r^* = \alpha/(\alpha+\beta) \tag{4.9}$$

4.4 Stability of cosexuality

When should individuals be both male and female (cosexual), i.e. when is an intermediate r-value stable? Taking the second derivative with respect to r_m in Eq. (4.7) $d^2W_m/dr_m^{\ 2} < 0$ indicates when this r^* value is a fitness maximum:

$$\alpha + \beta > 2\alpha\beta \tag{4.10}$$

Equation (4.10) (Lloyd 1984) gives the conditions under which cosexuality is locally stable. This area lies under the thick line in Fig. 4.3A. If both the values of α and that of β are smaller than 1, the condition from Eq. (4.10) is always met. There exists an asymptote at $\alpha = 0.5$ and $\beta = 0.5$: if the exponent of either the male gain curve or that of the female gain curve drops below 0.5, cosexuality is locally stable, regardless of the value of the exponent of the other gain curve (Fig. 4.3A). There are, however, also areas in Fig. 4.3A in which the second derivative of mutant fitness with respect to r_m is positive; here a potential branching point exists where the population splits up in two types. For instance, if both exponents are greater than 1 dioecy will evolve.

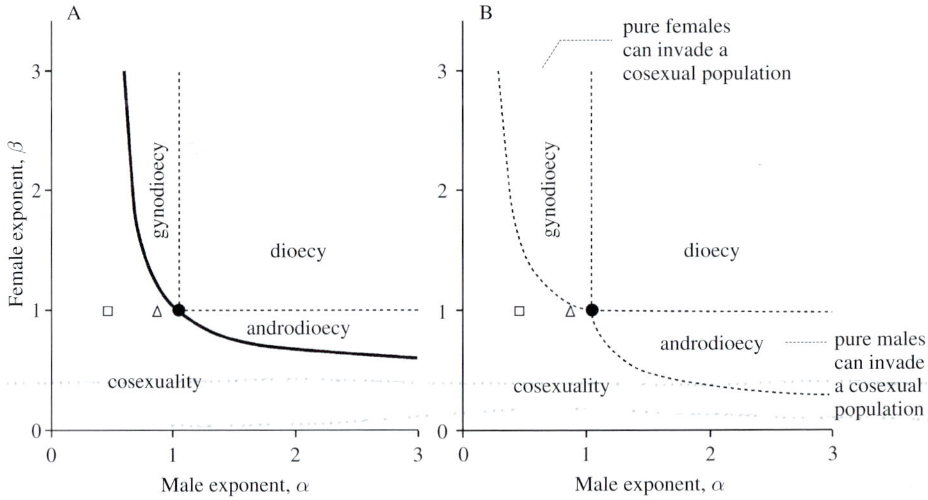

Fig. 4.3 Values of the exponents of the gain curves at which different sex systems are locally (A) and globally (B) stable. (A) The thick line indicates the boundary at which cosexuality is locally stable. Asymptotes are at $\alpha = 0.5$ and $\beta = 0.5$, so that cosexuality is always locally stable if one of the exponents is smaller than 0.5. (B) Pure males and females with full compensation can establish in cosexual populations through a large mutation step, in the areas indicated as androdioecy and gynodioecy, respectively. The points are: squares, *Ipomopsis aggregata* (Campbell 2000); triangles, *Echium vulgare* (Rademaker and de Jong 2000). Partly redrawn with permission from Seger and Eckhart (1996), *Proceedings of the Royal Society B*.

Seger and Eckhart (1996) asked under what conditions pure male or pure female mutants could invade into the population through a large mutation step. They assumed unisexuals reallocated all their resources completely to one function, which is the most optimistic estimate of the reproductive output of the unisexual plant. The absolute fitness W of a common cosexual type at the ESS is $2(1-r^*)^\beta$. We already know that at the ESS, $r^* = \alpha/(\alpha+\beta)$ (Eq. 4.9). The fitness of a female mutant with $r_{\mathrm{m}} = 0$ is found to be 1 by substituting $r_{\mathrm{m}} = 0$ in Eq. (4.7). The absolute fitness of a male mutant is found to be $(1-r^*)^\beta/(r^*)^\alpha$ by substituting $r_{\mathrm{m}} = 1$ in Eq. (4.7). By asking when the absolute fitness of the mutants exceeds that of the common type ($W_{\mathrm{m}} > W$) we can calculate, with some algebra, the boundary conditions:

$$\alpha = \beta(2^{1/\beta} - 1) \quad \text{and} \quad \beta = \alpha(2^{1/\alpha} - 1) \tag{4.11}$$

for invasion of the male and female, respectively. Males and females can invade in a slightly larger part of the parameter space than mutants, which are only slightly different from the cosexual resident (compare Fig. 4.3A and B). In Fig. 4.3 two points are indicated for two outcrossing species for which estimates of α were given in the literature and since the female curve is not well known in these cases, we set $\beta = 1$. Both points are within the range where cosexuality is stable.

4.5 Effects of local competition on the gain curves

We explained in the previous chapter that gain curves could be levelling off, due to efficiency factors, which are independent of the strategy of other plants. Frequently local mate competition (LMC) between pollen grains and local resource competition (LRC) between seeds are invoked as causes of the deceleration of gain curves. These processes do depend on the interaction between different plants. LMC originally referred to brothers competing in a local area for mating (Hamilton 1967). In a similar way LMC occurs when competition between pollen grains from the same parent (sib competition) leads to decelerating returns from male investment. For instance, if one pollen grain competes on a stigma with 99 pollen grains from other plants for fertilizing a single seed, the chance of its winning is 0.01. Doubling the pollen production may put two sib pollen into the race, which raises the chance of siring a seed from 0.01 to 2/101, which is almost a doubling. However, if the focal pollen grain is competing with only nine pollen from other plants, the gains increase from 0.1 to $2/11 = 0.18$ when pollen production is doubled. When all pollen grains on a stigma are sibs and there is sufficient pollen to fertilize all seeds, no further benefits exist from making more pollen. LRC has a similar definition: it occurs when competing seeds are siblings. Sibs are in competition with each other and with unrelated seeds for occupying open sites in the vegetation. If all competitors are sibs, the site is claimed anyway and there are no benefits from making extra seeds. It is worth noting that competition and LRC (or LMC) are two distinct phenomena. Competition is required for LRC or LMC to occur but is not sufficient. For LRC to occur seeds have to be poorly mixed by dispersal so that sibs remain in each other's vicinity (Campbell 2000). Increased dispersal reduces LRC and linearizes the gain curve. In the next paragraph we include LRC and LMC in the general sex allocation model.

4.6 Local resource competition

We consider again the absolute fitness of a rare mutant that allocates r_{m} to male and $(1-r_{\mathrm{m}})$ to female function in a population in which the common type allocates r and $1-r$ to male and female function, respectively. To keep the argument as simple as possible we assume that $\alpha=\beta=1$. Lloyd and Bawa (1984) assumed that seeds from a mother plant are not perfectly mixed in the population but are dispersed over N neighbourhoods, so that in each of these neighbourhoods a fraction $1/N$ of the seeds derives from one mother plant. The mutant makes the equivalent of $1-r_{\mathrm{m}}$ seeds that compete with $(N-1)(1-r)$ other seeds. By changing $1-r_{\mathrm{m}}$ the mutant can change the total density of seeds in its own neighbourhood and hence it can affect how much LRC its seeds encounter. For the common type the chance of establishment is $1/N$ in all neighbourhoods where it grows without the mutant. In a neighbourhood in which the mutant occurs the fitness equation for the mutant is the sum of a female and male component:

$$W_{\mathrm{m}}=\frac{(1-r_{\mathrm{m}})}{(1-r_{\mathrm{m}})+(N-1)(1-r)}+\frac{r_{\mathrm{m}}}{r}\frac{1}{N} \tag{4.12}$$

We use Parker's method from Box 4.1. Differentiating with respect to r_{m} and setting $r^*=r=r_{\mathrm{m}}$ to find the candidate ESS yields:

$$\frac{\mathrm{d}W_{\mathrm{m}}}{\mathrm{d}r_{\mathrm{m}}}=-1+\frac{1}{N}+\frac{(1-r)}{r}=0 \tag{4.13}$$

Rewriting Eq. (4.13) gives for the ratio of male to female allocation at the candidate ESS:

$$\frac{r^*}{1-r^*}=\frac{N}{(N-1)} \tag{4.14}$$

By calculating the second derivative of absolute fitness (Eq. 4.12) with respect to r_{m}, one can show that this is a strong ESS. Hamilton (1967), Charnov (1982a) and Lloyd and Bawa (1984) formulated these models in terms of the number of competitors N. However, the amount of LRC the mutant encounters depends on its own frequency p in the neighbourhood; it does not really matter how many parents are producing other competing seeds $(1-p)$. If all N mothers produce the same amount of seeds in the neighbourhood, the frequency p equals $1/N$, and we can substitute p for $1/N$ in Eq. (4.14):

$$r^* = \frac{N}{2N-1} = \frac{1/p}{2(1/p)-1} = \frac{1}{2-p} \tag{4.15}$$

When seeds are always alone in a neighbourhood, a single seed is enough to claim the site and r^* approaches 1. When seeds compete with only one unrelated seed ($N=2$), $r^*=0.66$. When seeds are well mixed in the population and there are many competitors, r^* approaches 0.5. Campbell (2000) argued that dispersal reduces p and therefore reduces LRC. In addition, one can see from Eq. (4.15) that when density of flowering plants increases, more seeds will land in one site so that the proportion p of seeds from any one plant will decrease. A higher density reduces LRC and makes the female gain curve more linear. We have to be extra careful to distinguish between competition in general, which increases with density, and LRC, which can increase or decrease with density. Although this model still is an oversimplification, the Lloyd and Bawa model of LRC seems realistic, and is close to the type of data a plant demographer would collect. In addition to describing the relation between density and yield, which has been studied extensively (Harper 1977), the plant demographer would have to mark seeds on different maternal parents and follow their fate through life. This allows one to make the connection between seeds produced per parent and the number of surviving seedlings. Molecular markers may well be applied but painting the seed coat would seem an excellent low-tech alternative in species in which the seed coat is still visible shortly after germination.

A slightly different way to model LRC was used by Ellner (1985a, b) who started from the density–yield relationships that have been firmly established in plant demography (Harper 1977). The reciprocal yield law is based on a hyperbolic relationship between total yield per seed (dry mass) and seed density:

$$\text{yield per seed} = \frac{1}{C + \text{density}} \tag{4.16}$$

in which C is a positive constant. If fitness is proportional to yield, then in a neighbourhood with N plants of the common type that each produce $1-r$ seeds, the female component of fitness (F) is:

$$F = \frac{(1-r)}{C + N(1-r)} \qquad (4.17)$$

Hence for a rare mutant with female allocation $1-r_{\mathrm{m}}$, which competes with $(N-1)(1-r)$ seeds produced by plants of the common type, the female component of fitness is:

$$F_{\mathrm{m}} = \frac{(1-r_{\mathrm{m}})}{C + (1-r_{\mathrm{m}}) + (N-1)(1-r)} \qquad (4.18)$$

This is analogous to the female fitness component in Eq. 4.12 except that a constant C has appeared in the denominator. Equation (4.18) is more realistic as it allows for a more gradual relation between density and yield in the monoculture; in the Lloyd and Bawa (1984) version, producing a single seed can already be enough to claim the site and to maximize fitness.

Few studies have addressed LRC in the field. Rademaker and de Jong (1999) estimated LRC by transferring different sized flowering plants of *Cynoglossum officinale* to a constant environment and monitoring survival and growth of seedlings. Seedling density was higher around plants with many seeds, but this had no negative effects on survival or growth. Since no competition could be demonstrated, they concluded that LRC is unimportant in this species and that there is no need to include it in the calculations of optimal allocation. More studies along these lines are needed.

4.7 Local mate competition: the barnacle model applied to plants

How do we incorporate LMC into the model? A good way to start is with Charnov's (1982a) barnacle model. Barnacles are hermaphrodite; they release their sperm in the water in which it is well mixed locally, before it can fertilize one of the neighbouring barnacles within the same mating group. The barnacles cannot mate with themselves so that in a mating group of $K+1$ each barnacle has K mates. After fertilization the larvae are then mixed over the whole population, i.e. over a scale much larger than the local mating group. In the same paper in which they introduced LRC, Lloyd and Bawa (1984) applied the barnacle model to self-incompatible plants. In plants, related pollen grains from the same parent compete for access to ovules if they land on the same stigma.

Suppose that pollen dispersal is local and one mating group consists of a mutant and K other self-incompatible plants, so that the group contains $K+1$ plants, from which K plants are potential competitors. For the mutant, at most $K(1-r)$ ovules are available in a mating group. It has to compete for these ovules with $K-1$ plants of the common type (one plant of the common type drops out, because it cannot sire its own seeds). In the mating group the single mutant produces r_{m} pollen which competes with pollen from $K-1$ plants of the common type, each producing r pollen, so that the mutant's competitive share is $r_{\mathrm{m}}/[r_{\mathrm{m}}+(K-1)r]$. The pollen grains of the mutant compete for $K(1-r)$ ovules. Leaving conversion factors (Eq. 4.2) out, absolute fitness of the mutant is:

$$W_{\mathrm{m}} = 1 - r_{\mathrm{m}} + \frac{r_{\mathrm{m}}}{(K-1)r + r_{\mathrm{m}}} K(1-r) \tag{4.19}$$

Now for the fitness of the resident there are two possibilities.

First, we can just consider the local mating group and assume that all seeds remain in this group. The mutant is not rare in the local group and the common type can sire seeds on $K-1$ plants with the same strategy of making $1-r$ seeds and on one mutant with $1-r_{\mathrm{m}}$ seeds. The fitness of the common type within the mating group is:

$$W = 1 - r + \frac{r}{(K-1)r + r_{\mathrm{m}}}[(K-1)(1-r) + 1 - r_m] \tag{4.20}$$

As you can check with a spreadsheet by comparing Eqs. (4.19) and (4.20), the ESS allocation r^* is always 0.5 for any value of K greater than 1. In the local population selection favours equal allocation to both sexual functions. This conclusion remains unchanged if the plants are self-compatible.

Second, at the other extreme, after seed production all seeds are mixed over all local mating groups. The mutant is not rare in its own mating group but is rare when we consider the whole population. In most mating groups the mutant will not be present. For the calculation of the fitness of the common type over the whole population we can now neglect the mutant: fitness of the common type is simply $W-2(1-r)$. For calculating the ESS we can now use Parker's method of differentiating the absolute fitness of the mutant W_{m} with respect to r_{m}, setting the derivative equal to zero and setting $r_{\mathrm{m}}=r=r^*$. We find that:

$$\frac{\mathrm{d}W_{\mathrm{m}}}{\mathrm{d}r_{\mathrm{m}}} = -1 + \frac{(K-1)(1-r^*)}{Kr^*} = 0, \tag{4.21}$$

so that

$$\frac{r^*}{(1-r^*)} = \frac{K-1}{K} \quad \text{or} \quad r^* = \frac{K-1}{2K-1} \tag{4.22}$$

The mutant should have 50% male allocation when many males compete in one group (K large), and should reduce male allocation if few males compete (K small).

We can redo the ESS calculation for a self-compatible, instead of a self-incompatible, species. In the self-incompatible species each plant in a group of $K+1$ plants had K potential mates, but with self-compatibility all $K+1$ plants in the group are potential mates. For a self-compatible species we then find that:

$$r^* = \frac{K}{2K+2} \tag{4.23}$$

We used the same notation as Charnov (1982a) and Lloyd and Bawa (1984). If the plants are self-compatible it would be simpler to say that each plant mates in total (itself included) with n, instead of $K+1$, individuals: $K=n-1$. Equation (4.23) is then equivalent to $r^*=(n-1)/2n$. This is a rather famous equation first derived by Hamilton (1967). Hamilton considered parasites that lay their eggs within the same host (equivalent to a mating group). The offspring mate within the host after which the fertilized females fly away to find a new host and are at that stage well mixed over the population. Thus Charnov's barnacle model differs only in details (self-incompatibility instead of self-compatibility and cosexuality instead of separate males and females) from Hamilton's model. In all these models within-group selection favours equal allocation but female-biased groups send out more offspring and if, at some stage, there is mixing of all genotypes, this between-group component of selection favours a female bias in sex allocation. We return to the subject of within- versus between-group selection in Chapter 17.

4.8 Problems with the LMC model

There are some problems with this way of handling LMC for plants. First, the whole result in Fig. 4.4 hinges on the assumption that after mating in a small group, all seeds are well mixed in the whole population. This is far less likely for plant seeds than for Hamilton's parasites that can fly and actively search for hosts or for Charnov's barnacles that release their larvae in the sea. If the seeds are not mixed we need only consider selection within a group; Eq. (4.20) holds and the optimal male allocation is 0.5 for all values of K larger than 1. The effect in Fig. 4.4 was not very strong, except for one to three mates, and if

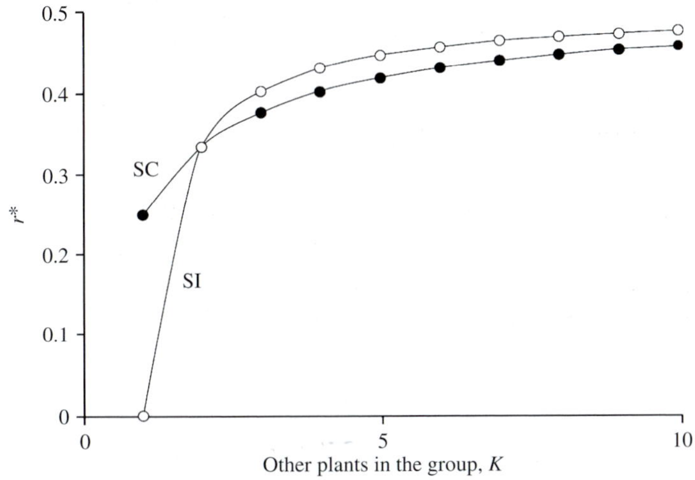

Fig. 4.4 Plot of ESS for male allocation r^* in Charnov's barnacle model for self-incompatible (SI) and self-compatible (SC) plants with $K+1$ plants in a mating group. The number of other plants in the group equals K, which is also the number of mates for the self-incompatible species. A self-compatible plant can also mate with itself, so that $K+1$ plants compete for access to ovules.

we change the assumption about seeds being well mixed in the whole population, the effect becomes even weaker.

Second, the fitness equation, Eq. (4.19), assumed that when a mutant doubles its pollen production, it doubles the amount of its own pollen that is competing with other pollen. This need not be the case; if a plant produces more pollen it may attract different pollinators who deliver the extra pollen produced on stigmas of other flowers of the same plant or even on different plants where the extra pollen is not competing with its sibs. In such a case LMC does not exist.

Third, the model is not tailored to plants and to the competitive arena in which pollen grains compete. Pollen grains on the same stigma compete strongly with each other for access to ovules. The number of fathers (pollen originating from different plants) in the seeds in a single fruit may then give a minimum estimate of K. However, developing fruits may also compete with each other for resources and in such a case more detailed models are necessary to describe the competition.

Without more detailed plant models it remains problematic to come up with a single estimate of K that describes LMC. Relevant data are, however, present in the literature on multiple paternity in fruits. Between two and eight fathers per fruit have been reported for animal-pollinated species (reviewed in

Bernasconi 2004). These numbers raise the question how strong the effect of LMC on allocation really is.

4.9 Measuring gain curves with paternity data

The theoretical quarrels on the sideline (Box 4.2) should not distract from the fact that gain curves can and have been measured using paternity analysis. Broyles and Wyatt (1990) plotted flower number versus number of seeds sired in a background population and concluded the relation was linear in *Asclepias exaltata* (poke milkweed). The regression line fitted to the data had a positive intercept with the y-axis, however, which shows that their statistical model is not realistic (Chapter 2). When we reanalysed the published data, we obtained a slope of 0.78 on the log–log plot; the gain curve is decelerating. Devlin *et al.* (1992) documented the relation between flower number and seeds sired in artificial *Raphanus sativus* (wild radish) populations. They plotted flower number against the logarithm of seeds sired, concluding diminishing gains of increasing flower production. This really confuses the issue as the deceleration

Box 4.2 Estimating exponents of the gain curve and mating group size from experimental data

Suppose you placed several plants with different allocation strategies, or simply with a different flower number, in a background population and measured their success on the surrounding plants. There are then two ways to estimate male fitness. The first method is based on observations of pollinators or using dye as a pollen marker to estimate reproductive success (Chapter 3). The dye estimates pollen delivery on the stigmas without taking into account LMC between pollen grains. The second method is based on paternity analysis. While in the first method LMC is assumed to play no role, in the second method sib competition between pollen grains for access to ovules is included in the estimate of reproductive success. Now how can we estimate the parameters from the sex allocation models?

For the first method (pollen or dye delivery) Eq. (4.7) is appropriate. If we plot logarithm of investment (r_m) on the x-axis against the male component of fitness, as estimated by the logarithm of dye delivered on stigmas ($M_m = cr_m{}^{\alpha}$, c being a constant independent of r_m), then α is the slope of the log–log regression. For this procedure the r-value of the background population does not matter. This population does not have to be at the ESS: the r-value of the population does not affect the slope of the regression. Therefore we can place rare mutants with different allocations r_m in any background population and obtain an estimate of the exponent α of the male gain curve.

The second method (paternity analysis) also includes LMC which adds the parameter K to the equations. If we write Eq. (4.19) in full so α and β

can take on values different from 1 we obtain for the male component of fitness of the mutant:

$$M_{\mathrm{m}} = \frac{r_{\mathrm{m}}{}^{\alpha}}{(K-1)r^{\alpha} + r_{\mathrm{m}}{}^{\alpha}} K(1-r)^{\beta} \tag{4.24}$$

If we now plot the logarithm of male investment (r_{m}) against the logarithm of the male component of fitness (M_{m}) the slope is no longer α but a function containing r, α and K. The consequence of this is that we may estimate the shape of the gain curve in the field but this does not easily translate to separate estimates of α and K. You cannot calculate the ESS from paternity data unless you have additional information on one of the parameters. Perhaps it would be good to combine dye with paternity analysis in a single experimental set-up. You would expect a stronger deceleration of the male gain curve estimated with paternity analysis than with dye because the former contains both efficiency factors and LMC and the latter only efficiency factors.

Ideally female fitness is determined after seeds have completed their life cycle and contributed to the next generation, i.e. after LRC has taken place. In practice this will seldom be the case and rather seeds produced and seeds sired are measured, before any LRC can occur. This suggests another restriction on using data from paternity analysis of seeds, before they are dispersed, for estimating gain curves. The mechanistic approach yields estimates of the relevant parameters. The genetic 'black box' method estimates fitness in the most appropriate way. Both methods have their advantages and complement each other.

is due to only taking the logarithm of the values on the y-axis. Re-examining their data for 1987 and 1988 populations on a log–log plot gives estimates of 0.64 and 1.74 for the slope, respectively, suggesting a decelerating curve in one year and an accelerating curve in the next. Campbell (1998) examined the relation between the relative allocation to pollen per flower and paternity in *Ipomopsis aggregata*. She found 0.48 as the best estimate of the slope on the log–log plot.

For species with abiotic pollination by wind or water, several authors have argued that the male fitness is often proportional to investment. Schoen and Stewart (1986) presented paternity analysis in a population of the wind-pollinated *Picea glauca* (white spruce) (Fig. 4.5A). The relationship is proportional, except for one large clone, which Schoen and Stewart (1986) suggested flowered simply out of synchrony. Bryozoans are moss animals that form large colonies in the sea. McCartney (1997) examined the outcrossing *Celleporella hyalina* by placing focal colonies of different sizes in the centre of four colonies. He then examined the male success of the central colony, which had a unique

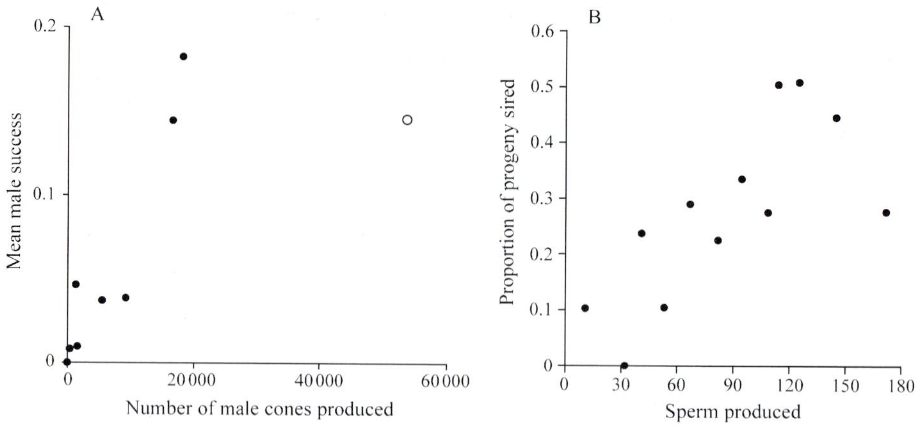

Fig. 4.5 Two examples of male gain curves in species with abiotic pollination. (A) Number of male cones produced versus mean siring success in the wind-pollinated *Picea glauca* (after Schoen and Stewart 1986). Schoen and Stewart (1986) considered the male curve as linear, since the low success of largest clone (open circle) was due to flowering out of synchrony with the rest of the population. (B) Male gain curve for the bryozoan *Celleporella hyalina* (McCartney 1997), established by placing colonies of different size in the centre of four background colonies and determining male success of the central colony through paternity analysis. Redrawn with permission from Schoen and Stewart (1986), *Evolution*, and McCartney (1997), *Evolution*.

isozyme marker, on the four surrounding colonies. The male success was proportional to the sperm produced (Fig. 4.5B). Apparently, there was intense sperm competition in this species, with sperm coming in over large distances, so that LMC was unimportant.

Plants with abiotic pollination may swamp their neighbours with pollen, but if part of the pollen is dispersed over very large distances it could land anywhere. This fraction of the pollen is free from LMC; making more pollen always contributes to fitness. In animal-pollinated plants, presenting more flowers with pollen will result in longer bout length, more geitonogamy and a decelerating male gain curve (Chapter 3). These processes play no role in plants with abiotic pollination. In an extensive literature review Hamrick and Godt (1990) compared isozyme variation within and between populations. They compared 258 outcrossing species of which 124 were animal-pollinated and 134 were wind-pollinated. The proportion of isozyme variation among populations was about half as low in plants with wind pollination as compared to plants with animal pollination. This indicates that pollen is better mixed in wind-pollinated species and suggests that LMC is of less importance in this group. At a finer scale Vekemans and Hardy (2004) found the same trend.

These results are suggestive but note that other factors like plant height are also important and one would have to control for these factors in order not to make comparisons between, for instance, an oak tree and *Arabidopsis thaliana* (wall cress).

With decreased costs and improved techniques of molecular methods, more male gain curves will undoubtedly be published in the near future. In time this will allow us to evaluate the importance of efficiency factors versus LMC for deceleration of gain curves and if indeed the male gain curve is less decelerating in species with abiotic pollination, as the present data suggest.

4.10 Data on gender allocation in outcrossing cosexual plants

Lloyd (1984) examined data on sex allocation in outcrossing plants. He first acknowledged the difficulties in assigning an exact value to r. The first problem is to identify which structures are male and which are female. Filament and anthers are clearly a male investment (A). Female investment includes style and stigma, but clearly also all the costs of fruits and seeds (B). However, it is not clear how to attribute costs of other flower parts like the calyx and petals (C) to the two sexual functions. The same holds for nectar production in animal-pollinated plants. It might be best to neglect C and write $r = A/(A+B)$ like Campbell (1998) did; there is a theoretical justification for this in Chapter 14. Lloyd (1984) attributed costs of attraction equally to the two sexual functions so that $r = (A+0.5C)/(A+B+C)$. The second problem concerns the units in which allocation is expressed. It is convenient to use dry weight but measuring limiting nutrients may be better and pollen and seeds may have different nitrogen or phosphorus contents. Considering these problems, also discussed by Goldman and Willson (1986), one should be careful about assigning one absolute value of r to a species. Nevertheless, as Lloyd and others have argued, if r is calculated in the same way in different species, highly informative comparisons can be made. The problem of estimating r can partly be solved with experimental data on trade-offs (Section 3.4) between male and female function, but such information is not readily available.

From the theory discussed in this chapter, one expects that the ratio of male to female allocation matches the exponents of the gain curves (Eq. 4.8) and is not necessarily equal; the sex allocation could be either male-biased or female-biased. One also expects (Section 4.9) that relative male allocation is higher in species with wind pollination than in species with animal pollination.

Data on exponents of both male and female gain curves are not available for a single species. In *Echium vulgare* and *Ipomopsis aggregata* estimates of $\alpha = 0.89$

Table 4.1 *Estimates of relative male allocation (r) in outcrossing cosexuals based on dry weight of flower parts, seeds and fruits*

	r	Pollination	Reference
Andropogon gerardii	0.60–0.89	Wind	McKone *et al.* (1998)
Bromus inermis	0.42	Wind	McKone (1987)
Microlaena polynoda	0.51	Wind	Lloyd (1984)
Oryza perennis[a]	0.16	Wind	Charnov (1979)
Plantago coronopus	0.30–0.40	Wind	Koelewijn and Hundscheid (2000)
Sorghastrum nutans	0.69–0.81	Wind	McKone *et al.* (1998)
Zizania palustris	0.16–0.33	Wind	Lloyd (1984)
Echium vulgare	0.45	Insects	de Jong, unpublished data
Epilobium angustifolium	0.09	Insects	Cruden and Lyon (1985)
Geranium maculatum	0.19	Insects	Cruden and Lyon (1985)
Gilia achilleifolia[a]	0.09	Insects	Schoen (1982)
Impatiens biflora	0.10	Insects	Lloyd (1984)
Impatiens pallida	0.09	Insects	Lloyd (1984)
Ipomopsis aggregata[b]	0.40	Hummingbirds	Campbell (1998)
Sisyrinchium campestre	0.13	Insects	Cruden and Lyon (1985)
Smyrnium olusatrum	0.04	Insects	Lloyd (1984)
Turnera ulmifolia[a]	0.11–0.12	Insects	Belaoussoff and Shore (1995)

Note: Unless indicated otherwise $r = (A + 0.5C)/(A + B + C)$, in which $A =$ mass male parts, $B =$ mass female parts and $C =$ shared costs such as petals and sepals.

[a] Species partly self-compatible, extrapolation towards plants with 0% selfing.
[b] $r = A/(A + B)$ at the flowering stage, before fruit set.

(dye delivery on stigmas: Rademaker and de Jong 2000) and $\alpha = 0.48$ (paternity analysis: Campbell 1998), respectively, have been published for the exponent of the male gain curve. With $\beta = 1$ this corresponds to theoretical estimates of r^* of 0.47 for *Echium* and 0.32 for *Ipomopsis*, in reasonable agreement with the measured allocations (Table 4.1). Starting from the idea that in wind-pollinated species the male gain curve is not decelerating and that with poor seed dispersal the female curve is likely to level off due to LRC, McKone *et al.* (1998) chose to measure carefully allocation to seeds and pollen in two grass species with poor seed dispersal. Theory would then predict more than 50% allocation to male function and this is indeed what they found for biomass in both *Andropogon gerardii* (bluestem grass) and *Sorghastrum nutans* (indian grass) (Table 4.1). Also for other currencies (N, P, Mg), allocation in *Sorghastrum nutans* was male-biased, while there was no significant difference from 50% for nutrients in *Andropogan gerardii*.

Lloyd (1984) listed six species and suggested that wind-pollinated plants typically have higher male allocation. We find the same result in an extended data set with 17 species (Table 4.1). Since seed dispersal does not differ systematically between wind- and animal-pollinated species, the difference between species must be due to the stronger deceleration of the male gain curve in animal-pollinated species. It is striking from Table 4.1 that estimates of relative male allocation for many of these outcrossers are far below 0.5. This is consistent with sex allocation theory provided that the exponent of the male gain is much lower than that for the female curve. Whether this is indeed the case should become clear when more gain curves are measured. Of course the value of the exponents cannot be inferred from the allocation data.

To sum up: ESS sex allocations for outcrossing plants can be calculated from the exponents of the gain curves. The problem becomes more complex if pollen or seeds from the same parent compete with each other (LMC or LRC). LMC is still poorly understood in plants and pioneering work can still be done. A strong argument was made that the male gain curve is more strongly decelerating in species with animal pollination than in species with abiotic pollination. Relative male allocation was higher in wind-pollinated species, consistent with this argument.

Box 4.3 A genetic model for sex allocation

In deriving the ESS for allocation to male function we counted the number of successful gametes. It is also possible to write a full genetic model for sex allocation for a panmictic population. The final outcomes of the models are the same. Assume sex allocation is determined by one gene with two alleles A and a. A is dominant over a. There are three genotypes AA, Aa and aa with initial frequencies f_1, f_2 and f_3, respectively. The male allocations are r_A, r_A and r_a, respectively and the matching female allocations are $1 - r_A$, $1 - r_A$ and $1 - r_a$. The frequency of A in the outcross pollen pool is $p_A = f_1 r_A + 0.5 f_2 r_A$, the frequency of a pollen is $p_a = 1 - p_A$. Each AA genotype produces 100% female gametes of type A, genotype aa produces only type a gametes and the heterozygote Aa produces 50% gametes of type A and 50% of type a. Now genotype frequencies in the next generation, denoted by the prime, are:

$$AA\text{: } f_1' = f_1 p_A(1 - r_A) + 0.5\, f_2 p_A(1 - r_A) \tag{4.24a}$$

$$Aa\text{: } f_2' = f_1 p_a(1 - r_A) + 0.5\, f_2 p_a(1 - r_A) + 0.5\, f_2 p_A(1 - r_A) + f_3 p_A(1 - r_a) \tag{4.24b}$$

$$aa\text{: } f_3' = f_3 p_a(1 - r_a) + 0.5\, f_2 p_A(1 - r_A) \tag{4.24c}$$

With a computer program or spreadsheet you can check that for any combination of two alleles A and a, the allele which is closest to 0.5 male allocation wins against the other one. If small mutations occur frequently the population evolves towards equal sex allocation. You can see that this genetic approach is more explicit, but also more complex than Eq. (4.3). The final result (the population will evolve towards 50% male allocation) of the two approaches is the same. The speed at which one allele replaces another is determined by the genetic details (degree of dominance).

Questions

(1) What conditions must be met before we call a strategy an ESS?
(2) What are local resource competition (LRC) and local mate competition(LMC)? What is the difference between LMC and sib mating? Is there a difference between competition and LRC?
(3) What is the difference between a locally versus a globally stable strategy?
(4) What is the difference between a fitness maximum and a fitness minimum?
(5) If you estimate the exponent of male and female gain curve using paternity analysis, can you use these data to calculate the ESS?
(6) Does the Shaw–Mohler equation include LRC and LMC?

5

Size at flowering

5.1 To flower or not to flower?

When to reproduce and how much to invest in each reproductive event are the two classical questions of life-history theory. In this chapter we focus mostly on the timing of flowering in short-lived monocarpic plants (a monocarp is a plant that dies after reproduction), using demographic models. In the next chapter we study optimal reproductive effort, using allocation models in which the plant chooses to allocate to current reproduction or to save resources for later.

The short-lived monocarps are the simplest example for illustrating the theory of when to reproduce. Well-studied examples include *Senecio jacobaea* (ragwort), *Verbascum thapsus* (mullein), *Echium vulgare* and *Cynoglossum officinale*. When these plants flower after a number of years, they are quite conspicuous with up to 100 simultaneously open flowers. The long-lived monocarps produce even more impressive inflorescences. Examples are some species of *Agave* and *Yucca*, the giant lobelias on Mount Kenya, *Echium wildpretii* (tower-of-jewels), which is the symbol of Tenerife and, of course, the mainland Chinese bamboo, discussed in Chapter 1. In this chapter we first define the fitness of monocarps with different generation times, next we list the theoretical predictions for the size at which plants should flower and finally compare this with the field data.

5.2 How to pick a winner

5.2.1 Population growth versus expected lifetime reproductive success

Consider two genotypes of an annual species. If genotype A produces more seeds than genotype B and the types were similar in all other respects, one would predict that genotype A will eventually replace B in the population. It would be correct to say that A has higher fitness than B and that seed production is the appropriate fitness criterion; we can be confident that the type with the highest seed production wins. However, when generation times differ it is no longer straightforward to pick the winner. Consider an annual that produces 30 seeds and a biennial that produces 65 seeds. Which type wins? Even with the additional information that survival in the second year is 50%, the answer to this question is not obvious. We will see later in this chapter that the answer depends on how density dependence (DD) operates. During DD the two types affect each other and the type that plays the evolutionarily stable strategy (ESS) with respect to generation time will win. This means that we have to search for a strategy that, when it is common, cannot be invaded by a rare mutant with a different strategy.

We soon return to this question that was solved by Mylius and Diekmann (1995) and Metz *et al.* (1996), but first we need to define some terms. Consider one type on its own, in a virgin environment in which density is still so low that there is no DD. We census all plants each year at the moment the seeds (N) have just been formed. These seeds need to escape from seed predators, keep free from pathogenic microorganisms in the soil, germinate, make it through the vulnerable seedling stage and survive as a vegetative and flowering plant. The probability that all goes well in the juvenile stage is denoted by C_j. A surviving annual individual produces S_a seeds. We assume that all seeds germinate. Now in the virgin environment the number of seeds in year $t+1$ depends on seeds in year t as:

$$N_{t+1} = N_t C_j S_a \tag{5.1}$$

The rate of increase per year (λ) or population growth of the annual when it grows on its own in a virgin environment is then:

$$\lambda_{\text{virgin}} = N_{t+1}/N_t = C_j S_a \tag{5.2}$$

Similarly, if a monocarpic genotype delays flowering until year g, and in each year that flowering is delayed the plant survives with constant probability C, the rate of increase over a year period of g years is:

$$N_{t+g}/N_t = C_j C^{g-1} S_g \tag{5.3}$$

and the rate of increase of a monocarp that flowers in year g equals:

$$\lambda_{\text{virgin}} = (C_j C^{g-1} S_g)^{1/g} \tag{5.4}$$

Now consider the number of seeds produced per generation by this monocarp, Z. In the animal literature R_0 is frequently used for notation and alternative names for this quantity are expected lifetime reproductive success, basic reproductive rate, basic reproductive ratio, the population growth rate per generation or mean lifetime offspring production. For a monocarp that reproduces once at age g in a virgin environment without any DD, Z_{virgin} is the number of seeds produced at that age multiplied with the probability that the plant survives to reproduce. For instance, if a plant always produces 100 seeds when it survives until flowering and has 95% chance of dying beforehand, $Z_{\text{virgin}} = 0.05 \times 100 = 5$. When we start with one seed of a monocarp that flowers after g years in the virgin environment, the expected number of seeds produced per generation equals:

$$Z_{\text{virgin}} = C_{\text{j}} C^{g-1} S_g \tag{5.5}$$

The parameter Z_{virgin} depends only on generation time, survival and seed production; the parameter does not depend on the frequency of other types in the population. Comparing Eq. (5.4) and Eq. (5.5) shows that $\lambda_{\text{virgin}} = Z_{\text{virgin}}{}^{1/g}$; λ is the population growth per year, Z is the population growth per generation, both measured in a population without any DD. Population growth λ_{virgin} depends much more strongly on generation time. An annual producing 10 seeds that all survive to reproduce and a biennial producing 10 seeds would have the same value for Z_{virgin} but the population growth of the annual would be a factor of 10 higher than that for the biennial. To attain the same population growth, the biennial would either have to become annual and flower in the first year or increase its seed production from 10 to 100. In growing populations there is strong selection for early flowering, only in stable populations are the two fitness measures equal: $\lambda_{\text{virgin}} = Z_{\text{virgin}} = 1$.

5.2.2 Fitness

Invasion fitness is the rate of increase of a type (or an allele) in a stable environment set by another common type (or allele). When all plants employ the ESS with respect to generation time, no mutant with a different generation time can invade into the population. Let us first assume that all DD effects occur in a single stage in the life cycle, for instance during seedling recruitment, and can be described by a term E. This term E is set by the strategy of the common type, from the point of view of the mutant E is a constant, as long as the mutant is rare. Absolute fitness of a mutant with generation time g_{m} in a population in which all plants have a slightly different generation time g is:

$$W_{\text{m}} = \left(E C_{\text{j}} C^{g_{\text{m}}-1} S_{g_{\text{m}}}\right)^{1/g_{\text{m}}} \tag{5.6}$$

We know that the population of the common type is stable so that its fitness W equals 1. Invasion fitness of the mutant is then $w_m = W_m/W = W_m$. Note that the term $C_j C^{g_m-1} S_{g_m}$ equals the expected seed production (Z_{virgin}) of type with generation time g_m growing in a virgin environment (Eq. 5.5). We can now write for invasion fitness:

$$w_m = (EZ_{virgin})^{1/g_m} \tag{5.7}$$

We calculate the candidate ESS by using Parker's method of differentiating invasion fitness with respect to the evolutionary character of the mutant (g_m in this case):

$$\frac{dw_m}{dg_m} = (EZ_{virgin})^{1/g_m}\left[-\frac{\ln(EZ_{virgin})}{g_m^2} + \frac{dZ_{virgin}/dg_m}{g_m Z_{virgin}}\right] = 0 \tag{5.8}$$

This is a complex equation, which you may want to check using the program Maple (Appendix 1). The next step in Parker's method (Box 4.1) is setting $g_m = g$. We can now simplify the equation because E in the resident population is known. We know that the population of the common type is constant so that $W = (EC_j C^{g-1} S_g)^{1/g} = 1$, so that $E = 1/(C_j C^{g-1} S_g)$. Using Eq. (5.5) we have $E = 1/Z_{virgin}$. Substitute this in Eq. (5.8) and we obtain:

$$\frac{dw_m}{dg_m} = \frac{dZ_{virgin}}{dg_m} = 0 \tag{5.9}$$

Differentiating invasion fitness of the mutant to calculate the ESS is equivalent to finding the optimum of the mean lifetime offspring production of a type growing on its own, without any DD, in a virgin environment. Under the assumption that DD works during a restricted period in the same way for all types (as in Eq. 5.6) mean lifetime offspring production in a model without DD is the appropriate fitness criterion; it correctly predicts the ESS. Under this assumption about DD, we can now solve the example we started out with: the annual had $Z_{virgin} = 30$ seeds and the biennial had $Z_{virgin} = 0.5 \times 65 = 32.5$ seeds. The biennial wins.

Isn't this a very complex way of saying something completely trivial and doesn't the biennial always win? No. Imagine DD affects rosette survival in each year. Survival is then lowered from C in the virgin environment to EC in the competitive environment. We assume that the first-year survival C_j is also reduced by a factor E to EC_j with competitors present, but this assumption is not essential. Invasion fitness of a rare mutant with generation time g is now:

$$w_m = (E^{g_m} C_j C^{g_m-1} S_{g_m})^{1/g_m} \tag{5.10}$$

which equals (using Eq. 5.5): $w_\mathrm{m} = E(Z_\mathrm{virgin})^{1/g_\mathrm{m}} = E\lambda_\mathrm{virgin}$.

$$\frac{\mathrm{d}w_\mathrm{m}}{\mathrm{d}g_\mathrm{m}} = E\frac{\mathrm{d}\lambda_\mathrm{virgin}}{\mathrm{d}g_\mathrm{m}} = 0 \tag{5.11}$$

Therefore under these new assumptions about DD, the type that wins in competition is the type with the highest λ in a virgin environment. This leads to much shorter generation times as when plants maximize Z. If DD affects rosette growth or mortality throughout life, the annual easily outcompetes the biennial in our example. The fitness criterion λ for the annual (30 seeds) is 30, the criterion for the biennial (65 seeds, 50% mortality in the second year) is $\sqrt{32.5} = 5.7$. Now the annual wins.

Mylius and Diekmann (1995) pointed out that there is no special reason for these two fitness criteria and emphasized that for more complex situations it will be impossible to write down a fitness criterion, i.e. it will not be possible to predict the ESS from measurements on each type growing alone in a virgin environment without DD. The point to remember is that invasion fitness is a general fitness concept (Metz *et al.* 1992); it helps us to predict the winner when two types (or two alleles) compete. In special situations there exists a simple fitness criterion: if you know how types compare for this criterion when growing on their own, you can predict the winner of the evolutionary contest.

Before Mylius and Diekmann (1995) and Metz *et al.* (1996) came up with their clear explanation when λ or Z measured in a virgin environment was the appropriate fitness criterion, some models had already anticipated their result. For instance, Klinkhamer and de Jong (1989) used the safe site model from Skellam (1951). In this model safe sites occur at random places in an environment, which is otherwise unsuitable for establishment. A safe site could be a molehill or a gap in the vegetation after the death of an established plant. Seedlings within the safe site compete for space, with a single winner per site. A population of a single genotype will either go extinct or reach an equilibrium at which a fraction of the safe sites, not necessarily 100%, is occupied. In this example DD occurs only during recruitment and Z_virgin is the appropriate fitness criterion. Indeed the simulations showed that finally the biennial (65 seeds, 50% mortality in year 2, $Z = 32.5$) outcompeted the annual (30 seeds, $Z = 30$) (Fig. 5.1). Note however that DD occurred only in 20% of the safe sites (Fig. 5.1) and exclusion of the annual was a slow process; after 20 years the annual still outnumbered the biennial. If a bulldozer were to come along every 20 years, after which new populations were established, then the annual would win. We will return to this issue when we discuss metapopulations in Chapter 17.

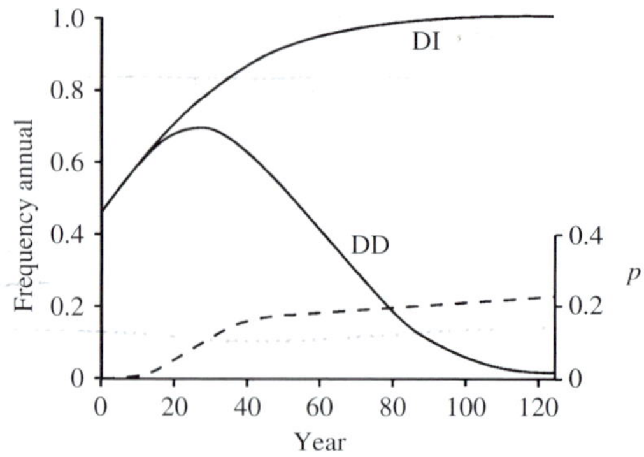

Fig. 5.1 Relative frequency of an annual type growing together with a biennial under density-independent (DI) and density-dependent (DD) selection. DD occurs only in the phase of seedling establishment. Seed productions: $S_a = 30$, $S_b = 65$. Rosette survival in the second year is 0.5. Both types started at low density. The proportion of individuals suffering from DD (more than one seed landing in a safe site) is denoted by p on the axis on the right-hand side (broken line). Redrawn with permission from Klinkhamer *et al.* (1989), *Acta Botanica Neerlandica*, Blackwell, Royal Botanical Society of the Netherlands.

Several authors (for instance, Rees and Rose 2002) have embraced Z_{virgin} as the most appropriate fitness criterion. These authors pointed out that DD occurred mostly in the seedling stage, after which established plants did not interfere with each other any more. While the distinction of λ vs. Z is critical in the theoretical models we discussed, it is not important when we analyse field data. Demographic data collected in a stable population in the field include DD that occurs in the population as it is. In a model based on field data we can introduce a rare mutant with a different generation time and examine whether the mutant increases in frequency *at the level of DD set by the common type*. Because over a sufficiently long time, the common type has both $\lambda = 1$ and $Z = 1$, it is of no consequence whether we take the mutant's λ or Z as a fitness criterion. In such models one tests the hypothesis that the population is at the ESS. If all mutants do worse than the common type the hypothesis is confirmed. If one of the mutants does better than the common type, the hypothesis is rejected. In the latter case it is, in general, not possible to predict the ESS from the demographic data. If a mutant takes over this may greatly affect the DD in the population. However, if DD effects take the form of a multiplier, as in Eq. (5.6), there is no problem and the generation time that produces the highest expected number of offspring Z in a non-ESS population is also the ESS.

With these concepts clearly in mind we now examine early life-history models and field studies of monocarps.

5.3 Maximizing λ: the large advantages of short generation time in growing populations

Early studies of age at first reproduction have focused on population growth λ in a virgin environment as a measure of fitness. As we argued in the previous section this corresponds to certain assumptions about DD. The method for comparing phenotypes for λ was pioneered by Cole (1954) and developed by others in the 1970s (for instance, Charnov and Schaffer 1973). Following these early life-history models, Hart (1977) wrote an article 'Why are biennials so few?' Of all species listed in the *Flora of North America* 21.3% are annual, 77.3% are perennial and only 1.4% are biennial. We drop the subscript for the virgin environment. The equations for the rates of increase are then:

$$\text{Annual:} \qquad \lambda_{\mathrm{a}} = N_{t+1}/N_t = C_{\mathrm{j}} S_{\mathrm{a}} \tag{5.12a}$$

$$\text{Biennial:} \qquad \lambda_{\mathrm{b}} = (C_{\mathrm{j}} C S_{\mathrm{b}})^{1/2} \tag{5.12b}$$

$$\text{Perennial:} \qquad \lambda_{\mathrm{p}} = C_{\mathrm{j}} S_{\mathrm{p}} + C \tag{5.12c}$$

$$\text{Monocarpic perennial:} \qquad \lambda_g = (C_{\mathrm{j}} C^{g-1} S_g)^{1/g} \tag{5.12d}$$

The extra factor C for the perennial in Eq. 5.12c indicates the survival of plants to the next year after reproduction. Because $Z = \lambda^g$ genotypes can also easily be compared for their value for Z (de Jong and Klinkhamer 1987), but we will not elaborate on this. Combining Eqs. (5.12a) and (5.12c) shows that the annual always has to produce more seeds to outcompete the perennial. The disadvantage is smallest if the perennials have a high probability of dying anyway after reproduction, which could happen under very stressful conditions such as in a desert or under conditions of regular disturbance. The perennial strategy is vastly superior if adult survival C is high. One would expect that with high adult survival there is more competition, reducing the conditions for juvenile survival. Annuals are then expected in environments with high juvenile and low adult survival and perennials in environments with low juvenile and high adult survival.

Biennials increase faster than perennials when $\lambda_{\mathrm{b}} > \lambda_{\mathrm{p}}$, i.e. when $S_{\mathrm{b}} > C/C_{\mathrm{j}} + 2S_{\mathrm{p}} + (C_{\mathrm{j}}/C)(S_{\mathrm{p}})^2$. Apparently the ratio of juvenile to adult survival (C_{j}/C) is the critical parameter here. By differentiating the right-hand side with respect to C_{j}/C we find that for a given S_{p} the right-hand side of the above inequality is minimized at $C_{\mathrm{j}}/C = 1/S_{\mathrm{p}}$. At this ratio of C_{j} and C the biennial stands the best chance against the perennial. At $C_{\mathrm{j}}/C = 1/S_{\mathrm{p}}$ the term $C/C_j + 2S_p + (C_j/C)(S_p)^2$ equals $S_p + 2S_p + S_p = 4S_p$. Even at the most profitable ratio of juvenile and adult survival, the biennial needs to produce four times as many seeds as the perennial! At other ratios it needs to produce even more than four times as many seeds to keep up with the perennial. Biennials increase faster than annuals if $\lambda_b > \lambda_a$. If $S_b > 4S_p$ and $C_j/C = 1/S_p$, then $S_b > (S_a)^2/(S_b/4)$ or $S_b > 2S_a$. At this specific ratio of juvenile and rosette survival the biennial needs to produce twice as many seeds as the annual. Hart (1977) went on to argue that only in a limited range of C_j and C do biennials come close to competing with the other life-history strategies. While annuals are typical of stressful open or disturbed habitats and perennials dominate closed vegetations, biennials are found in semi-open and intermittent habitats.

The so-called demographic approach, of which Hart (1977) is a prime example, was applied by Young (1990) to the lobelias growing on Mount Kenya. *Lobelia keniensis* is perennial and *Lobelia telekii* is monocarpic with 8–16 years between reproductive events. With high survival rates of rosettes (above 0.98 per year), Young (1990) calculated that the monocarp needed to have an unrealistically high seed production to keep up with the perennial. With lower rosette survival the monocarp could win if it produced three to four times as many seeds. Such differences in seed production were observed between the two species. Young (1990) argued that differences in rosette survival between the habitats of *L. keniensis* and *L. telekii* are critical for which type wins.

Early studies on demography of biennials by Werner and Caswell (1977), van der Meijden and van der Waals-Kooi (1979) and Gross (1981) showed that the plants rarely flowered in the second year, but delayed flowering to the third, fourth or fifth year or rather to the year when a certain minimal size for flowering was reached. The plants were only biennial under fertile conditions as in gardens and 'monocarpic perennial' or 'facultative biennial' is a more accurate description of this species group than biennial. With population growth in a virgin environment as a fitness criterion, these life histories are even more difficult to explain than the biennial habit.

5.4 Calculating ESS generation time from maximizing expected lifetime reproductive success

The monocarpic perennial species may be few in numbers, but some representatives are not rare and can even be noxious weeds. It was thus a puzzle how the monocarpic life history can be explained and the 'Letters to the Editor' column of the *American Naturalist* has received a number of responses on Hart's theme why biennials are few and sometimes not so few. If DD acts during a restricted period, rather than throughout life, Z is a more appropriate fitness criterion than λ. The fitness criterion of a common monocarpic type with generation time g can be written as $Z_g = C_j C^{g-1} S_g$. A type with generation time $g+1$ has $Z_{g+1} = C_j C^g S_{g+1}$. The type with generation time g wins if $Z_{g+1} < Z_g$. Consequently, further delay of flowering becomes unfavourable when $C_j C^g S_{g+1} < C_j C^{g-1} S_g$ or (de Jong and Klinkhamer 1987):

$$\frac{S_{g+1}}{S_g} < \frac{1}{C} \tag{5.13}$$

Plants should delay flowering until the increase in seed production no longer matches the chance of dying in the extra year of vegetative growth (see also Kelly 1985). Metcalf *et al.* (2003) compared monocarps to rational gamblers making a bet. One option is to collect \$10 today, the other option is to wait until tomorrow, when the amount could be increased to \$20, or could be reduced to nothing. What you should do depends on the odds of getting \$20 and losing everything. If the expected gain tomorrow, which is \$20 multiplied by the probability that you can collect your bet, is larger than \$10, you should wait.

5.5 Plant size affects survival, growth and flowering

So far we have considered optimal generation time for plants as a problem of flowering at the right age. While this was appropriate for the Chinese bamboo flowering once every 120 years, this is not how it works in most short-lived monocarps. Under fertile garden conditions a short-lived monocarp usually flowers in the second year, because it requires winter cold for flowering. However, under less fertile field conditions flowering is typically delayed for several years until the plant has reached a certain size. Also the probabilities of dying as a rosette and the relative growth of the rosette typically depend on the plant's size rather than age (Fig. 5.2).

The physiology of flowering is also relevant for the life-history models. While actual flowering occurs at a certain time of year, during a period earlier in life the plant makes a 'choice' to induce flowering or to remain vegetative. For instance, tulips flower in spring but the decision to flower, which involves the apical meristem switching from producing leaves to producing a flower, is made some time in advance. One can observe the primordial flower in large tulip bulbs in autumn. The plants are following a certain growth trajectory and at some census moment the question is to flower or to delay flowering with at least one year.

So far we have assumed rosette survival C is constant, but since old/large rosettes usually have higher survival, it is more reasonable to write survival as a

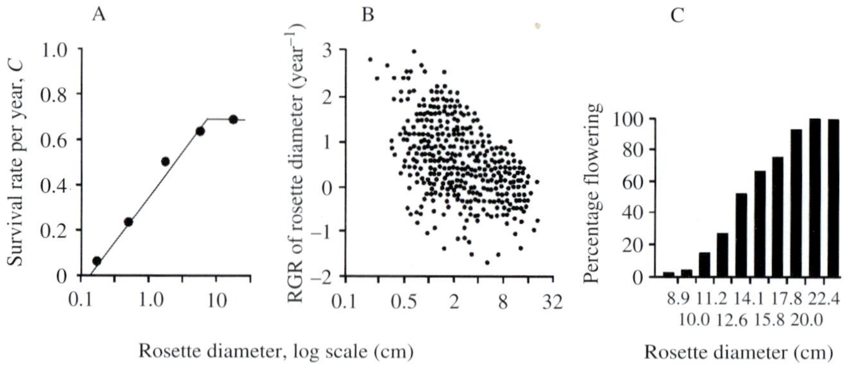

Fig. 5.2 Typical life-history data for a monocarpic perennial, *Oenothera glazioviana* (Kachi and Hirose 1985). (A) Rosette survival C increases with plant size until a level of 0.7 per year. (B) Relative growth rate (RGR) decreases with rosette size. (C) Flowering probability is a sigmoid function of plant size. Redrawn with permission from Kachi and Hirose (1985), *Journal of Ecology*, Blackwell.

function of age or size. The survival rate C_1 refers to the survival of a one-year-old rosette until the next year (Fig. 5.3). Further delay of flowering is unfavourable when if $Z_{g+1} < Z_g$ or $C_j C_1 C_2 \ldots C_{g-1} C_g S_{g+1} < C_j C_1 C_2 \ldots C_{g-1} S_g$. In the end we are left with the same condition as Eq. (5.13) for delay of flowering to be favourable: $S_{g+1}/S_g < 1/C_g$. Plants should look one year ahead to decide to flower (Rees *et al.* 2000). It is convenient to apply this approach to monocarps. A growth trajectory is sketched in Fig. 5.4. In year 1 the growth expectation exceeds $1/C$ which is $1/0.5 = 2$ in the example of Fig. 5.4. Only in the fourth year does the growth expectation drop below 2 and it becomes favourable to flower. If rosette survival C becomes larger than 0.5, so that 1/C decreases, the plant should flower in year 5 and with a further increase in survival in year 6. We denoted g as the age of the plant and survival was considered a (constant) function of age. Several studies, starting with Werner and Caswell (1977), have shown that probabilities of flowering and surviving and the growth of the plant are determined by the size or weight of the plant (Fig. 5.2) rather than its age. It is therefore preferable to draw the curves in Fig. 5.4 with respect to weight. This also makes data collection easier: we only need to estimate rosette weight by an indirect method (for instance, by measuring rosette diameter) and determine the fate of the plant over a one-year period, instead of following individual plants of different ages through time.

Once we know approximately the time of year in which plants initiate flowers, we can collect data on size-dependent growth and survival curves of rosettes, and plot these (Fig. 5.4) to calculate the optimal size for flowering.

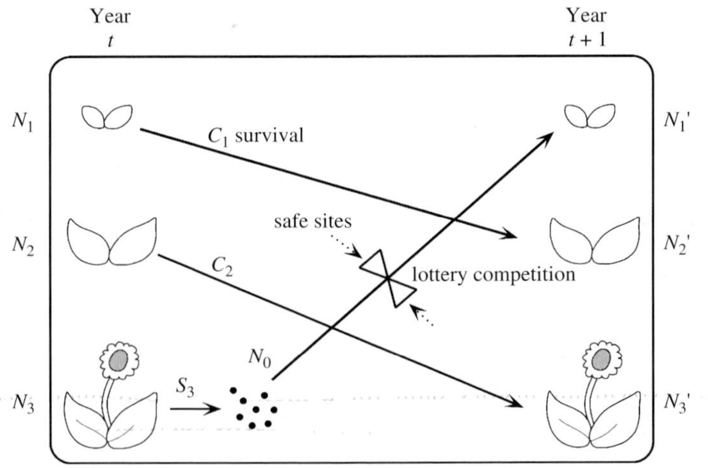

Fig. 5.3 Diagram showing various transitions for a triennial. N_i is the number of plants at age i, C_i is survival of i-year-old plants to the next year, S_3 is the seed production of a three-year-old plant when it survives until reproduction. All competition is assumed to occur in the stage between N_0 (seeds) to N_1 (established rosettes).

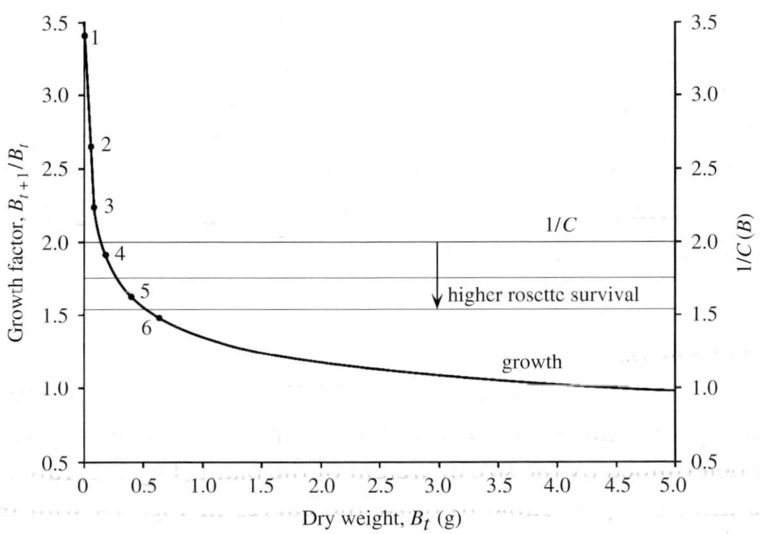

Fig. 5.4 Relative increase in weight (B_{t+1}/B_t) from age t to t+1 as a function of rosette dry weight at time $t(B_t)$. $C(B)$ is rosette survival until the next year, as a function of dry weight. The dots denote the growth path of vegetative plants in six subsequent years. The graphical method, also know as the one-year-look-ahead approach, predicts the optimal size for reproduction. As long as the relative increase in weight, corresponding to seed production, is higher than $1/C$ then plants should delay flowering. Yearly survival C was chosen constant, independent of plant weight or age.

Independently from those data collected by measuring vegetative plants we could examine the sizes (or ages) at which the plants actually flower. This allows testing of the hypothesis that the population is at the ESS with respect to the size for flowering.

5.6 Predictions of threshold size for flowering in one habitat

One of the first tests of the one-year-look-ahead approach (Fig. 5.4) was by Kachi and Hirose (1985), who examined *Oenothera glazioviana* (large-flowered evening primrose) on sand dunes in Japan (Fig. 5.2). Maximizing Z_{virgin} yielded the prediction that plants should flower at 7.2 cm rosette diameter. In the field the probability of flowering increased gradually with rosette diameter, with 50% flowering at 14 cm diameter. This size at which 50% of the plants flowers, is usually referred to as the threshold size (TS) for flowering. Rees and Rose (2002) reanalysed the Japanese data using variation in growth between plants within the same year; Fig. 5.2B shows that this is more realistic. The variation around the average growth curve pushed optimal TS towards higher values; Rees and Rose (2002) predicted that *Oenothera* plants should flower at 16–17 cm, in good agreement with the data.

The models of maximizing Z have provided fair predictions in a number of other species. Other examples include *Cirsium vulgare* and *Cynoglossum officinale* on sand dunes in the Netherlands (de Jong *et al.* 1989), *Onopordum illyricum* (cotton thistle) in France (Rees *et al.* 1999) and *Carlina vulgaris* (carline thistle) in English chalk grasslands (Rose *et al.* 2002). The conclusion of these studies was that the field population was at the ESS or close to it. The models of Mark Rees and co-workers have evolved from the one-year-look-ahead approach (Fig. 5.4). State-of-the-art models now include temporal variations, with growth and survival varying from year to year, and integral projection, rather than discrete transitions between different size classes (Rees and Rose 2002; Metcalf *et al.* 2003). While flowering sizes of *Carlina vulgaris* in the UK could be predicted with these models, the same species studied on the dunes of the Netherlands, where it grows in a rather specific habitat at the edge of scrubs of *Salix repens* (creeping willow), seems to defy the theory (de Jong *et al.* 2000). Plants grew exponentially, multiplying their weight annually by a factor 2.32. Rosette survival was 0.7 per year for both small and large rosettes. In three years during which plants were monitored growth was always higher than $1/0.7 = 1.42$ so that the model of maximizing Z predicts that plants should postpone flowering indefinitely. Instead, 50% of the *Carlina vulgaris* plants had already flowered at a weight of 0.8 g (de Jong *et al.* 2000). Rosettes were quite rare and widely spaced, so the assumption that all DD occurred in the first year and plants should maximize Z seems reasonable. Jong *et al.* (2000) found that locally many *Carlina* populations go extinct and suggested that selection should be considered in the context of a metapopulation, which would select for earlier flowering (see Chapter 17).

5.7 Predictions for different habitats

The predictions for a single habitat may differ, depending on the year in which data were collected. Comparison between habitats that differ consistently with respect to the demographic parameters is therefore a more robust way of testing the model. This comparison has been made for a number of species.

Cynoglossum officinale was studied by Renate Wesselingh in both the dune area of Meijendel in the Netherlands and the dunes at Holkham in the UK. Growth at Meijendel was lower and levelled off more quickly with plant size than at Holkham (Wesselingh *et al.* 1997). In addition, at Meijendel plants suffered from larvae of the beetle *Mogulonus cruciger* which fed in the top of the root. Survival was higher at Holkham where this beetle did not occur. The model from Fig. 5.4 then predicts selection for a larger size at flowering at Holkham. This was observed in the field: 50% of the plants flowered at a dry weight of 2.33 g at Meijendel and 5.13 g at Holkham. Moreover, this difference was maintained when seeds of the two populations were transferred to uniform conditions in a growth room and grown for different periods (Fig. 5.5). This shows that genetic differences in size for flowering exist between Meijendel and Holkham and that these differences are in the direction in which we expected natural selection to work in the field.

Lacey (1988) carried out reciprocal transplant experiments with *Daucus carota* (wild carrot) in Northern America. She compared populations in Greensboro, North Carolina (36° N), Ann Arbor, Michigan (42° N) and Ottawa, Canada (45° N) at different sowing densities. Rosette survival was typically highest in the northern populations; at the lowest density 50%, 85% and 95% survival (from south to north) are typical values. No difference in plant weight existed between the three locations, so the three habitats seemed equally suitable for growth. In the south the growing season may be longer but some dry periods may also be too extreme for growth. Lacey found that the plants from the southernmost populations produced the highest fraction of annuals at the three different planting sites (Fig. 5.6). Generation time is thus shortest for the genotype from Greensboro, the site with the highest rosette mortality. This is consistent with the model predictions. The carrot was introduced into North America in the seventeenth century so that the latitudinal cline in generation time must have evolved in the last 300 to 400 years (Lacey 1988).

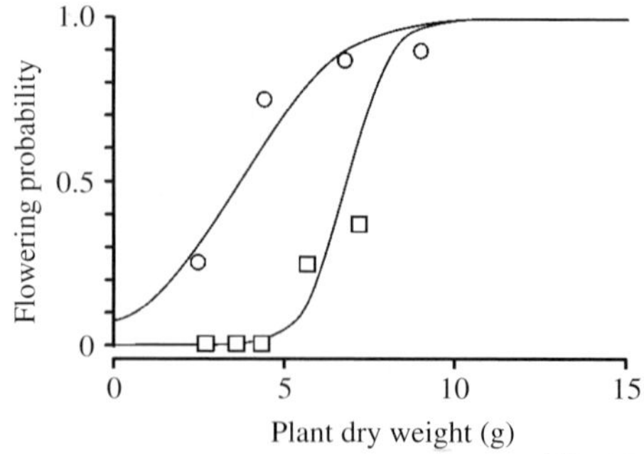

Fig. 5.5 Comparison of size-dependent flowering under laboratory conditions for *Cynoglossum officinale* plants, grown from seed collected at Meijendel, the Netherlands (circles) and Holkham, UK (squares). The points indicate the mean dry weight and the fraction of flowering of subgroups of eight plants. Redrawn with permission from Wesselingh *et al.* (1993), *Acta Botanica Neerlandica*, Blackwell, Royal Botanical Society of the Netherlands.

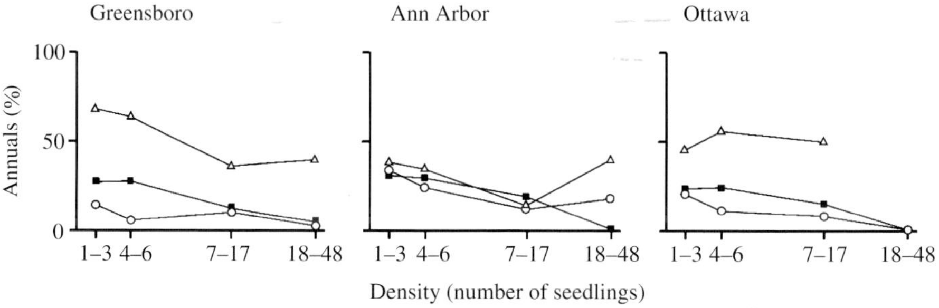

Fig. 5.6 Percentage annuals in *Daucus carota* at different planting densities in a reciprocal transplant experiment. Plants from seeds collected in Greensboro (open triangles), Ann Arbor (solid squares) and Ottawa (open circles) (arranged from south to north) were planted in each others' habitat. Redrawn with permission from Lacey (1988), *Ecology*.

Reinartz (1984) transferred plants of *Verbascum thapsus* from different areas across North America (southern Texas/Georgia, North Carolina and Canada) to a common garden in Durham, North Carolina. Many northern genotypes delayed flowering until the third year and annuals were only present among the southern genotypes. Based on climatic data Reinartz argued that in the south rosette survival was more uncertain and this favoured shorter generation times.

Wild beets (*Beta vulgaris* ssp. *maritima*) are short-lived perennial plants that have been studied along the Atlantic and Mediterranean coasts of Europe. Wild beets are not monocarpic but because they have many similarities to biennials, like the rosette stage and the requirement of cold for flowering, we

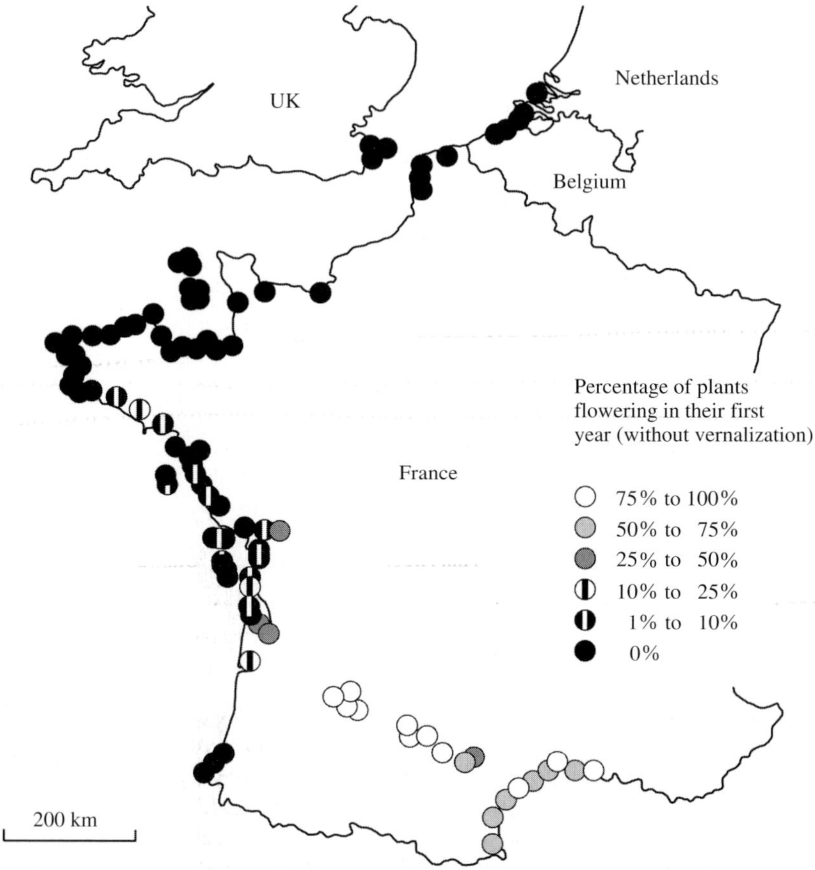

Fig. 5.7 Percentage flowering in the first year of *Beta vulgaris* ssp. *maritima* plants grown from seeds collected at different places in France, the UK, Belgium and the Netherlands, in a greenhouse at the University of Lille. Redrawn with permission from van Dijk *et al.* (1997), *Acta Oecologica*, Elsevier.

include them here. Henk van Dijk collected seeds of wild beets all over western Europe and grew seedlings under nutrient-rich conditions at the University of Lille; it was then recorded what fraction of the plants flowered in the first year (van Dijk *et al.* 1997). None of the plants from the northern population flowered in year 1, while a considerable fraction of the southern plants did so (Fig. 5.7). It is not simply a matter of south versus north: for instance, none of the seeds collected from plants on the Atlantic coast near the Bay of Biscay flowered in the first year, whereas many seeds taken from inland plants or from the Côte d'Azur did. The plants that do not flower in the first year need cold for flowering (vernalization). In beets it is known that a single gene is responsible for the vernalization requirement. Boudry *et al.* (2002) argued that populations in the south of France are more disturbed, so that annuals are the only plants that produce seeds in this habitat. In the north survival is higher and annuals are outcompeted by longer-lived plants that produce more seeds over their extended life (see also Hautekeete *et al.* 2002).

5.8 Genetic variation in generation time

The examples above suggest variation in size at flowering upon which selection has acted. We have already noticed that the probability of flowering increases gradually with plant size (Figs. 5.2, 5.5) (Werner and Caswell 1977; van der Meijden and van der Waals-Kooi 1979; Gross 1981). In *Cynoglossum officinale* a steep increase of percentage flowering with plant size has only been found in seed samples obtained from botanical gardens, in which the genetic variation is presumably small (Wesselingh *et al.* 1993). Is the gradual increase in flowering with size due to genetic or environmental variation? We noted that plants do not flower as soon as they reach a certain size, but rather that plants larger than a certain minimal size are induced to flower at a certain time of the year. These induced plants may have just reached the minimum size at the moment of flower induction or already passed this size some time ago and are much larger. To take this into account Wesselingh *et al.* (1993) referred to threshold size (TS) of plants as the minimum required size for flower induction. If we take a point x at the X-axis in Fig. 5.5 then all plants that had a TS smaller then x flowered. If we then move slightly to the right from x to $x + \Delta x$, the percentage flowering rises because the plants with TS between x and $x + \Delta x$ are also included in the sample. Consequently, we can obtain the frequency distribution of TS in the population by differentiating the sigmoid curve with respect to size x. If the frequency distribution is symmetric, the peak of the distribution of TS coincides with the size at which half the plants in the population flower. Next Wesselingh and de Jong (1995) selected plants that flowered at small sizes and crossed these among each other (Fig. 5.8). They also selected plants that had not flowered at a large size. These plants were grown to an even larger size and after another cold period they flowered and could be crossed. They compared the frequency distributions of flowering sizes before and after selection. While average TS was 3.30 g in the original seed sample, after selection for flowering at small or large size, TS had shifted to 1.85 g and 5.41 g, respectively. It is apparently easy to select for a different TS for flowering, even within a single generation. Wesselingh and de Jong (1995) calculated the heritability of this character in the greenhouse to be 0.32–0.35. In a similar selection experiment carried out with *Lobelia inflata* (indian tobacco), Simons and Johnston (2000b) found a heritability of 0.42 for age at flowering. With such large

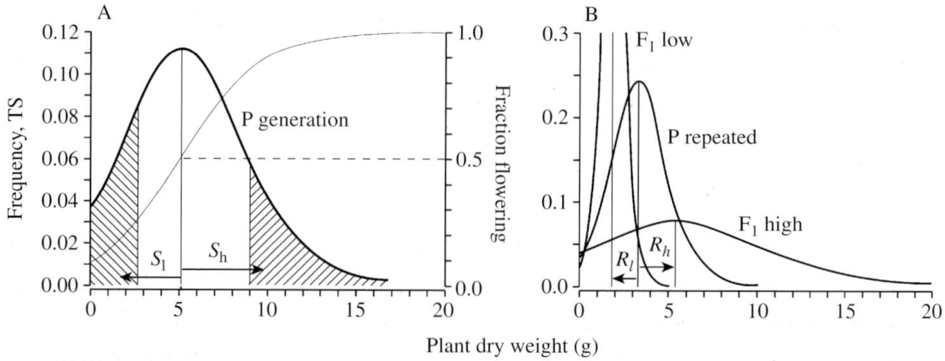

Fig. 5.8 Plants of *Cynoglossum officinale* were selected for size at flowering. (A) The thin line indicates the flowering probability versus plant size. The frequency distribution of threshold size for flowering in the parental (P) generation (left axis) can be found by differentiating with respect to size. The hatched areas under the curve represent the plants selected for flowering at low or high sizes. S_l and S_h denote the selection differentials for low and high TS, and the corresponding responses to selection are denoted R_l and R_h. (B) The response to selection R shows how much the mean of the frequency distribution of threshold size for flowering has shifted compared to a repeated P generation. The average threshold size for flowering (TS) is at the peak of the frequency distribution and is the size at which 50% of the plants in the sample flowered. Redrawn with permission from Wesselingh and de Jong (1995), *Heredity*, Nature Publishing Group, http://www.nature.com.

genetic variation for size at flowering, the average value of this character could easily change due to natural selection. What maintains this genetic variability is currently not known.

5.9 *r* versus *K* selection

The concept of r–K selection was introduced in MacArthur and Wilson's (1967) famous book *The Theory of Island Biogeography*. The r and K refer to the logistic equation of population growth:

$$\frac{dN}{dt} = rN\left(\frac{K-N}{K}\right) \tag{5.14}$$

When $K = N$, population size N no longer increases; K is known as the carrying capacity of the environment. When N is low $(K - N)/K$ equals 1 so that the population grows exponentially with a relative growth rate of r. The parameter r is referred to as the intrinsic rate of increase or the Malthusian parameter after the mathematician T. R. Malthus who wrote in 1798 a still-relevant essay about the exponentially growing human population. At low population density $dN/dt = rN$, so that $N_t = N_0e^{rt}$ and $N_{t+1}/N_t = e^r$. The term e^r thus corresponds to our familiar λ from Eq. (5.2) and $r = \ln\lambda$.

MacArthur and Wilson (1967) argued that when a species has reached an island, it first colonizes the island and then reaches a maximum density K due to intraspecific competition. They imagined that during the colonization phase other life-history characters were selected than when the species was at the carrying capacity. During colonization intraspecific DD will be low and the population is growing. Population growth certainly selects for high r (or λ) in a virgin environment, and genotypes with reproduction early in life and early investment into reproduction are favoured by selection (Eq. 5.12). After a stable population has formed DD will be high (Antonovics and Levin 1980) and the population does not grow any more. MacArthur and Wilson (1967) initially suggested that this would favour genotypes that convert resources more efficiently. In fact characters that allow plants to do better than their neighbours are favoured. Such characters could also include interference strategies such as growing tall or producing toxic litter and do not necessarily involve more efficient resource use. Nevertheless, MacArthur and Wilson's assertion that other life-history characters are selected during colonization than in a stable population is likely to be true: genotypes with a high $r(\lambda)$ are selected in a growing population without DD and genotypes with a high Z take over under DD selection (if DD works during a restricted time period; see also Fig. 5.1). In the first case there is strong selection for reproducing early, in the second case generation time is less important for fitness. It would be interesting to test this idea for plant species that have recently expanded their geographic range after being dispersed to other continents. Genotypes at the front of the wave should be under r selection, shortening generation times as compared to the original population (see also Brown and Eckert 2005). Selection on size for flowering may be complex in such cases because colonizers on new continents may also have escaped from their specialist herbivores (Müller-Schärer *et al.* 2004). When this results in higher survival and better growth, selection favours later reproduction, an effect that counteracts r selection.

As Dawkins (1982) noted: 'ecologists enjoy a curious love/hate relationship with the r–K concept, often pretending to disapprove of it while finding it indispensable'. Certainly r–K selection was immensely popular in the 1970s. Its fame faded after it became clear that the concept was rather broad and gave rise to a wealth of interpretations. For instance, in Stearns's (1992) book r–K selection is no more than a footnote. Reviews by Boyce (1984) and Reznick *et al.* (2002) cautioned to keep in mind what exactly the theory predicts and how this matches the observations. For instance, Gadgil and Solbrig (1972) compared reproductive effort, size and survivorship of three dandelion populations. The population they considered as *r*-selected was frequently trampled and mowed and thus thought to be experiencing high, DI mortality. The so-called *K*-selected population was in a less disturbed site with high competition from other species. Gadgil and Solbrig (1972) found the highest reproductive effort in the disturbed population, and concluded that this was consistent with *r–K* selection. However, the dandelion density was highest on the disturbed site, so perhaps in this situation intraspecific competition was highest, leading to more *K* selection according to the brief summary of the theory above. It is unclear whether Gadgil and Solbrig's observation, interesting as it is, supports the theoretical expectation based on DI versus DD selection.

Grime (1977) divided plant strategies into three major categories. Plants are adapted to stressful (*S*), ruderal (*R*) or competitive (*C*) conditions. Grime suggested that vegetation succession is similar to island colonization. At the start of succession ruderals are present, later competitors come in, while towards the end of succession plants dominate that can withstand the stress induced by their neighbours. This would mean that ruderals are *r*-strategists, stress-tolerant species are *K*-strategists and competitors are in the middle of the *r–K* continuum. While there is certainly merit to the idea of comparing island colonization to vegetation succession there are problems also. Early-successional species may germinate massively from a seed bank, thus encountering high levels of intraspecific competition, contrary to the situation for the colonizer on the island. The *r–K* and *C–S–R* concepts are useful metaphors as long as they do not cloud the underlying assumptions.

To sum up: we have sketched life-history models that predict optimal size for flowering in monocarpic perennials. There exists considerable genetic variation in generation time within species and high rosette mortality appears to select for smaller sizes for flowering and consequently shorter generation times, both when we compare habitats at the same latitude (*Cynoglossum officinale*) and along a latitudinal cline. This trend is consistent with the model predictions, but there is still a need for more studies comparing genotypes between habitats with clear differences in demographic parameters.

Questions

(1) Why are age-structured models often inappropriate for plants?
(2) What is meant by threshold size for flowering?
(3) What is expected lifetime reproductive success (Z) for a monocarpic plant?
(4) When are short generation times especially advantageous?
(5) What is a fitness criterion and when is Z or λ preferred?

6

Reproductive effort

How much resource should a plant invest in each reproductive event? Reproductive effort (RE) is one of the key elements determining an organism's life history because it not only determines the fitness gains in the current year but also influences the survival probability and the size and growth in the following years. For instance, a high RE is associated with small adult sizes and short lifespans. Usually the reproductive effort is expressed as the ratio of reproductive to vegetative biomass (RE_1) or as the proportion of total biomass invested in reproductive structures (RE_2). Plant species show a wide variation in lifetime reproductive schedules ranging from monocarps that reproduce only once and then die, to perennials that reproduce repeatedly with or without years with unusually high fruit production, mast years (Fig. 6.1). Plant ecologists usually refer to monocarpic and perennial whereas animal ecologists are more familiar with the terms semelparous and iteroparous. In monocarpic plants lifespan varies from short-lived annuals to long-lived monocarpic perennials such as bamboos that reproduce after a juvenile lifespan of up to 120 years and then put all resources into reproduction and die. In perennial plants lifespan varies from short-lived herbs such as *Plantago lanceolata* that live for a few years to trees that grow to reach adult size and may reproduce over periods of hundreds of years.

Many researchers have modelled optimal lifetime reproductive schedules to explain the evolution of different life histories. Particularly the evolution of monocarpic versus perennial life histories has received much interest. Basically there are two types of model: demographic models like the one by Hart (1977) that were discussed in the previous chapter, and reproductive effort models. The approach in demographic models is to compare different fixed life histories and to find demographic parameters that favour one over the other. Mostly these models use assumptions such as constant adult survival, constant RE in the perennial species or a fixed ratio of seed production of the

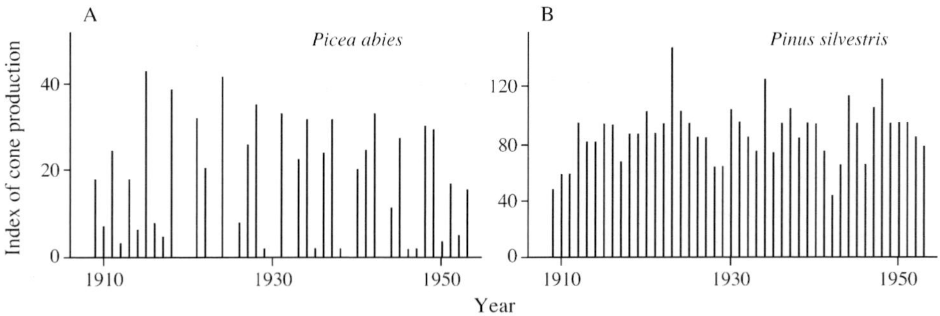

Fig. 6.1 Fluctuations in the cone crop of *Picea abies* (Norway spruce) and *Pinus sylvestris* (Scots pine) in south Sweden. Redrawn from Fig. 22.8 in Harper (1977) with permission from Elsevier.

monocarpic and perennial species. Demographic models may go a long way to explaining why a monocarpic species is found in a different environment than its close perennial relative (see the *Lobelia* example in the previous chapter). These models do not, however, optimize allocation and they miss the link between variation in RE and its effect on survival and growth etc. They cannot, therefore, explain why a monocarpic species diverts all its available resources to reproduction instead of, for instance, 90%, thereby gaining the possibility of surviving and reproducing again. Neither can they predict the lifetime reproductive schedule. To explain monocarpy we need models that focus on optimal reproductive effort taking investment–fitness gain relations and consequences for future growth and survival into account.

6.1 Trade-offs between RE and survival and growth

Before discussing the trade-offs and models, we need to spend a few words on the currency that we use to measure RE. It can be argued that RE should be measured in terms of the particular resource that limits plant growth. Thompson and Stewart (1981) therefore suggested that RE should be measured in terms of nutrients rather than biomass, especially because reproductive structures are capable of photosynthesis themselves. In contrast, Reekie and Bazzaz (1987a) argued that carbon, because of its important structural and energetic role in plants, will tend to reflect the allocation patterns of other resources, in particular the one that is most limiting. Reekie and Bazzaz (1987a) confirmed their prediction using *Agropyron repens* (couch grass) by comparing RE calculated in terms of total carbon with RE measured in terms of nitrogen and phosphorus along gradients of availability of the latter two. The allocation of carbon tended to be biased towards that of the most limiting resource and they concluded that carbon is an adequate measure of phosphorus allocation.

All reproductive effort models are based on the assumption that investment in reproduction reduces growth and survival. Most models on animals focus on the trade-off with survival because in many animals size at maturity is fairly constant. In plants, because of their plastic growth, both trade-offs are important. Many authors have attempted to detect costs of reproduction by comparing, within natural populations, future growth and survival of individuals that differ in current investment in reproduction (Silvertown and Dodd 1999 and references therein). Law (1979), using *Poa annua* (annual bluegrass), was among the first to demonstrate a cost of reproduction in this way. Flowering individuals showed reduced reproduction in the next year compared to non-flowering ones. In woody dioecious species females generally invest more in reproduction than males and often this is associated with reduced growth and/or survival (Antos and Allen 1999; Delph 1999; Nicotra 1999; Turcotte and Houle 2001; Obeso 2002). However, surprisingly often no costs were detected by comparing flowering individuals with different RE (Lord 1998 and references therein). In some cases the level of natural variation in RE may simply be too low to detect any costs (Ehrlén and van Groenendael 2001). Sometimes even a positive correlation between RE and vegetative growth was found (Cruz and Moreno 2001). Such a positive correlation may be explained by the house–car problem from Chapter 3. Any variation in how many resources individuals have will obscure a trade-off (van Noordwijk and de Jong 1986). This problem can partly be avoided by controlling for resource levels in combination with comparing genotypes that differ in allocation pattern (Reekie and Bazzaz 1992). The best way to detect costs of reproduction, however, is probably experimentally by manipulating flowering or fruit set (Antonovics 1980). This can be done for instance through photoperiod (Reekie and Bazzaz 1987b; Reekie 1998), bud or flower removal (Ehrlén and van Groenendael 2001) or setting different levels of pollination. In response to reduced levels of pollination and seed set *Ipomopsis agregata* shifted from the normal monocarpic mode of reproduction to a perennial life history (Paige and Whitham 1987). Primack and Stacey (1998) found that in the perennial orchid *Cypripedium acaule* control plants only rarely made fruits and repeated flowering resulted in no detectable reduction in either plant size or probability of flowering in subsequent years. A cost of fruit production became evident after additional hand-pollinating plants. These hand-pollinated plants showed reduced flowering and smaller leaf area in subsequent years, but mortality was not affected. The effect of experimental treatment that was repeated during successive years reached maximum values at three to seven years, indicating that short-term studies may be too brief to reveal experimental effects (Primack and Stacey 1998). Based on the available experimental evidence we can conclude that often costs in terms of reduced growth, survival or future reproduction are associated with reproduction. However, costs may go unnoticed if studies do not control for resource levels or if reproductive investments are low as in the case of *Cypripedium acaule*. Costs of reproduction in terms of future growth may be more evident in resource-poor than in resource-rich environments (Bell and Koufopanou 1986). A further complication, that is not included specifically in the models discussed below, is that the costs of reproduction depend often on size or age (Silvertown and Dodd 1999).

6.2 The reproductive effort model

To address the problem of optimal RE we will start by using the approach outlined by among others Gadgil and Bossert (1970), Schaffer (1974) and Stearns (1976). We will use expected lifetime seed production (Z) as the criterion to pick the winner of the evolutionary contest. This is justified for a population where density dependence is restricted to a short period (like seedling survival in safe sites; Chapter 5). In other situations a different criterion may be more appropriate. Many authors used population growth rate (λ), the seed production in a particular year is then discounted by λ^{-x}, with x being the age at which the offspring is produced. For expected lifetime seed production all seeds are equally valuable irrespective of the year of production. We will discuss discrete time models and consider a plant that grows and gathers resources during the season. At the end of the season part of the resources can be channelled into reproductive structures, the other part is available for growth next year. For some species this may not be a realistic assumption because they flower and grow simultaneously during a longer period within the year. In such species reproduction may impair an immediate cost through reduced photosynthesis (Saulnier and Reekie 1995). For convenience we will not make a distinction between vegetative parts and storage and we will assume that the plant has an optimal allocation schedule during the season.

6.2.1 Effects of reproductive investment on future survival

Expected lifetime reproductive success is defined as:

$$Z = \sum l_x S_x \tag{6.1}$$

in which l_x gives the survival until age x ($l_x = C_j C_1 C_2 \ldots C_{x-1}$) and S_x denotes the seed production at age x. Although it is often forgotten or implicit in the models, it is important to realize that for hermaphrodite plants S_x is the sum of the number of seeds produced and the number of seeds sired. The latter is usually left out, more for practical than for scientific reasons. Expected lifetime reproductive success, Z, is maximized if at every age x the reproductive value, V_x, is maximized. The reproductive value of an organism of age x is its expected reproduction for the rest of its life, counting from that age. The reproductive value, V_x, equals the sum of seed production in year x and the future reproductive value $C_x V_{x+1}$. If we single out a particular age x we get:

$$V_x = S_x + C_x V_{x+1} \tag{6.2}$$

If the impact of increased reproductive investment on future reproductive value is mainly through smaller survival probabilities, we can use a graphical method to study the impact of the relationship between seed production S_x and survival C_x on life-history evolution. From Eq. (6.2) it can be seen that in the $S_x - C_x$ plane each line with slope $-V_{x+1}$ represents a fitness isocline (Fig. 6.2). The higher the intercept with the S_x axis the higher V_x is. The point where such a line just touches the $S_x - C_x$ curve therefore maximizes V_x. From Fig. 6.2 it can be seen that the shape of the trade-off between

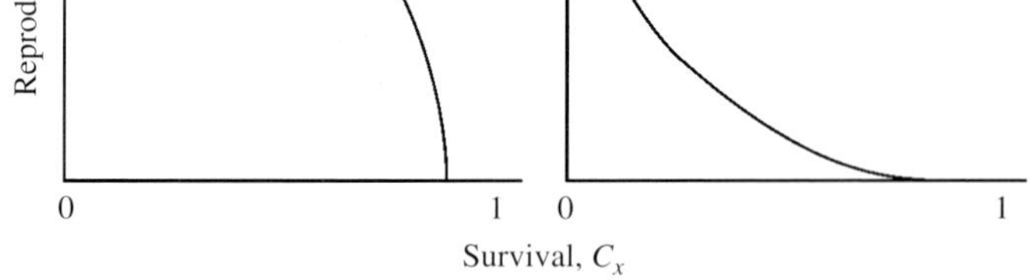

Fig. 6.2 Current year's seed production, S_x, plotted against survival, C_x. Selection for a perennial life history or monocarpy depends on the shape of the reproductive success–survival trade-off. The point where a line with slope $-V_{x+1}$ touches the S_x–C_x curve maximizes V_x and represents optimal S_x and thus optimal investment in reproduction (see text). (A) The convex curve selects for a perennial life history. (B) The concave curve selects for monocarpy.

reproductive success and future survival is important in determining the optimal life history. If the trade-off curve is convex, bowed outward, we expect selection for a perennial life history, if the curve is concave, bowed inward, we expect selection for monocarpy. We started with the assumption that the impact of increased S on future reproductive value is mainly through smaller C. This assumption may be more appropriate for animal than for plant species because in many animal species adult size is often constrained to a certain size while many plant species have the capacity to continue growth after reproduction. For instance in birds it is well known that increased brood size leads to reduced parental survival (Daan *et al.* 1990). There is less evidence for such a trade-off in plants (Lord 1998), although indirectly survival may be reduced because survival increases with plant size (Fig. 6.2) and plants with large investments in reproduction attain smaller sizes.

Direct effects of reproduction on future survival may occur when flowering makes the plant more attractive for herbivores or more vulnerable to diseases. Anther smut fungus on species of the Caryophyllaceae is a well-known disease transmitted by pollinators. The fungus only infects flowering plants and makes the plant sterile so that it is dead in an evolutionary sense. In theory we expect that pollinator-transmitted fungal or virus diseases can lead to concave relationships between investment in reproduction and survival because a pollinator visit depositing a single spore can be fatal to the plant. More detailed studies on this are needed, however. Many butterfly species use apparent flowering stems to detect their hosts (Wiklund 1984). In *Senecio jacobaea* the percentage of plants with egg batches of *Tyria jacobaeae* was twice as high in flowering plants as in non-flowering ones even when corrected for size differences between flowering and non-flowering plants (van der Meijden 1976). Courtney (1985) argued that if size differences among plants are relatively large as compared to the distance at which they can be detected by herbivores larger flowering stems are disproportionately more attacked by herbivores (see also Chapter 3). This disproportionate attack would yield a convex trade-off curve and select for a perennial life history. On the other hand if size differences among plants are small relative to the distance at which they can be detected by herbivores, large reproductive stems may get disproportionately fewer attacks. In *Cynoglossum officinale*, a monocarpic species from calcareous, sandy soils, the probability of being attacked by the root weevil *Ceutorhynchus cruciger* increased from 15% in small flowering plants to 100% in large plants while non-flowering plants were hardly ever attacked. Because plants become infected relatively late during seed set, the effect of the weevil on seed production is relatively small (Prins *et al.* 1992) while its effect on future

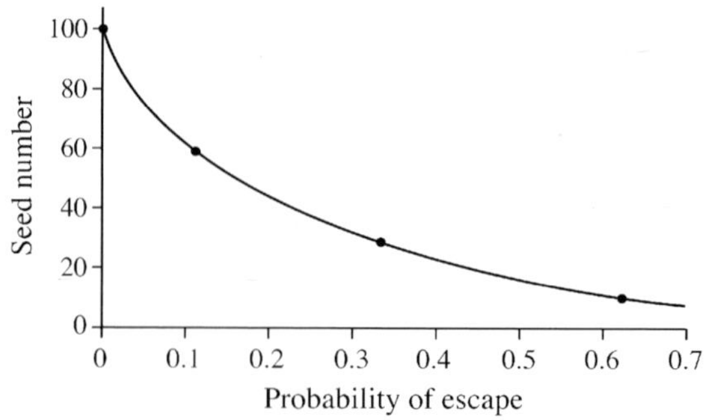

Fig. 6.3 The relationship between seed production and the probability of escaping from the root weevil *Ceutorhynchus cruciger* in *Cynoglossum officinale*. Calculated from Prins *et al.* (1992).

reproduction is probably disastrous. The weevil eventually destroys the root of the plant so that the survival of infected plants is close to zero. The probability of escaping attack as function of the reproductive investment is given in Fig 6.3. The shape of the function corresponds to Fig. 6.2B and could potentially explain monocarpy in this species.

6.2.2 *The relationship between investment and reproductive output*

We can rewrite Eq. (6.2) by taking seed production, S, as a function of reproductive investment. We will use i instead of RE_2 to denote the fraction of all biomass that can be invested in reproduction:

$$S = a(iB)^b \tag{6.3}$$

Here a is a scaling parameter and b is the exponent of the gain curve. Biomass is denoted as B. If we again single out a particular age x we get:

$$V_x = a(i_x B)^b + C_x V_{x+1} = aB^b i_x^{\ b} + C_x V_{x+1} \tag{6.4}$$

Optimal i is found by summing current year's seed production (S_x) and expected future seed production ($C_x V_{x+1}$). Equation 6.4 shows that the shape of the gain curve determines the value of i that maximizes V. If this shape is concave, i.e. there is a more than proportionate increase in seed production with increasing RE, we may expect a disproportionate decrease in residual reproductive value as well. A small decrease in growth and subsequent absolute investment in reproduction in future years would then lead to

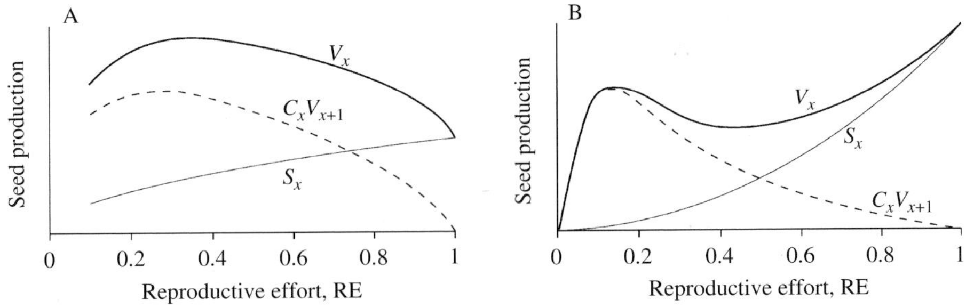

Fig. 6.4 Lifetime seed production is given by the sum of current year's seed production, S_x, and residual reproductive value, C_xV_{x+1}. (A) Two convex functions select for a perennial life history. (B) Two concave functions select for monocarpy. In both graphs a constant RE during life is assumed; this implies that lifetime seed production is zero if RE is zero.

a relatively large decrease in seed output. This would lead to selection for monocarpy as depicted in Fig. 6.4B. Similarly, a smaller than proportional increase of S_x with RE (a convex, decelerating curve) would select for a perennial life history (Fig. 6.4A). The shape of the gain curve has been discussed in Chapters 3 and 4. In general, we would expect the gain curves in animal-pollinated plants to be decelerating because of, for instance, geitonogamy (male gains) and local resource competition (female gains). Perhaps this explains why perennial life histories are far more common than monocarpic ones (Hart 1977). However, a number of processes may lead to accelerating gain curves. In relation to monocarpy Schaffer and Schaffer (1977, 1979) emphasized the importance of pollinator preference. They found that in monocarpic species of *Yucca* and *Agave* pollinators preferentially visited larger inflorescences and percentage fruit set was correlated with inflorescence height. However, later studies on these species (Udovic 1981; Aker 1982) and on the monocarpic *Lobelia telekii* (Young 1990) indicated that fruit set was limited by resources and not by pollinators. Young found that in the semelparous *L. telekii* the number of seeds per flower increased with plant size while this was not the case in the perennial *L. keniensis*. He argued, however, that this difference did not reveal selective forces favouring either monocarpy or iteroparity but was caused by the fact that taller inflorescences of the monocarpic species grew on more favourable sites. He concluded therefore that the reproductive effort model did not explain the evolution of monocarpy in Mount Kenya *Lobelia* species. This is of course only half the story because, as said before, we should also look at the male function of reproduction, pollen export. In addition, more detailed studies of pollinator visitation and pollen movement than that by Schaffer and Schaffer (1977, 1979) showed that the male gain curve is likely to be decelerating in animal-pollinated plants (Chapter 3) and is unlikely to play any role in selection for monocarpy.

Another suggestion leading to accelerating gain curves is that supportive tissues such as flowering stems have to be formed before any seed can be produced. The production of the first seed would then be extremely expensive but additional seeds would be relatively cheap. The most clear-cut example of such economies of scale comes from animals. Semelparous salmons migrate at great cost to their spawning areas. Because of this travel, producing the first offspring is extremely costly while producing further offspring is cheap, leading to an accelerating gain curve. Studies on the relationship between plant size and reproductive output in eight monocarpic species showed that in only two of them seed production increased more than proportionally to biomass (Klinkhamer and de Jong 1997). Because in most of these species the seed-to-flower ratio increased with plant size, this reflected a shift in sex allocation rather than an accelerating female gain curve. The evidence for economies of scale for seed production in plants are therefore weak but more data would certainly be welcome.

Factors after seed production may also lead to accelerating female gain curves and have been proposed as candidate selective forces for monocarpy. These factors include fruit removal (Chapter 3) and seed predator satiation (the bamboo example of Chapter 1: Janzen 1976). The latter factor has also been invoked to explain mast years (Janzen 1976; Kelly and Sork 2002). It is rather unlikely, however, that seed predator satiation provides a general explanation for the evolution of monocarpy, when individuals do not flower in synchrony between years. Instead of satiating the predator population an isolated plant producing many seeds may just provide the best place for seed predators to forage.

6.3 Residual productive value

So far we have only qualitatively discussed the way in which investment i in one year influences reproductive value in the next year. In the next section we will look at residual reproductive r value in somewhat more detail. Suppose biomass in the next year B_{x+1} depends on biomass in the current year (B_x) and on growth, while growth itself is a decelerating function of the resources left after reproduction:

$$B_{x+1} = B_x(1 - i_x) + d[B_x(1 - i_x)]^z \tag{6.5}$$

with d being a scaling factor and z being the exponent of the growth curve.

Substituting Eq. (6.5) in Eq. (6.4) gives:

$$V_x = a(i_x B)^b + C_x\left[a(B_x(1 - i_x) + d(B_x(1 - i_x))^z)^b + C_{x+1}V_{x+2}\right] \quad (6.6)$$

We can continue with substituting the formula for the biomass of year $x+2$, year $x+3$ etc. The second part of Eq. (6.6) is often denoted as the residual reproductive value, RRV, so that we can write:

$$V_x = a(i_x B)^b + \text{RRV} \quad (6.7)$$

The important thing here is that reproductive investment in year x has consequences for the biomass and optimal i in all following years, and the optimal investment in year x is itself determined by investments in all previous years. Survival from one year to the next is in most species also dependent on i directly or through size dependence (Fig. 6.2). Because the investment in one year has consequences for the optimal reproductive schedule for the rest of the plant's life it is impossible to work out the optimal lifetime reproductive schedule analytically. It is also nearly impossible to work it out by simulation if we start in year 1 and then work forward to the end of the life. If we simulate i with 1% accuracy, we have 100 possibilities in the first year. Starting with $i = 0.01$, and assuming an initial biomass we have to calculate the corresponding seed production, survival and growth in year 1 then increase i with 0.01 and repeat the procedure for all possible values of i. In year 1 seed production directly contributes to lifetime reproductive success. For each possible value of i in year 1 we then have to repeat the procedure for year 2. For year 2 we have to discount seed production for (size-dependent) survival, which itself depends on i in year 1. We need to continue this until the year in which the plant dies. We only know the optimal solution for a given year if we have written down the possibilities for all later years so that we have to consider all possible combinations until the end of the plant's life. For a plant that can live for 10 years this would already yield 100^{10} or 10^{20} possibilities. Only for a small number of years we can do the simulation this way. For a larger number of years, the problem is too hard to handle. Fortunately there is a trick. We can work our way backwards if we assume a maximum age (or size). If we know the plant will die in year 30, we know for all possible plant sizes what it should do in year 29: it should put all resources it can in reproduction. Once we know the reproductive value, V_{29}, of year 29, we can, for all possible sizes, work out the value of i_{28} that maximizes V_{28} using Eq. (6.4). Working backwards we end in year 1 for which we know B and the corresponding optimal i. We can then work forward because for each value of B_x we have already found the corresponding optimal i_x and thus B_{x+1}. Working in this way the number of possible

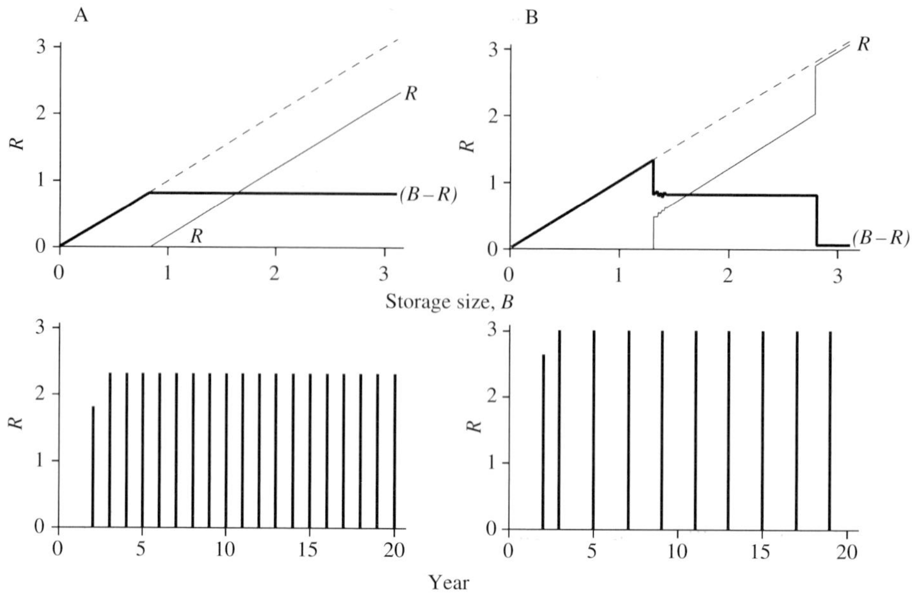

Fig. 6.5 Simulated optimal investment in reproduction, $R = Bi$, as a function of biomass at the end of the growing season, B. $B - R$ gives the amount of resources invested in growth for next year. In the simulation, the growth curve was assumed to be decelerating and survival was constant over the years and independent of R. (A) $S = R$ (i.e. a linear gain curve) selects for a constant investment in reproduction over the years. (B) $S = R^9$ (i.e. a strongly accelerating gain curve) selects for mast years. Redrawn with permission from Klinkhamer *et al.* (1997), *Journal of Evolutionary Biology*, Blackwell.

solutions we have to consider is greatly reduced because for all years they don't multiply but add up. Suppose we consider 1000 biomass classes we have $1000 \times 100 = 10^5$ possible cases in each of the years 28 to year 2. This way of programming is called dynamic programming. For a very small number of years we can use the trick of reasoning backwards to find an analytical solution for the optimal reproductive schedule (Kozlowski and Uchmanski 1987; Kozlowski and Wiegert 1987). In most other cases we have to rely on simulation models. Iwasa and Cohen (1989) and Klinkhamer *et al.* (1997b) showed that with a decelerating growth curve and a linear gain curve the optimal solution is either annual or perennial but never monocarpic perennial. The optimally growing plant grows until a maturation size is reached and diverts all extra resources to reproduction. As a result it stops growing and reproduces with the same seed output every year thereafter (Fig. 6.5A). A decelerating gain curve doesn't change the optimal life history qualitatively and the perennial life history is still the optimal solution. With an accelerating gain

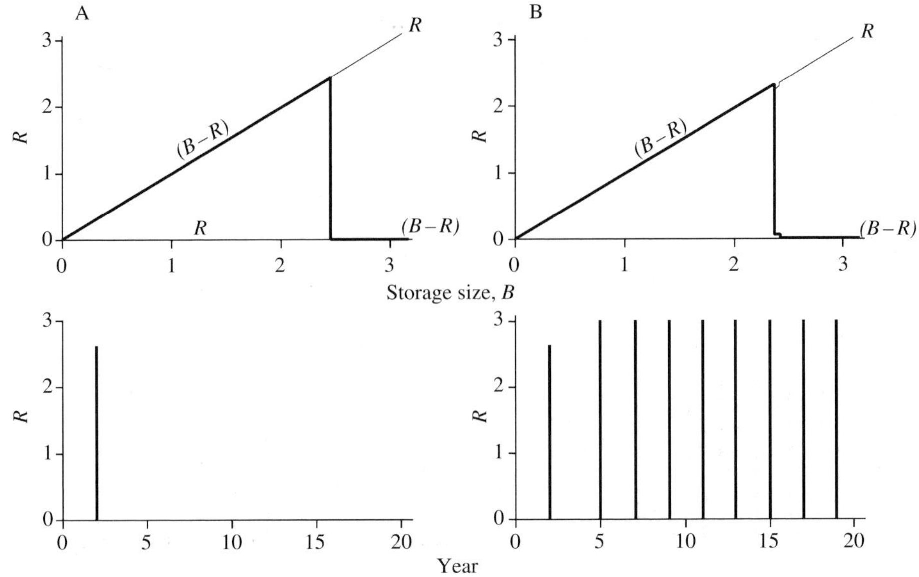

Fig. 6.6 Simulated optimal investment in reproduction, $R = Bi$, as a function of biomass at the end of the growing season, B. $B - R$ gives the amount of resources invested in growth for next year. In the simulation, the growth curve was assumed to be decelerating and survival was constant over the years. $S = R$. Reproduction-related herbivory reduces plant survival, $C_x = C_0 e^{-cR}$. (A) $c = 2$ selects for monocarpy. (B) $c = 1$ selects for mast years. Redrawn with permission from Klinkhamer *et al.* (1997), *Journal of Evolutionary Biology*, Blackwell.

curve mast years is the most likely outcome. Note that in this context mast years refer to the reproductive schedules of an individual plant and not a whole population. Even very high values of the exponent of the gain curve do not select for monocarpy (Fig. 6.5B). Only when survival is extremely low, $C_x = 0.1$, and the efficiency with which resources at the end of the season can be used for next year's growth is very low monocarpy can be selected for. Klinkhamer *et al.* (1997b) especially looked at the effects of reproductive investment on the attraction of herbivores and the subsequent consequences for survival. They assumed that the absolute investment in reproduction, iB, was important for attracting herbivores. Survival, C, is then a decreasing function of both plant size and investment in reproduction. The latter was modelled as: $C(i) = C(0)e^{-cBi}$ with $C(0)$ being the survival when reproductive investment $i = 0$ and c being a scaling constant. With a strong effect of reproductive investment on survival ($c = 2$) monocarpy is the optimal solution (Fig. 6.6A). More moderate effects of reproductive investment on survival ($c = 1$) again select for mast years (Fig. 6.6B).

The case of the monocarpic *Cynoglossum officinale* was modelled in detail. Recall the reproduction-related mortality of Fig. 6.3. The idea was that if a herbivore is attracted the plant always dies irrespective of its size. The root weevil that attacks flowering plants of *C. officinale* enters the root crown and destroys the apical meristem. Often such an attack is associated with infestation by fungi or bacteria leading to further infection of the thick tap root which eventually completely rots away. Furthermore a linear relationship between investment and reproductive output was assumed, $b = 1$. Biomass at the end of the season was considered a decelerating function of biomass at the beginning of the season. With the parameters specific for *C. officinale* the optimal solution was close to monocarpy. Reproduction should be in the third year and nearly all resources (99%) should be invested in reproduction. Only very few resources would be left for future growth and reproduction and the survival probability would be less than 1% so that over 99% of all individuals would be effectively monocarpic. The modelled optimal solution comes close to the actual life history of *C. officinale*. Taking reproduction-related mortality into account provides a good hypothesis for the evolution of monocarpy in this species. This explanation is, however, rather specific and it remains to be investigated if other monocarpic species have herbivores that select plants in a similar way and also cause such damage that the survival of the plant becomes problematic.

The most striking result from the dynamic modelling approach is that monocarpy is not easily selected for. Even steeply accelerating gain curves select for mast years rather than monocarpy. In part this may be an artefact from the model caused by the way size-dependent growth and survival are included. Growth and survival in these models depend on how much resources are available at the beginning of the season. If this amount is small the relative growth rate is very high making it profitable not to put everything into reproduction but to leave a small amount for future growth. A perennial plant that just reproduced and that has only 0.1 g of usable carbohydrates left has the same growth rate and survival as a juvenile plant with 0.1 g usable carbohydrates. This, most likely, does not comply with reality. It seems reasonable to assume the plant that has just flowered and put most of its resources in reproduction will have a much lower growth rate and survival than the juvenile because it uses a degenerated root and stem system. Introducing this difference into the model will make it much easier to select for monocarpy. Unfortunately lifetime reproductive schedules are not available in combination with data on (size-dependent) growth and mortality, and the shape of the gain curves for many species. It is therefore difficult to test the reproductive effort model as described above for specific cases.

To sum up: selection for different life histories depends on the shape of the gain curve and on the relationships between reproductive expenditure and future survival and growth. If the relationship between reproductive expenditure and survival is convex we expect selection for a perennial life history, if this relationship is concave we expect selection for monocarpy. Likewise, if the gain curve is decelerating we expect selection for a perennial life history, if the gain curve is accelerating we expect selection for monocarpy. Because decelerating gain curves are more likely than accelerating ones this explains why monocarpy is relatively rare. Although the hypothesis of seed predator satiation to explain monocarpy in long-lived bamboos is well known there is little experimental data to test the reproductive effort model. More detailed studies on lifetime reproductive schedules are needed.

Questions

(1) Can demographic models explain the evolution of monocarpy?
(2) What is meant by residual reproductive value?
(3) Why do reproductive effort models only in a very restricted parameter range show selection for monocarpy?
(4) Give two possible explanations for selection for monocarpy.

7

Size and number trade-offs: the evolution of seed size

Seed size is highly variable between species. The dust seeds of orchids and the giant seeds (up to 27 kg) of *Lodoicea seychellarum* (double coconut) are two extreme examples and even related species sometimes differ considerably in seed size (Fig. 7.1). This variability may indicate a lack of constraints on seed size. At the same time, however, seed size is sometimes extremely constant within species. A famous example is *Ceratonia siliqua*, the carob tree, which produces seeds so remarkably similar in size that they were originally used as the unit of weight for diamonds (193.3 mg = 1 carat).

In this chapter we will first discuss a straightforward optimization model in which the plant optimizes seed mass without interacting with other plants in the population. This optimization model was first presented by Smith and Fretwell (1974), in a paper that later became a citation classic. Later in this chapter we introduce an ESS approach to the problem. In this ESS model seeds of different sizes compete with each other. Finally we will look at the interplay between risk-spreading strategies (dormancy and seed dispersal) and seed size.

7.1 The relationship between seed mass, seed number and seedling success

All seed size models are based on two essential assumptions.

First, seed size trades off against seed number. With a given amount of resources a plant can produce many small seeds or fewer big ones. It is generally accepted that this trade-off exists (Salisbury 1942; Harper 1977; Coomes and Grubb 2003). It may be difficult, however, to demonstrate this trade-off in comparisons between species because of confounding factors; bigger plants may have bigger seeds and phylogeny also plays a role (Kelly and Purvis 1993). After controlling for both these factors, Jakobsson and Eriksson (2000) found a strong negative relationship between seed

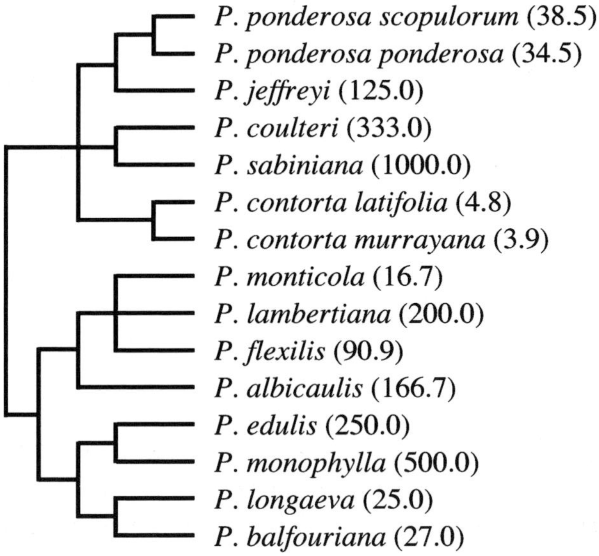

Fig. 7.1 Phylogeny of a number of *Pinus* species. Seed mass (mg) can be highly variable, even between closely related species. Redrawn with permission from Veech *et al.* (2000), *Evolutionary Ecology Research.*

mass and seed number (Fig. 7.2). Likewise, Dalling and Hubbell (2002) found a negative relationship between mass and density of seeds of different species in the surface 0–6 cm of the soil of a moist tropical forest (Fig. 7.3).

Second, seedling success increases with seed size. Germination and seedling recruitment were positively associated with seed mass in studies that compared among species (Figs. 7.2, 7.3) or within species (Figs. 7.4, 7.5). Seedlings from big seeds survived better (Shaukat *et al.* 1999; Simons and Johnston 2000a; Dalling and Hubbell 2002), although the effect may be restricted to the very early life stages (Winn 1988; Saverimuttu and Westoby 1996). Surviving seedlings from big seeds also grew larger (Figs. 7.2, 7.4) (Schaal 1980; Stanton 1984; Shaukat *et al.* 1999; Jakobsson and Eriksson 2000; Simons and Johnston 2000a), although this effect may disappear later in life (Meyer and Carlson 2001). Effects of seed size on survival and plant size may be indirect, through differences in timing of germination (Simons and Johnston 2000a), through differences in metabolic rate under deep shade (Saverimuttu and Westoby 1996), or through seed predation (Fig. 7.6) (Reader 1993). Although the way seed size influences seedling success may differ among species, the examples above show that there is ample evidence, both within and across species, for a positive correlation between seed size and success.

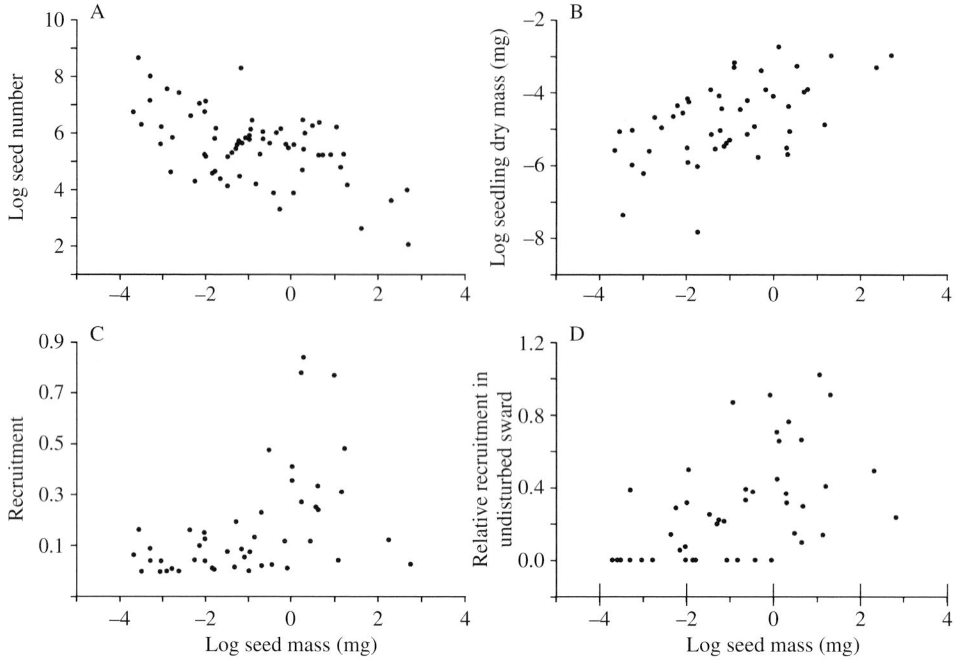

Fig. 7.2 The relationship between log seed mass and (A) log seed number per individual, (B) seedling weight (after three weeks of growth in pots in the lab), (C) recruitment and (D) relative recruitment. Recruitment = number of recruits/seeds sown. Relative recruitment = recruitment in undisturbed plots/recruitment in disturbed plots; this measures the impact of disturbance on recruitment. Small seeds benefit more from disturbance than big ones. All correlations are significant ($P < 0.05$). Redrawn with permission from Jakobsson and Eriksson (2000), *Oikos*.

7.2 Seed mass and environmental conditions

The benefits of producing bigger seeds can be more pronounced when seedlings are grown under harsh conditions, which include competition from established vegetation, deep shade, defoliation, nutrient shortage, drought, the presence of litter or deep burial in the soil (reviewed in Westoby *et al.* 1997). Because of this interaction with environmental conditions one expects strong habitat effects on seed mass. However, a comparison between species of dry vs. wet and nutrient-poor vs. nutrient-rich habitats revealed no difference in seed mass (Mazer 1989; Westoby *et al.* 1997). The only clear relationships were with shade and the age or maturity of the vegetation (Salisbury 1942; Mazer 1989; Hodkinson *et al.* 1998). Fenner (1985), based on data of Salisbury (1942), showed that species in more mature, forest vegetation produced bigger seeds

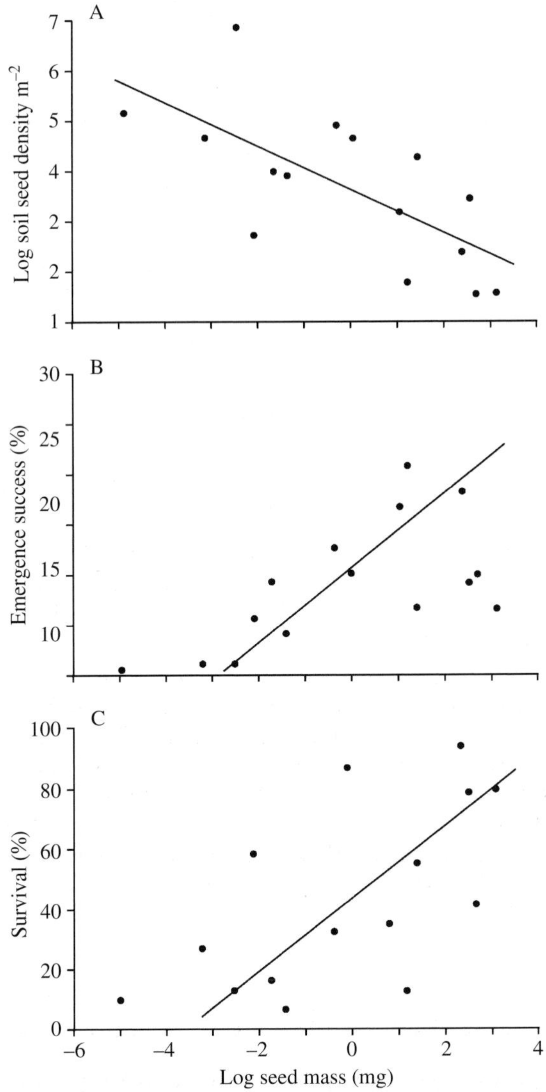

Fig. 7.3 The relationships between log seed mass and (A) log seed density (m^{-2}) in the top 0–6 cm of the soil, (B) the percentage of seeds emerging as a seedling from October to April and (C) the percentage of surviving seedlings, for 15 abundant species in secondary forest at the Barro Colorado Nature Monument in Panama. Redrawn with permission from Dalling and Hubbell (2002), *Journal of Ecology*, Blackwell.

than those in immature, open vegetation. Many other factors, including phylogeny, height, longevity, growth form and mode of dispersal, are correlated with seed size and can confound the picture (Kelly and Purvis 1993; Leishman *et al.* 1995; Grubb and Metcalfe 1996; Rees 1997).

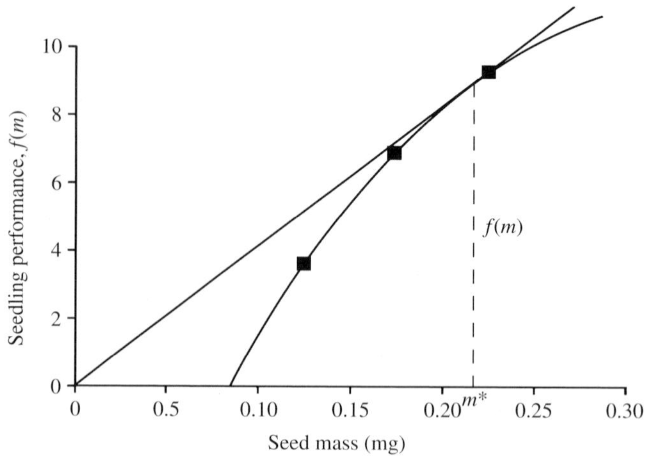

Fig. 7.4 The relationship between seed mass and seedling performance (the product of fraction germination, fraction seedling survival and final seedling biomass) in a growth room experiment for the tree *Acacia nilotica* ssp. *indica* (after Shaukat *et al.* 1999). Squares indicate the data points, m^* indicates the optimal seed mass as predicted by the Smith and Fretwell (1974) model.

7.3 The Smith–Fretwell model

The Smith–Fretwell model predicts the optimal investment in a single offspring based on the two assumptions (size–number trade-off and big seeds are more successful) discussed above. The model is straightforward and provides a simple graphical tool to determine the optimum, given the relationship between seed size and offspring success.

Suppose the total amount of resources available for reproduction equals T. The investment in each seed equals m, so that T/m seeds can be produced. The fitness that a parent gains by producing offspring is the product of the number of seeds and the fitness gain per seed. The fitness gain per seed, $f(m)$, is an increasing function of mass m. Fitness $W(m)$ as a function of seed mass m is:

$$W(m) = T(f(m)/m) \tag{7.1}$$

We can conveniently analyse this graphically (Fig. 7.4). For a given amount of resources, T, fitness W is maximized at the highest value of $f(m)/m$ (Eq. 7.1). The slope of a line through the origin that touches or crosses the $f(m)$ curve at point m equals $f(m)/m$. The line that just touches the curve gives the highest slope. Analytically this can be shown as follows. We find the optimal seed mass by differentiating $W(m)$ with respect to m and setting the derivative equal to zero. Thus $\mathrm{d}W(m)/\mathrm{d}m = T[\mathrm{d}f(m)/\mathrm{d}m]/m - Tf(m)/m^2 = 0$, so that:

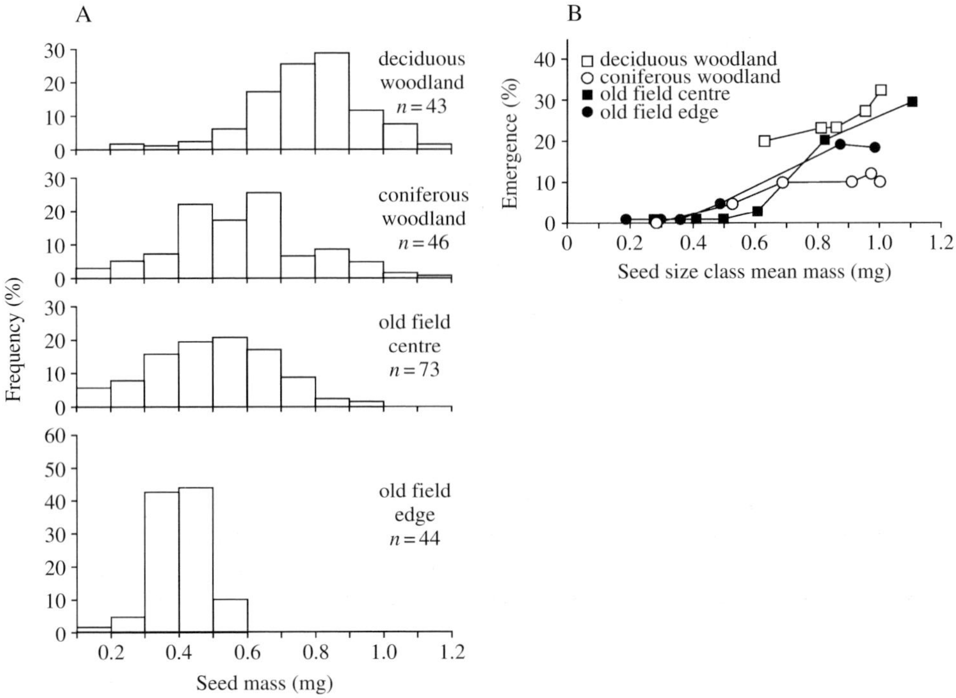

Fig. 7.5 (A) Frequency distributions for seed size based on all seeds produced in each of four populations of *Prunella vulgaris* (selfheal). The number of individuals from which seeds were collected is denoted as n. All population means were significantly different from each other. (B) Mean percentage emergence for each of up to six seed size classes, sown into its home population. The data in (B) can be used to estimate the optimal seed size according to the Smith and Fretwell (1974) model from Fig. 7.4: average seed size observed is below the Smith–Fretwell optimum in three out of four populations. Redrawn with permission from Winn (1988), *Ecology*.

$$\frac{\mathrm{d}f(m)}{\mathrm{d}m} = \frac{f(m)}{m} \tag{7.2}$$

Optimal seed mass is found at the point where the tangent of the $f(m)$ curve $[\mathrm{d}f(m)/\mathrm{d}m]$ is a straight line through the origin with slope $f(m)/m$ (Eq. 7.2). This optimum is independent of T, the amount of resources available to the plant.

To date there are only few empirical studies that have tested the predictions of the Smith–Fretwell model. Data for *Acacia nilotica* (Fig. 7.4) and *Prunella vulgaris* (Fig. 7.5) suggested that actual seed size was somewhat smaller than predicted. A possible explanation is that these studies did not take seed predation into consideration. Bigger seeds are more frequently eaten and this could lead to an overestimation of the optimal seed size. It should be kept in

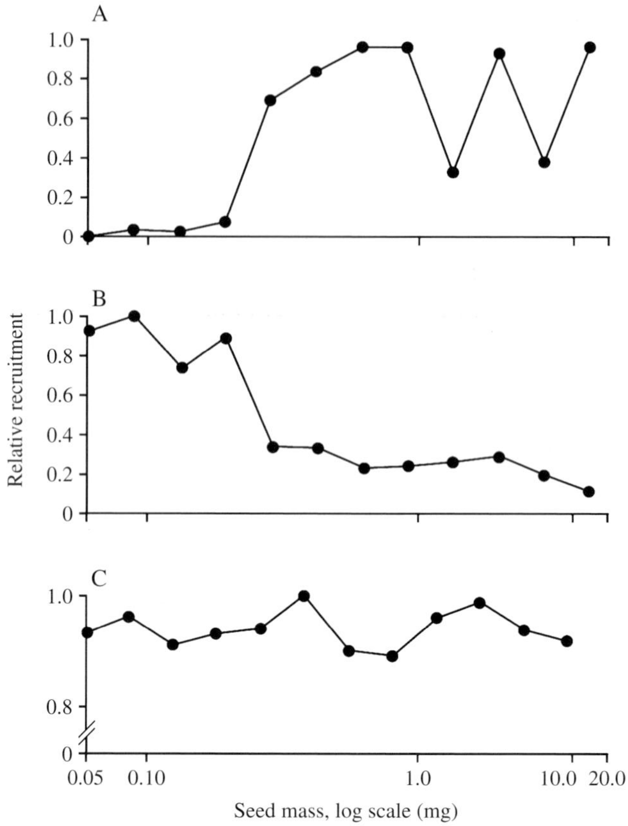

Fig. 7.6 Relationship between seed mass of different species and relative recruitment in an old field after (A) removal of ground cover, (B) addition of a cage to prevent seed predation and (C) a combination of both. Relative recruitment = recruitment in control plots/recruitment in treatment plots. The presence of ground cover strongly reduced germination of small seeds, whereas big seeds suffered most from predation. Redrawn with permission from Reader (1993), *Journal of Ecology*, Blackwell.

mind, however, that although intuitively appealing, it is perhaps difficult to obtain a reliable estimate of the $f(m)$ curve in any single species. There are several reasons for this. Some species lack the variation in seed mass that is necessary to draw the curve accurately. Furthermore, seeds should be as genetically similar as possible. When seed mass depends on the genetic quality of the seed, then seeds that gather more resources due to their genotype may also be more vigorous later in life. This is another effect than the effect that Smith and Fretwell (1974) had in mind: a mother plant can produce seeds of different size but similar in other aspects, including genetic quality. If there is indeed some degree of control by seed genotype over its mass, this violates the assumptions of the Smith–Fretwell model. In such a situation parent–offspring conflict arises, a situation that we will deal with in Chapter 15. The Smith–Fretwell model leads to three predictions, listed below.

7.3.1 Prediction 1

Plants should produce seeds of the same size. At first sight this prediction seems to be supported by empirical data. Seed size variation within species has typically been considered to be low (Harper *et al.* 1970), with *Ceratonia siliqua* as the prime example. However, significant variation within species, populations and even individuals has also been reported (Fig. 7.5) (Michaels *et al.* 1988). Within plants, seed mass often depends on the position in the inflorescence (Cavers and Steel 1984; Klinkhamer and de Jong 1987; Simons and Johnston 2000a) with the first seeds produced being heavier. This may simply be a matter of declining resource status but a functional explanation has also been suggested. Geritz (1995) pointed out that the Smith–Fretwell model implicitly assumed that additional costs for reproduction are independent of seed size or number and this assumption may not always hold. The production of a seed requires additional costs for making supporting structures such as flowering stems, flowers, fruits etc. If these costs are proportional to seed number (a constant cost c for each seed) we can rewrite Eq. (7.1) as:

$$W(m) = f(m)R/(m + c) \tag{7.3}$$

By differentiating fitness with respect to seed size m and setting this to zero we find that the optimal seed size is found at:

$$\frac{\mathrm{d}f(m)}{\mathrm{d}m} = \frac{f(m)}{m + c} \tag{7.4}$$

This is at a point m where the tangent is smaller than without additional costs ($c = 0$) (see Fig. 7.4). A smaller tangent means that the optimal seed size has shifted to the right. In other words, with additional costs bigger seeds should be produced. When additional costs are not the same for each seed, but decline with each extra seed produced, optimal seed size decreases over the flowering season (Geritz 1995).

With a conceptually different model Begon and Parker (1986) reached the same conclusion. They considered that plants first gather resources and then start producing offspring at a certain rate. During the period of offspring production the parent may die. Consequently, later-initiated offspring have a higher probability of not being produced at all and contribute less to fitness. The parent optimizes its fitness by allocating more resources to the early-initiated offspring.

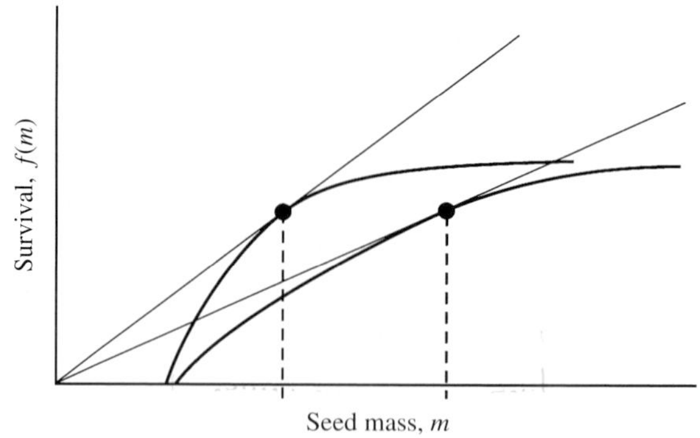

Fig. 7.7 The less decelerating the curve of seedling performance as a function of seed mass, the higher the optimal seed size. The $f(m)$ curve is expected to decelerate less under harsh conditions or strong competition.

The prediction that it is optimal to produce seeds of a single size depends on the assumption that seedlings do not compete with each other. As we will show in Section 7.4 ESS models, based on competition within safe sites, with seedlings from heavy seeds outcompeting smaller ones, predict that plants should produce a range of seed sizes.

7.3.2 *Prediction 2*

If a plant has more resources available it should produce more seeds instead of producing bigger seeds. The data on *Cynoglossum officinale* (Klinkhamer *et al.* 1987) seem to support this conclusion: a 40-fold increase in plant size was not associated with a significant increase in mass per seed. However, in a number of species big plants do produce bigger seeds (Thompson and Rabinowitz 1989; Venable 1992). Venable (1992) suggested that, in contrast to the Smith–Fretwell model, the shape of the $f(m)$ curve might depend on the total number of seeds produced. Producing more seeds can result in DD reduction of survival or growth of siblings in the seed shadow of their maternal parent (Augspurger 1983; but see Rademaker and de Jong 1999 for *Cynoglossum officinale*). When only a limited number of sibs can survive from a very large number of seeds, the parent attains the highest fitness when big seeds survive (they each perform better and there is hardly any disadvantage of producing fewer seeds). In other words, in plants with many resources (T large) big seeds have an additional advantage as compared to plants with few resources. This makes the curve of $f(m)$ less decelerating for plants with more resources (Fig. 7.7) and the Smith–Fretwell approach predicts a larger optimal seed size for big plants. For a mathematical treatment of this model we refer to Venable (1992).

7.3.3 Prediction 3

The more strongly the $f(m)$ curve decelerates, the smaller the optimal seed size becomes. In Venable's (1992) example above, the shape of the curve changed with resource status of the plant, but environmental factors may also have an effect. Big seeds have more reserves to make up for lack of resources in the environment just after germination. This helps to overcome the hazards of seedling survival and to win competition with other seedlings. Grime and Jeffrey (1965) placed saplings of North American trees in the shade and found after 12 weeks that seedling mortality was inversely related to seed mass. Similarly, Saverimuttu and Westoby (1996) reported that under deep shade seedlings of species with larger seed mass tended to live longer. Big seeds are advantageous in stressful environments, in nutrient-poor soils or with intense competition. For the Smith–Fretwell model this means that we expect the $f(m)$ curve to be less decelerating in stressful environments and as a consequence we expect seed size to be higher in such environments. For example, in the Australian shrub *Hakea sericea* the smallest seeds were produced when plants were grown in nutrient-enriched soil. This counterintuitive result fits in well with prediction 3 and suggests an adaptive, conditional response of the plant to nutrient status (Haig and Westoby 1988).

7.4 Including competition: an ESS model for seed size

So far, we have explained variation in seed mass within or between species by referring to the shape of the $f(m)$ curve. This approach cannot explain the large differences in seed mass that sometimes occur between species from the same habitat, nor can it explain why some plants produce seeds of different sizes at the same time. Recently, game-theoretical models have been made for the evolution of seed mass (Geritz 1995; Rees and Westoby 1997). In these models the outcome of competition for available space depends on seed size, with seedlings from big seeds outcompeting the smaller ones. The outcome of these ESS models is, in general, that no single seed mass strategy can be evolutionarily stable.

We will discuss the model of Stefan Geritz in more detail here because his approach shows how, without complicated mathematics, several interesting conclusions can be drawn based upon reasoning about ESS conditions. Think of an environment in which seedling establishment is confined to small favourable spots such as little gaps in the vegetation, so-called safe sites. Within each safe site competition for space allows only one seed to establish itself and to reproduce. Silvertown and Bullock (2003) showed that such a situation was realistic for grazed grasslands with DD survival in gaps smaller than 3 cm diameter. For such an environment Geritz (1995) considered two distinct periods with corresponding survival functions depending on seed size. The first function gives the number of seedlings that survive until the moment competition takes place, $N(m)$. During this period survival is independent of the seed size strategies of other plants in the population, as in the Smith–Fretwell model. The second function gives the probability of winning the competition within safe sites, $U(m)$, so that the fitness contribution of seeds of size m can be written as:

$$W(m) = N(m)U(m). \tag{7.5}$$

The probability of winning the competition depends on the seed size strategies of the plants in the population. Assume that competition within safe sites is asymmetric, so that a seed with slightly higher mass always wins when competing against a smaller seed. Then, any single seed size strategy can easily be invaded by a strategy producing heavier seeds because the latter seeds always win the competition. On the other hand, if plants produce just a few, very big seeds, then at equilibrium many safe sites will not contain any seeds; plants with small seeds can invade the population because they produce many more seeds and occupy vacant sites. Clearly the probability of winning the competition depends on the seed-setting strategies of the plants in the population and producing seeds of only one size cannot be an ESS.

Eventually we want to calculate an ESS for the frequency distribution of seeds of different sizes, but we start in a simple way by examining how a mutant with only one seed size would perform in a population in which the common type produces this ESS frequency distribution. For the first period, before seeds enter the competition phase, we denote seedling survival as $f(m)$, as in the Smith–Fretwell model. For a single seed size strategy, the number of seeds entering the competition, $N(m)$, is the number of seeds produced, T/m, multiplied by their survival $f(m)$:

$$N(m) = f(m)T/m \tag{7.6}$$

The chance of winning the competition, $U(m)$, is constant or increases with seed mass, m. The relationship between $U(m)$ and m is determined by the seed size strategy in the resident population. For convenience, think of the range of seed sizes produced as three separate size classes: when a continuous range of seed sizes is produced, $U(m)$ always increases with m because with each additional small increase in m a seed moves up in the competitive hierarchy. If however in the middle size class no seeds are produced, the place in the competitive hierarchy is the same for the smallest seed in the middle size class and the smallest seed in the highest size class. In other words $U(m)$ is constant over the middle size class. In more general terms we can say $U(m)$ is constant over a size class if, within that size class, no seeds are produced and vice versa. Therefore, if we know $U(m)$ in an ESS population (this chance is denoted as $U^*(m)$), we can derive which seed masses the common type produces. Now we will show how $U^*(m)$ can be constructed graphically, using three rules.

Rule 1 Any ESS for producing seeds of different sizes should be stable against invasion by any mutant that makes all seeds of size m. Mutant fitness is: $W_m = N(m)U(m)$ (Eq. 7.5). When the population is at equilibrium in the safe site model, one plant replaces another plant after reproduction, so that the absolute fitness W for the common type is 1. Because the mutant cannot do better than the ESS, the fitness of a mutant producing all seeds of size m in a population at the ESS is:

$$W_m = N(m)U^*(m) \leq 1 \tag{7.7a}$$

so that

$$U^*(m) \leq 1/N(m) \tag{7.7b}$$

In words, at the ESS the chance of winning the competition for seeds with mass m is equal to or less than the reciprocal of the number of seeds entering the competition if only this seed size were produced. Furthermore we know that if $U^*(m) < 1/N(m)$ mutant fitness W_m is smaller than 1. In that case the resident type at the ESS produces no seeds of size m, because such a strategy could be invaded by a strategy allocating to seed sizes for which fitness does equal 1. In turn this implies that $U^*(m)$ is constant. Therefore, if we plot the curves of $U^*(m)$ and $1/N(m)$ as a function of m, the first curve either coincides with the second curve or is smaller and constant (Fig. 7.8B). Note that the curve of $1/N(m)$ is known (Eq. 7.6) and can be drawn once the shape of the survival curve, $f(m)$, has been measured (Fig. 7.8A,B).

Rule 2 The function $U^*(m)$ cannot decrease with m. In other words, bigger seeds always have the same or a higher chance of winning the competition, but never a smaller chance.

Rule 3 For the largest possible seed size (T), $U = 1$, because the biggest seed can never be outcompeted.

Starting with the largest seed size possible T, which is the only value for which we know $U^*(m)$ in advance (rule 3) and working our way backwards using rules 1 and 2 we can derive the shape of $U^*(m)$. At the right-most end of the graph (i.e. at $m = T$ in Fig. 7.8C), $U^*(m)$ is equal to 1 which is typically below the graph of $1/N(m)$. According to rule 1 these seeds should not be produced. For smaller seeds $U^*(m)$ then remains constant until it meets the graph of $1/N(m)$ at m_4. Between m_4 and m_3, $U^*(m)$ and $1/N(m)$ necessarily coincide,

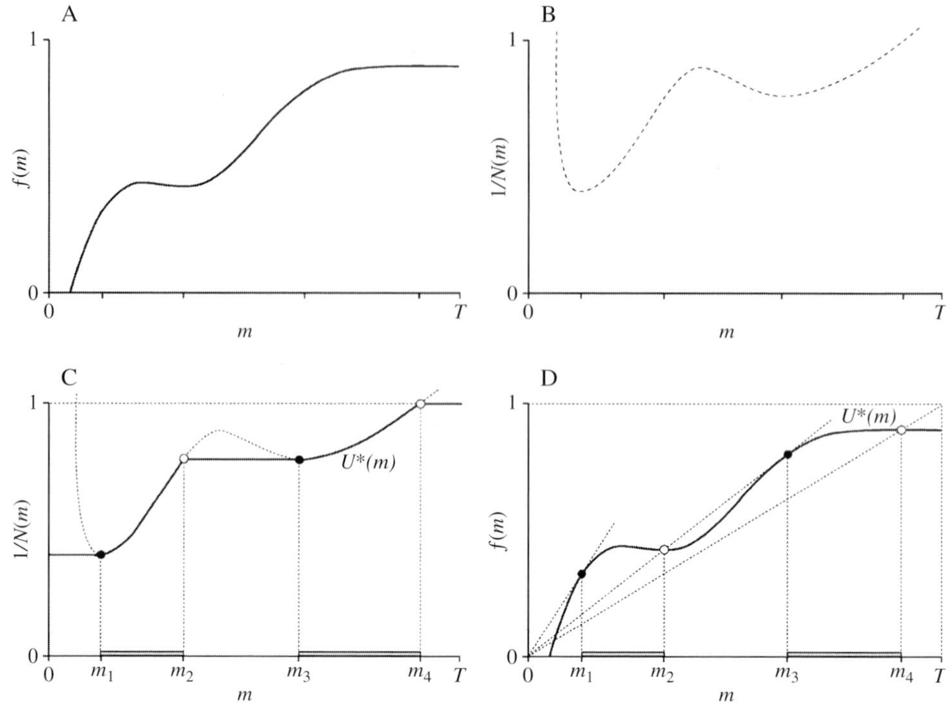

Fig. 7.8 The relationship between seed mass (m) and (A) survival until the moment competition takes place ($f(m)$), (B) $1/N(m)$, with $N(m)$ being the number of seeds of size m entering the competition phase ($1/N(m) = m/[f(m)T]$, Eq.7.6). (C, D) Construction of the ESS for producing seeds of different sizes. Thick line shows $U^*(m)$, the probability that a seed with size m wins the competition in a population at the ESS. At the ESS seeds with sizes between m_1 and m_2, and between m_3 and m_4 are produced as indicated by the horizontal bars (see text for explanation). Redrawn with permission from Geritz (1995), *American Naturalist*, University of Chicago Press.

because if continued as a constant function, $U^*(m)$ would end up above the graph of $1/N(m)$. Since $U^*(m)$ is a non-decreasing function of seed size (rule 2), it can no longer follow the graph of $1/N(m)$ once m_3 has been reached, and therefore $U^*(m)$ must be continued as a constant function until the graph of $1/N(m)$ is joined again at m_2. From m_1 downwards $U^*(m)$ remains constant. Horizontal bars indicate where $U^*(m)$ increases and these seed sizes should be produced. We conclude that at the ESS seeds between m_1 and m_2 and between m_3 and m_4 are produced.

There is a quicker way to derive the same conclusion directly from the graph of $f(m)$ (Fig. 7.8A, D). First, note that when reading the graph in Fig. 7.8C from left to right we see that for all seed sizes that should be produced $1/N(m)$ is smaller than for any other possible seed size. Because $N(m) = f(m)T/m$, this means that for any seed size that is produced at the ESS, $f(m)/m$ is larger than for all bigger seeds. In Fig. 7.8D it can be seen that the smallest seed size that should be produced is the one for which the slope of the line connecting the graph of $f(m)$ with the origin has a local maximum (m_1). Seed sizes between m_1 and m_2 have a lower value of $f(m)/m$, while this is no longer true for seed sizes between m_2 and m_3. Seed sizes between m_2 and m_3 are therefore not produced at the ESS. From m_3 onwards the above condition is again met and seeds of this size and larger sizes are produced.

Second, we have to find the largest seed size that is produced at the ESS. For the biggest seed, U is by definition 1, because it will never lose the competition. It then follows from $W(m) = N(m)U^*(m) = 1$ that $N(m) = 1$. Because $N(m) = Tf(m)/m$ (Eq. 7.6), this means $f(m)/m = 1/T$ for seeds of the largest size. For all points on the diagonal in Fig. 7.8D this condition holds. The largest seed size that must be produced is where the $f(m)$ curve intersects the diagonal, which is at m_4.

We have calculated the range of seed sizes that plants produce at the ESS and the next step is to calculate how much resource should be invested in each size category. For these further calculations the reader is referred to Geritz (1995). The predictions of this ESS model are fundamentally different from the predictions of the Smith–Fretwell model. The ESS model predicts that (1) seeds of different sizes should be produced, (2) the range of seed sizes produced should be extended to larger seed sizes in populations where plants are able to gather more resources and (3) the range of seed sizes should increase with the largest seed size produced (Geritz 1995). The first and the third prediction seem to be supported by data on *Prunella vulgaris* (Fig. 7.5).

7.5 Dormancy, dispersal and seed size variation

So far we have treated the environment as constant in space and time and have considered the adaptation of seed size to one type of habitat. If the environment is unpredictable and consists of two or more sets of conditions, each with a different optimal seed size, this may give rise to seed size variation or even several distinct seed sizes (somatic seed heteromorphism) if environmental variation is large and intermediate seed strategies have low fitness (Harper 1977). In a single environment, success of seed germination and seedling survival may strongly differ among years. A number of authors have suggested that seed heteromorphism represents a form of bet-hedging in the face of temporal variation in environmental suitability (Harper 1977; Westoby 1981; Venable 1985). We will discuss this in more detail later but first we will set out some basic principles about bet-hedging.

7.5.1 Cohen's model: optimization of the germination fraction without DD

The full text of one of the shortest introductions ever written to a scientific paper (Cohen 1966) reads:

> Most living organisms are faced with a considerable risk of failure when attempting to reproduce. One obvious way to survive and reproduce in a risky environment is to spread the risk that one failure will not be decisively harmful.

Cohen's paper was the starting-point for a whole series of papers on risk-spreading strategies. Two risk-spreading strategies are of particular interest in relation to seed size variation: seed dormancy and dispersal. Here we will only detail seed dormancy but many of the basic ideas of the two risk-spreading strategies are similar. We first outline the simplest case discussed in Cohen's model in which each individual optimizes its germination fraction on its own without regard to the strategies of other plants in the populations. In Cohen's model no DD occurs and selection maximizes population growth λ. Next we extend the analysis to an ESS model, in which individuals do interact.

Consider an annual plant growing in an environment without DD mortality. We count the number of individuals, N_t, at a time of year when they are all in the same stage (in this case the number of seeds just prior to germination). It is assumed that all seeds are equal, so that seeds that have remained in the soil for years have the same quality as seeds from last year. Each year a fraction G of the seeds germinates, and of the fraction $1-G$ of the seeds remains dormant in the soil. A fraction D of the dormant seeds dies every year. Each seed that germinates produces Y_t seeds, with Y_t being a random variable. The population growth λ between year t and year $t+1$ (Fig. 7.9) is:

$$\lambda = N_{t+1}/N_t = (1-D)(1-G) + GY_t \tag{7.8}$$

We only discuss the simplest case in which Y_t can take on two values, 0 in a bad year and Y in a good year. Then in a good year: $N_{t+1}/N_t = (1-D)(1-G) + GY$, and in a bad year: $N_{t+1}/N_t = (1-D)(1-G)$. If good years occur with probability P, we can write for population growth over a very large number of years:

$$\lambda = [(1-D)(1-G) + GY]^P[(1-D)(1-G)]^{(1-P)} \tag{7.9}$$

When Y or D are large enough so that $GY \gg (1-D)(1-G)$ Eq. (7.9) can be simplified and after taking logarithms we obtain:

$$\log\lambda = P\log(GY) + (1-P)\log(1-D)(1-G) \tag{7.10}$$

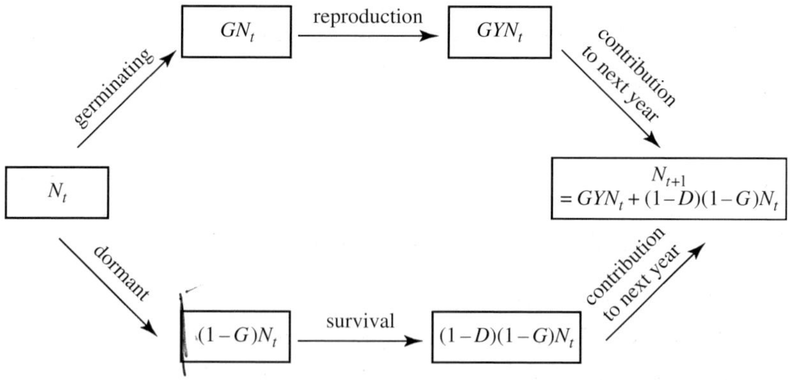

Fig. 7.9 Graphical representation of the year-to-year changes in the number of seeds (N) of an annual plant in the soil just before germination (Eq. 7.8). The germination fraction equals G; germinating seeds reproduce and produce Y new seeds. A fraction $1 - G$ of the seed does not germinate, these seeds remaining dormant in the soil; a fraction D dies each year.

We want to find the value of G that maximizes $\log \lambda$ for a given P. Taking the derivative with respect to G and setting it to zero gives $\mathrm{d}\log\lambda/\mathrm{d}G = P/G - (1-P)/(1-G) = 0$ so that the optimal germination fraction is:

$$G^* = P \tag{7.11}$$

This surprisingly simple equation states that the optimal germination fraction equals the probability of encountering a good year.

7.5.2 Ellner's model: ESS for germination fraction in an environment with DD

In Cohen's model a low germination fraction helps to 'average' reproduction over good and bad years. On the other hand, dormant seeds may not survive and when seeds are in the soil for many years reproductive events are fewer. Dormancy reduces the arithmetic mean of reproductive success over the years while it increases the geometric growth rate by limiting variance in reproductive success. Note that the geometric growth rate of a strict annual without a seed bank is zero as soon as the population encounters a year with total reproductive failure. In growing populations the disadvantage of fewer reproductive events in time is substantial (see Chapter 5). Any DD process that limits population growth diminishes this disadvantage of dormancy. In a well-mixed population with constant population size and with dormant seeds surviving with probability 1, it no longer matters in which year seeds germinate as there are no mortality costs and no costs of missing out on reproductive events. Additional advantages arising from dormancy may come from avoiding sib competition among seedlings and reducing inbreeding effects among the offspring of a single parent, because sibs germinate and reproduce in different years (Venable and Brown 1988 and references therein). In either case dormancy would be favoured as a way to escape crowding because most seeds are produced at times of high density and dormancy reduces the average density that seeds encounter when germinating (Ellner 1985a, b). Ellner (1985a, b) extended Cohen's model to include DD, using the reciprocal yield law (Chapter 4). If the yield parameters are constant in time the population will grow to some equilibrium density; at this point the only disadvantage of dormancy is the lower survival in the seed bank until germination. At the equilibrium there is no longer a premium for fast reproduction as in Cohen's model. Predictions of Cohen's model and Ellner's ESS model differ. While in Cohen's model increasing the frequency of bad years always favours a higher germination fraction, this is no longer true in the ESS model. With DD included, dormancy is favoured by the occurrence of superlative years in which the population density increases sharply, even if conditions are always favourable for reproduction. The reason for this is that a low germination fraction limits the reduction of number of individuals after a bad year and for a longer period retains the positive effects of an exceptionally good year. Another striking difference between the models is the effect of seed mortality (D). In Cohen's model the prediction of Eq. (7.11) does not depend strongly on the survival of seeds in the soil. In Ellner's ESS model, however, the optimal germination fraction goes to zero if most dormant seeds survive. The reason for this is as follows. In the ESS model an increase in survival leads to more seeds in the soil. When the population becomes more crowded, this reduces both the variance and the average of the success of germinating seeds for any germination fraction. As a result there is no longer a penalty for not germinating and the optimal germination fraction goes to zero.

7.5.3 Experimental tests for dormancy as a bet-hedging strategy

Clearly, the relationship between demographic parameters and the ESS with respect to the germination fraction is not as straightforward with DD as in Cohen's model. A thorough test would require long-term demographic data on germination fractions, survival of dormant seeds and germination and reproductive success. Such data are mostly absent and as a substitute several researchers have related germination fractions in

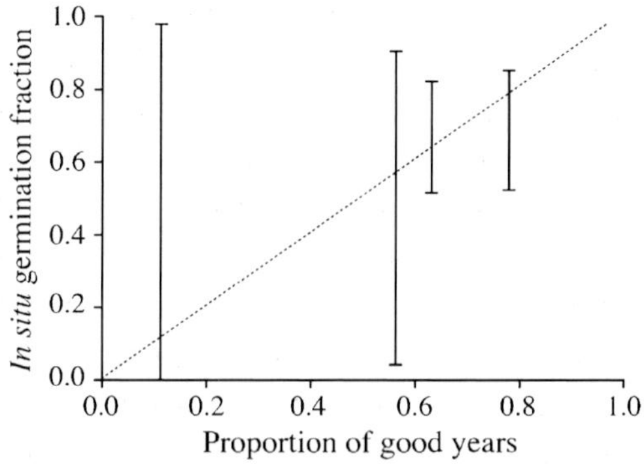

Fig. 7.10 Germination fractions of four populations of a desert annual *Plantago insularis* (woolly plantain), along a moisture gradient. Bars indicate the range observed over four years. The dotted line indicates the optimal germination fraction predicted for different proportions of good years (Eq. 7.11). Redrawn with permission from Clauss and Venable (2000), *American Naturalist*, University of Chicago Press.

different habitats to environmental factors, especially rainfall data. Using Cohen's prediction that the germination fraction should increase with the proportion of good years they tested for a positive correlation between the probability of sufficient rainfall and the germination fraction across different populations. Some studies indeed found such a positive correlation (Fig. 7.10) (e.g. Hacker 1984; Hacker and Ratcliff 1989; Philippi 1993; Clauss and Venable 2000). The general pattern is, however, rather unclear and some studies have even reported a decrease in germination fraction with the probability of good years (Jain 1982; Platenkamp 1991). Given the more complex relationship between demographic data and dormancy predicted by the ESS model this is perhaps not surprising. In addition, Clauss and Venable (2000) stressed that germination fractions should be measured within the natural population itself and not, for instance, in Petri dishes in the lab because the seed genotype that germinates most readily in the lab may germinate at the smallest rate when placed in its natural (dry) environment.

7.5.4 Dispersal, dormancy and seed size

Optimal seed size and germination rates do not solely depend on the probability of survival and establishment. Seed size also affects dispersal; in wind-dispersed species heavy seeds drop faster to the ground and are dispersed over smaller distances (Fenner 1985). Likewise, seed mass will affect germination probabilities and survival of dormant seeds in the soil. On the one hand, a plant species with small seeds has reduced predation and easier transport of seeds into deeper soil layers. On the other hand, the surface-to-volume ratio of small seeds is larger, which leads to better diffusion of water into the seed and germination inhibiting hormones out of the seed. Small seeds face greater problems emerging after deep burial and therefore often evolved a light requirement for germination, which, as a by-product, leads to a long-lived seed bank (Cook 1980). Venable and Brown (1988) drew attention to the fact that seed size, dispersal and dormancy are alternative ways to deal with variation in reproductive success. Big seeds reduce yearly variation in success because seedlings produced by such seeds are less likely to suffer from shading or drought. Likewise seed size, dormancy and dispersal share two other population dynamic functions: escape from high densities and escape from sib competition. Which of these traits is most strongly favoured will depend on the particular spatial and temporal structure of the habitat. For instance, if drought is the crucial factor for seedling establishment in a large homogeneous area dormancy will be a more efficient bet-hedging strategy than dispersal. Nevertheless, in many ways the three traits are complementary and we may expect negative covariance between them. A negative relationship between dormancy and dispersal has been reported in several data sets (Venable and Lawlor 1980; Klinkhamer *et al.* 1987; Rees 1997). Cook (1980) speaks of a continuum of species with at one end species with a small seed bank and a high capacity to disperse and at the other end species with a large seed bank and poor dispersal. Analysing existing data sets, Klinkhamer *et al.* (1987) found that composites with a special dispersal device (pappus) had higher germination rates and were less likely to have a persistent seed bank than composites without such a device. Rees (1997) analysed data from Roberts (e.g. Roberts and Boddrell 1983) and the Sheffield database (Hodgson *et al.* 1995) and found evidence that dormancy was negatively correlated with both adult longevity (temporal risk-spreading) and lateral spread (spatial risk-spreading).

There is increasing evidence for a negative relationship between seed mass and dormancy in the field, even though Klinkhamer *et al.* (1987) could not demonstrate this. In particular plants with small seeds can build up large seed banks in the soil (Cook 1980). Rees (1997) concluded that seed mass was positively correlated to plant height, longevity and animal dispersal and negatively related to lateral spread and seed bank score. Hodkinson *et al.* (1998) studied data of the British flora and found, after controlling for phylogeny and life history, that seed mass was positively correlated with habitat shade and negatively with seed longevity.

7.6 Seed heteromorphism

Imbert (2002) provided a recent overview of seed heteromorphism. In many species, individuals produce two types of seeds with discrete differences in size, shape, colour or dormancy (Fenner 1985; Venable 1985; Venable *et al.* 1995; Imbert 2002; Chapter 11). Such somatic seed heteromorphism occurs in families such as the Asteraceae, Poaceae, Brassicaceae, Caryophyllaceae, Polygonaceae and Boraginaceae. Well known are the dimorphisms in composites that have two types of florets in each flowerhead; ray florets on the perimeter of the flowerhead and disk florets in the centre. Often also two types of seeds are produced, which differ in size and germination rate. Ray seeds usually are bigger and more dormant than disk seeds (Fig. 7.11) (Venable *et al.* 1995; Imbert 2002). Venable and Lawlor (1980) interpreted this dimorphism in terms of bet-hedging. In a heterogeneous environment, the variance in temporal success for the most dispersing type (small seed) is less reduced by dormancy compared to the non-dispersing type (big seed) because the smaller seeds already sample over more patches and so reduce the variance. In *Chenopodium album* (lamb's quarters) four different kinds of seed are produced on the same plant (Imbert 2002). Seeds are brown or black and reticulate or smooth. Thin-walled big brown seeds, which constitute only 3% of the seeds of a plant, germinate more quickly than black seeds. Seedlings arising from these seeds have a high probability of dying over winter but if they survive they develop into very large plants with high seed production.

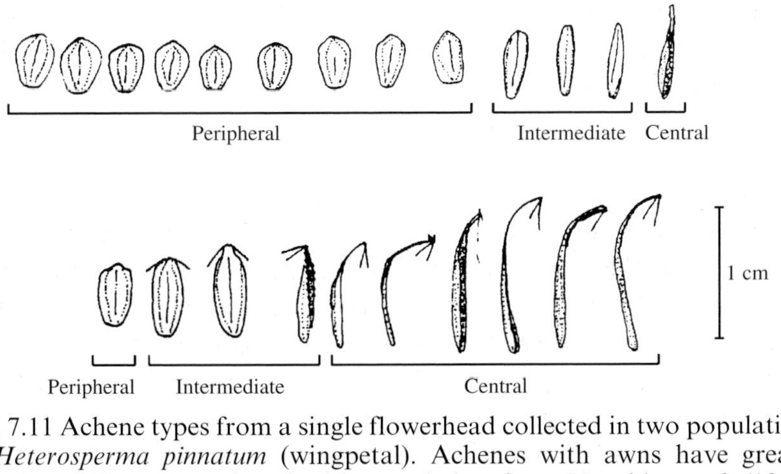

Fig. 7.11 Achene types from a single flowerhead collected in two populations of *Heterosperma pinnatum* (wingpetal). Achenes with awns have greater adhesive ability. Reproduced with permission from Venable *et al.* (1995), *American Journal of Botany*.

Harper (1977) and Venable (1985) discussed the evolution of seed heteromorphism. If all seeds on a plant are under the same evolutionary constraints, seed heteromorphism can evolve if intermediate adaptations experience low fitness, for instance because two types of years (wet and dry) occur and intermediate adaptations do very poorly in both.

Two types of heteromorphism may occur depending on the level of similarity of the evolutionary constraints. If the morphs differ substantially each morph may specialize in a different year type or habitat type, so that both types experience a high (arithmetic) mean reproductive success but also a high variance (which reduces the geometric growth rate over the years). Venable called this the high-risk–high-risk (HRHR) heteromorphism. If both morphs do best in the same year type but differ somewhat in their constraints one morph may be selected for dormancy or dispersal while the other is not. For instance, if seeds in some positions at the plant are less effective nutrient sinks and therefore smaller they may more easily specialize in dispersal than bigger seeds at other positions (Venable 1985). The dormant or dispersed seeds will have a lower (arithmetic) mean with low variance between years while the non-dormant type will have a higher (arithmetic) mean with high variance between the years. This is called the high-risk–low-risk heteromorphism (HRLR). We already discussed that such HRLR seed heteromorphisms are common in many plant families. The fact that seed heteromorphisms are especially common among annual plants in arid environments underlines the fact that bet-hedging is indeed a major factor in the evolution of such traits (Harper 1977; Venable *et al.* 1995; Imbert 2002; Chapter 11).

To sum up: seedling emergence and recruitment are positively associated with seed mass both across and within species. The classical Smith–Fretwell model does not take competition among seedlings into account. The model predicts that plants should adopt a single seed size strategy. ESS models that do take competition among seedlings into account predict that individual plants should produce a range of seed sizes. Plants face a considerable risk of failure when reproducing. Dormancy and dispersal are two ways to spread the risk of failure. For a very simple model of an annual plant with a seed bank, Cohen showed that the optimal germination fraction equals the probability of a good year. Dormancy reduces the arithmetic mean of the reproductive success over the years while it increases the geometric growth rate. ESS models that include DD generate different predictions. Dispersal and dormancy are two ways of dealing with environmental variance. A negative correlation between the two was predicted and found across species.

Questions

(1) What are the two assumptions that underlie the Smith–Fretwell model?

(2) Describe, in terms of the Smith–Fretwell model, why larger seeds are predicted for shaded habitats.

(3) What are the predictions of the Smith–Fretwell model and the Geritz model for the effects of an increasing investment in seeds on optimal seed size?

(4) Consider a genotype of an annual in a fluctuating environment such that fitness per germinated seed equals 10 in good years and 0.1 in bad years. We consider the number of seeds just before germination and start with 10 seeds. What is the number of seeds of a genotype without dormancy after a good and a bad year? What is the number of seeds of a genotype with 50% dormancy after a good and a bad year if 50% of the dormant seeds die each year?

8

Sex allocation theory for partially selfing plants

Some plant species combine outcrossing with self-pollination. This has some interesting consequences for sex allocation. In this chapter we first ask how selfing quantitatively affects allocation strategies. Next we ask how females can be maintained in combination with cosexuals in gynodioecious species and at what frequency. We have already asked this question for outcrossers in Chapter 4, but as we will see, in selfing species there exists an additional outbreeding advantage for females. This outbreeding advantage helps to maintain females in the population. Finally we ask when selfing cosexual species can evolve towards dioecy.

8.1 The effect of prior selfing on sex allocation

Consider two genotypes with different alleles coding for selfing or outcrossing. Each seed of the selfer contains two copies of the corresponding 'selfing' allele, whereas an outcrossed seed contains only a single, maternal copy of the 'outcrossing' allele. Twice as many copies of the 'selfing' allele are passed on through the seeds. A selfed seed, however, may perform worse than an outcross seed, through a combination of lower germination and survival, poorer growth or a lessened ability to produce seeds and viable pollen. This reduced performance is called inbreeding depression (δ) and is defined as:

$$\delta = 1 - \frac{\text{performance self seed}}{\text{performance outcross seed}} \tag{8.1}$$

If the performance of selfed seeds equals that of outcross seeds, inbreeding depression equals zero. If the selfed seeds do very poorly, inbreeding depression approaches 1. The old-fashioned way to calculate δ is to take an average over the population for the performance of all selfed and outcrossed seeds. This gives a single value for population δ, which we can use in the models. Note, however, that when data were collected by crossing plants, it is usually more informative to calculate inbreeding depression for individual plants (Holsinger 1992). This then yields a value per plant, rather than a single value for the population.

Now suppose there is some fixed selfing rate in the population. This could occur with prior selfing, in which a fraction of the ovules are selfed even before the flower opens so that there is never competition between outcross and self pollen grains for the selfed ovules. A partially selfing plant transfers alleles to the next generation in three ways. First, all seeds that are outcrossed contain a single copy of the allele. Second, all seeds that are selfed contain two copies. The chance that they transfer these alleles to the next generation is, however, reduced by a factor $(1-\delta)$. Third, through the pollen the plant can be the father of seeds on other plants; outcross pollen can fertilize the non-selfed ovules on other plants.

Fitness W is the number of allele copies that is transmitted to the next generation, for a certain sex allocation strategy. In analogy to Eq. (4.2) for an outcrossing species, absolute fitness of a rare mutant with allocation strategy r_m to male function and $1-r_m$ to female function in a partly selfing population of the common type that allocates r to male and $1-r$ to female becomes:

$$W_m = (1-S)(1-r_m) + 2S(1-\delta)(1-r_m) + \frac{r_m}{r}(1-S)(1-r) = 0 \quad (8.2)$$

The terms correspond to allele copies transferred through outcrossed seeds, selfed seeds and siring seeds on other plants. The fitness of the common type follows from substituting r for r_m. With some algebra this yields $W = (2-2S\delta)(1-r)$. For convenience linear gain curves are assumed. Now how does selfing affect the ESS value r^* for male allocation? We use Parker's method to differentiate the absolute (or invasion) fitness of the mutant with respect to the evolutionary variable r_m. This yields $\mathrm{d}W_m/\mathrm{d}r_m = -(1-S) - 2S(1-\delta) + (1/r)(1-S)(1-r) = 0$, so that

$$r^*/(1-r^*) = (1-S)/(1+S-2S\delta) \quad (8.3)$$

or

$$r^* = (1-S)/(2-2S\delta) \quad (8.4)$$

The second derivative of W_m with respect to r_m is equal to zero; therefore r^* is a weak ESS. Allocation to male function r_m declines steadily with the selfing rate S (Fig. 8.1). This result is due to the assumption of a fixed, prior selfing rate which 'guarantees' that a certain fraction of the seeds will be selfed regardless of the allocation strategy of the plant. High inbreeding depression selects for a higher value of r^*, pushing male allocation back to 0.5 (Fig. 8.1).

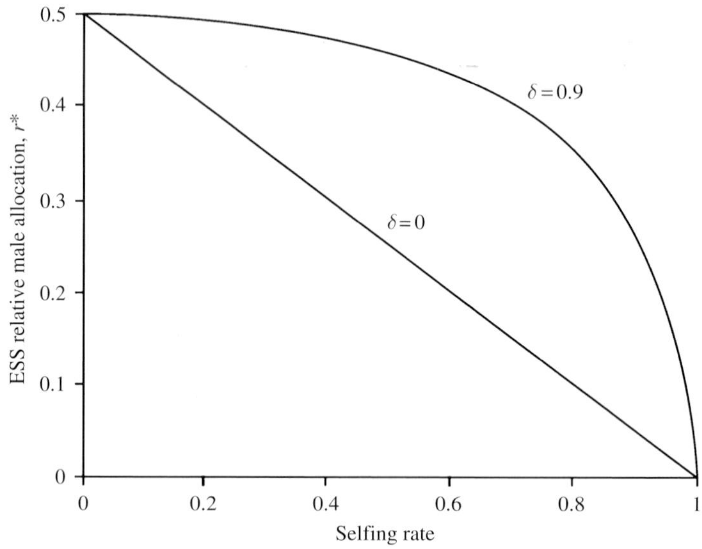

Fig. 8.1 Selfing selects for a female bias in sex allocation. If the selfing rate is constant, inbreeding depression (δ) has the opposite effect, pushing r^*, the ESS for male allocation, back to 0.5 (see Eq. 8.4).

8.2 Effect of selfing on sex allocation if the selfing rate is not fixed

So far we have assumed a fixed selfing rate, as in prior selfing. If outcross pollen arrives at the same time as self pollen, the pollen grains will compete for access to ovules and the selfing rate is no longer a fixed parameter. Does more male allocation (a higher pollen production) result in a higher selfing rate? In plants with a single flower this is perhaps least likely. Very few pollen grains are necessary for fertilization, compared to the tens of thousands produced in a flower and selfing rate of the flower will depend more on synchronization of male and female function within a flower than on the number of pollen grains produced. In species with many flowers, selfing is more likely to increase with male allocation. Damgaard and Abbott (1995) found the selfing rate to increase with the pollen/ovule ratio of flowers in *Senecio vulgaris* (groundsel) and *Brassica napus* (rape) (Fig. 8.2). Stigmas could simply receive more self pollen if the pollen/ovule ratio is high. At a plant level, producing more flowers with pollen, but each with fewer seeds, induces pollinators to make longer foraging bouts, resulting in more geitonogamous pollination (Fig. 8.3) (reviewed by Snow *et al.* 1996; Chapter 4). For plants with abiotic pollination one would also expect higher selfing with increased pollen release: each pollen grain has a small chance of being captured on one of the plant's own stigmas

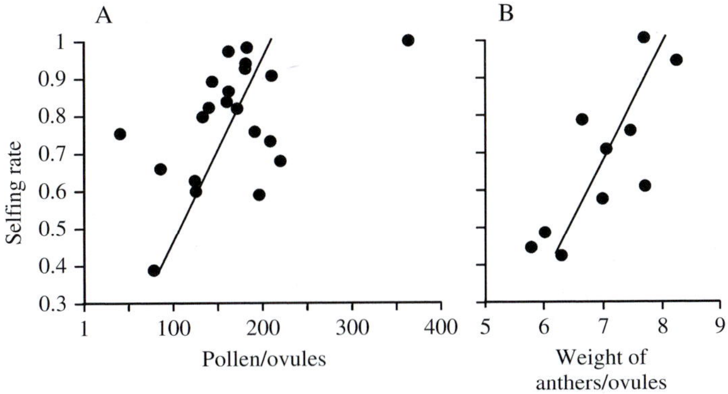

Fig. 8.2 Individuals with a high relative male allocation had a higher selfing rate in (A) *Senecio vulgaris* and (B) *Brassica napus*. Redrawn with permission from Damgaard and Abbott (1995), *Evolution*.

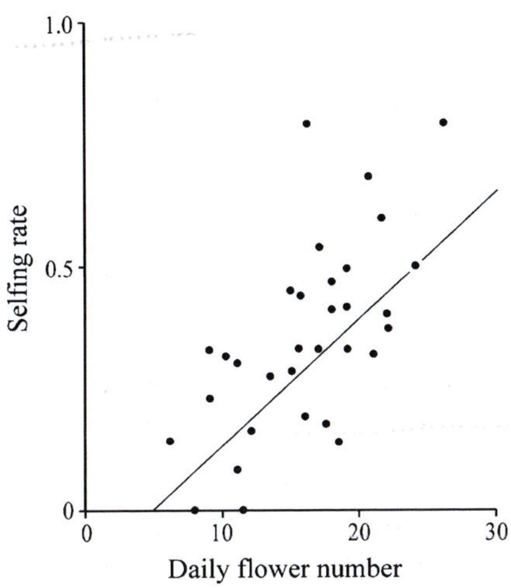

Fig. 8.3 Because bees make longer foraging bouts on plants with many flowers the rate of geitonogamous selfing usually increases with flower number. The example is *Malva moschata* (musk mellow). Redrawn with permission from Crawford (1984), Blackwell.

and the more pollen is locally released, the higher the rate of geitonogamous selfing. For plants with abiotic pollination (Gregorius *et al.* 1987), the shape of the curve of selfing rate versus pollen released is very similar to that for plants with biotic pollination (Eq. 3.11).

Charlesworth and Charlesworth (1981) examined how results are affected if we no longer consider the selfing rate constant but write S as an increasing function of r_{m}. Absolute fitness of a mutant is similar to Eq. (8.2):

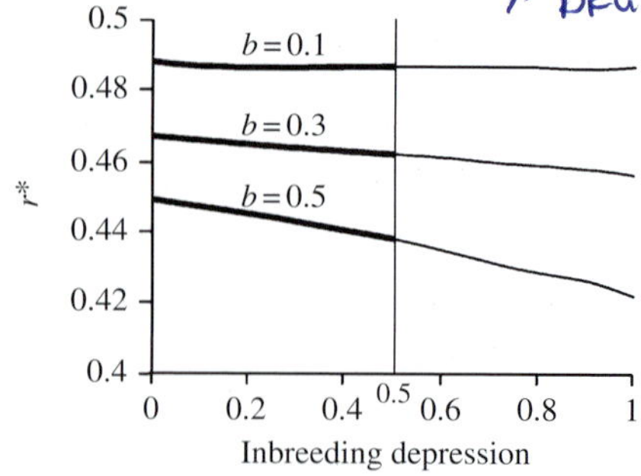

Fig. 8.4 The ESS r^* for allocation to male function when the selfing rate S is no longer constant but increases with male allocation r; $S = br$. More selfing (a steeper value for slope b) selects for a female bias in allocation. Contrary to earlier results shown in Fig. 8.1, inbreeding depression now operates in the same direction as selfing, selecting for more female-biased allocation. For $\delta < 0.5$ the second derivative of mutant fitness with respect to r_m is negative, so that r^* is a strong ESS. At the right-hand side of the graph, where $\delta > 0.5$, the second derivative is positive, so that r^* is no longer an ESS, but a branching point.

$$W_m = (1 - S_m)(1 - r_m) + 2S_m(1 - \delta)(1 - r_m) + \frac{r_m}{r}(1 - S)(1 - r) \quad (8.5)$$

Here S is the selfing rate of the common type with strategy r, and S_m is the selfing rate of the mutant with male allocation r_m. To find the ESS we use again Parker's method to differentiate the fitness of the mutant with respect to r_m. We abbreviate $dS_m/dr_m = S'_m$.

$$\frac{dW_m}{dr_m} = -(1 - S_m) + (1 - r_m)(-S'_m) - 2S_m(1 - \delta) + 2(1 - \delta)S'_m(1 - r_m) + \frac{1}{r}(1 - S)(1 - r) = 0 \quad (8.6)$$

By setting $r_m = r$ and $S_m = S$ it is still possible to write Eq. (8.6) as $r^* = \ldots$, but it is just as easy to specify some convenient function for the selfing rate like $S_m = a + br_m$ and compute r^* numerically for a given value of δ. Of course a and b can only take on values such that the selfing rate remains between 0 and 1. A way to compute r^* values is to use a spreadsheet: use the first column to generate r^* values between 0 and 1 with steps of 0.01, type Eq. (8.6) (with $r = r_m$) in the top cell of the next column and copy it down. The r_m-value at which Eq. (8.6) becomes zero gives the candidate ESS r^*. As before with a constant selfing rate, more selfing results in female bias (Fig. 8.4). However, inbreeding depression can now have the same effect as selfing, instead of working in the opposite direction as with a fixed selfing rate (compare Fig. 8.1).

8.3 Data on selfing and pollen/ovule ratios

For some situations one could equate r to relative allocation to pollen and $1-r$ to relative allocation to seeds. When P is the number of pollen grains, plants have T resources and C_p is the production cost per pollen grain: $rT = PC_p$, so that $P = rT/C_p$. Similarly if N denotes the number of seeds and C_s the cost per seed, then $(1-r)T = NC_s$ and $N = (1-r)T/C_s$. When only a fraction γ of the ovules (O) produces a seed, the number of seeds equals $N = \gamma O$. We neglect costs of ovules that do not produce a seed. The ratio of male to female allocation is then $r/(1-r) = TPC_p/(TNC_s) = PC_p/(\gamma OC_s)$, so that:

$$P/N = \frac{C_s}{C_p}\frac{r}{1-r} \quad \text{and} \quad P/O = \gamma\frac{C_s}{C_p}\frac{r}{1-r} \tag{8.7}$$

Cruden (1976) showed that pollen/ovule ratios differed greatly between plant species with different breeding systems, and found the ratio to be highest in outcrossing plants and lowest in cleistogamous plants in which the flowers do not even open and 100% selfing occurs (Table 8.1). Cruden's analysis did not take into account the costs per pollen grain or per seed and assumed that each ovule develops a seed. Taking this extra information into account, as in Eq. (8.7), we can be more precise about the shape of the predicted relationship (Charnov 1982a). At the ESS the allocation r^* is known (Eq. 8.3) and substituting this in Eq. (8.7) gives:

$$P/O = \gamma\frac{C_s}{C_p}\frac{r^*}{1-r^*} = \gamma\frac{C_s}{C_p}\frac{1-S}{1+S-2S\delta} \tag{8.8a}$$

or

$$\log(P/O) = \log\left(\frac{1-S}{1+S-2S\delta}\right) + \log\left(\frac{\gamma C_s}{C_p}\right) \tag{8.8b}$$

Table 8.1 *Pollen/ovule ratios differ strongly between species with different selfing rates (Cruden 1976), in accordance with theory*

Breeding system	Number of species	Pollen/ovule ratio (standard error)
Cleistogamy[a]	6	4.7 (0.7)
Fully selfing	7	27.7 (3.1)
Mostly selfing	20	168.5 (22.1)
Mostly outcrossing	38	796.0 (87.7)
Fully outcrossing	25	5859.2 (936.5)

[a] Cleistogamous flowers remain closed and full selfing occurs.

Table 8.2 *Data on pollen/ovule ratio (*P/O*), selfing rate (*S*) and costs of seeds*[a] *(*C_s*) and pollen (*C_p*) in* Astragalus

Species	P/O	S	C_s (mg)	C_p (mg 10^{-6})	$(C_s/C_p) \times 10^3$
A. edulis	715	0.324	34.82	21.71	1604
A. epiglottis ssp.*epiglottis*	196	0.625	4.19	6.21	674
A. hamosus	193	0.565	8.18	15.48	528
A. epiglottis ssp. *asperulus*	170	0.583	2.59	7.14	363
A. cymbicarpos					
chasmogamous flowers	151	0.201	2.82	22.23	127
cleistogamous flowers	84	0.676	9.60	18.06	531

[a] Cost of seeds were estimated by taking seed weight.
Source: Gallardo *et al.* (1994).

This is a full regression model for predicting pollen/ovule ratio as a function of selfing rate, pollen and seed weight, seed set per ovule and the level of inbreeding depression (Charnov 1982a). Charnov (1982a) showed that pollen grain volume, as an estimate of C_p, explained extra variation in P/O in addition to the effects of selfing rate shown in Table 8.1. Data for testing the model of Eq. (8.8b) were collected by Gallardo *et al.* (1994) for *Astragalus* species. These species have ingenious flowers with ten anthers, in which some of the pollen is ripe before the flower opens, causing prior selfing. By caging flowers, Gallardo *et al.* estimated the rate of prior selfing for five species (Table 8.2). Seed and pollen weight varied greatly between species, so these factors need to be taken into account for predicting the pollen/ovule ratio. An estimate of inbreeding depression δ and seed abortion is missing. We arbitrarily set $\delta = 0.5$ and $\gamma = 1$ for all species. We can combine all effects by computing the complete right-hand side of Eq. (8.8b), using the values in Table 8.2. We denote this as log X, so that $\log(P/O) = \log(X)$. Figure 8.5 shows that the correlation between $\log(X)$ and $\log(P/O)$ is positive, as predicted. Gallardo *et al.* (1994) concluded that the *Astragalus* data support the prediction from sex allocation theory that selfing species invest relatively less resource in pollen.

Many other studies have found that selfing reduces the pollen/ovule ratio. For instance, in *Plantago* this ratio ranges from 37 in the cleistogamous *P. patagonica* (woolly plantain), from 900 to 6500 in the predominantly selfing *P. major* (rat-tail plantain), *P. ovata* (desert indianwheat) and *P. drummondii* (dark plantain), from 4000 to 18 000 in the mostly outcrossing *P. coronopus* (buck's-horn plantain) and *P. lagopus*, and from 12 000 to 40 000 in the outcrossing *P. media* (hoary plantain), *P. maritima* (sea plantain) and *P. lanceolata* (ribwort) (Kuiper and Bos 1992; Sharma *et al.* 1999).

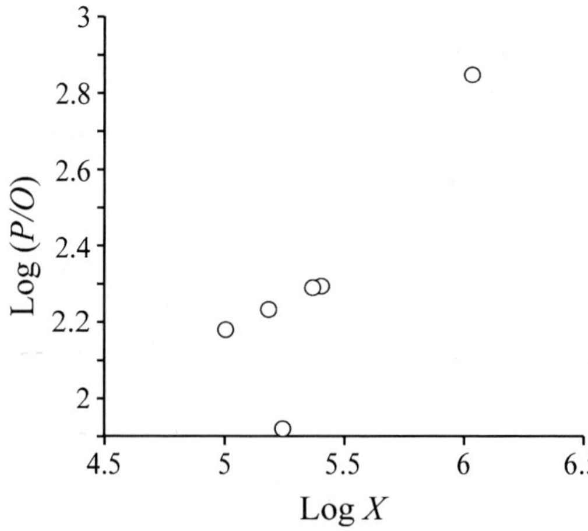

Fig. 8.5 A combination of selfing rate and costs of seeds and ovules explains variation in pollen/ovule (P/O) ratio in *Astragalus* species (Gallardo *et al.* 1994). Here, $\log(X) = \log[(1 - S)/(1 + S - 2S\delta)] + \log[\gamma(C_o/C_p)]$ (Eq. 8.8b) with $\delta = 0.5$ and $\gamma = 1$. Since the error in X is probably largest (γ and δ are unknown) we should consider $\log(X)$ as dependent variable in ordinary LS regression. This results in a slope of 1.01, and $r^2 = 0.77$.

The same trend was found within species for *Oryza perennis* (wild rice) (Charnov 1987) and *Gilia achilleifolia* (blue gilia) (Schoen 1982) (Fig. 8.6). Ramsey (1993) compared two populations of *Blandfordia grandiflora* (Christmas bells) in Australia. Pollen/ovule ratio was 11 500 in the outcrossing, coastal and 5600 in the selfing, inland population, consistent with theory.

8.4 Data on selfing and seed-to-flower ratios

In some species not every ovule produces a seed, even if pollination is assured (Chapter 13), and this can be regarded as a sex allocation problem. Similar to Eq. (8.7) and using Eq. (8.3) for sex allocation r^* at the ESS, we can write:

$$\text{seed-to-flower ratio} = \frac{C_f}{C_s}\frac{(1 - r^*)}{r^*} = \frac{C_f}{C_s}\frac{(1 + S - 2S\delta)}{1 - S} \qquad (8.9)$$

Without selfing ($S = 0$) the seed-to-flower ratio (Eq. 8.9) reduces to C_f/C_s, the ratio of costs per flower and costs per seed. Outcrossing plants should allocate equally to male and female function (Chapter 4), but if one function is more expensive than the other, the cheaper sex should be overproduced (Fisher 1930).

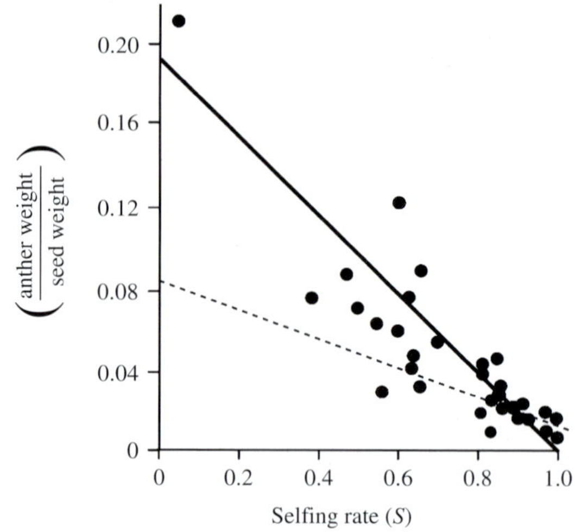

Fig. 8.6 The ratio of male (total anther weight) to female (total seed weight) allocation declined with selfing rate in varieties of *Oryza perennis* (Charnov 1987). The dotted line indicates a similar relation for *Gilia achilleifolia* (Schoen 1982). Redrawn from Fig. 1 in Charnov (1987), *Evolutionary Ecology*, with kind permission of Springer Science and Business Media.

Rademaker and de Jong (2000) calculated the ESS seed-to-flower ratio, using data on *Cynoglossum officinale* (Fig. 8.7) and *Echium vulgare*. Both plants belong to the Boraginaceae; they have four ovules per flower, so that each flower can produce at most four seeds. In the field the seed set is much lower: on average, only 0.62 seeds per flower are produced in *C. officinale* and 0.96 in *E. vulgare*. Average selfing rates were *c*.40% and *c*.10%, and best estimates for inbreeding depression were 0.16 and 0.5, for *C. officinale* and *E. vulgare*, respectively. The cost of a seed, relative to a flower, was estimated at 3.64 in *C. officinale* (Eq. 3.2) and 1.87 in *E. vulgare*. Equation (8.9) predicts an ESS seed-to-flower ratio of 0.59 for *C. officinale* and 0.53 for *E. vulgare*. Predictions are somewhat lower than observed values. Nevertheless the simple model clearly illustrates that sex allocation theory predicts low seed set per flower if seeds are more costly than flowers with pollen. This is the case in these boraginaceous species, but the effect should be even stronger in other species with very large fruits, like avocado or melon.

In conclusion, data strongly support the theoretical prediction that selfing promotes a female bias in allocation, either through lowering the pollen/ovule ratio or by increasing seed set per flower. The effect of inbreeding depression is less clear and since this factor helps to determine the shape of the relationship,

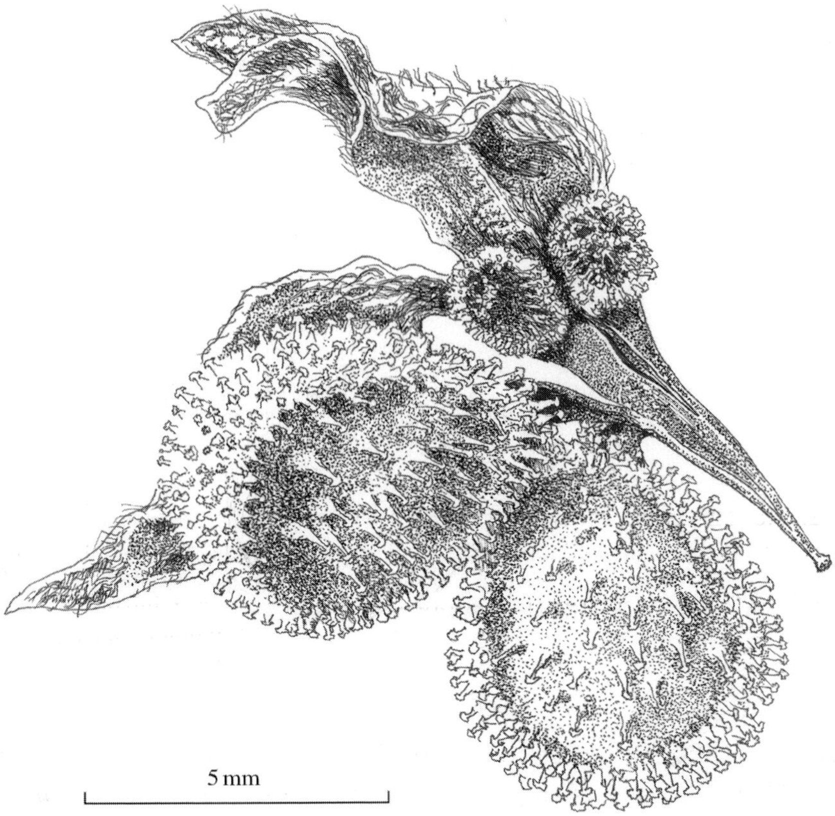

Fig. 8.7 Seed production in *Cynoglossum officinale*. Each flower has four ovules but, in the flower shown two fruits (seeds) were aborted. On average, in the field less than one seed per flower is formed, even with sufficient fertilization. Seeds weigh 20–30 mg and production costs of one seed were estimated to be 3.64 times higher than those of a flower, pollen included. Drawing by Frank van der Meer.

it is still an open question how well the shape of the relation between selfing and male allocation fits the theory.

8.5 Gynodioecy and androdioecy: when can females or males invade into a cosexual population?

We have already considered this problem for an outcrossing species (Chapter 4), but selfing makes a big difference. A female plant in a selfing population will produce 100% outcross seeds and if the selfed seeds on cosexuals are all of poor quality due to inbreeding depression, this represents an outcrossing

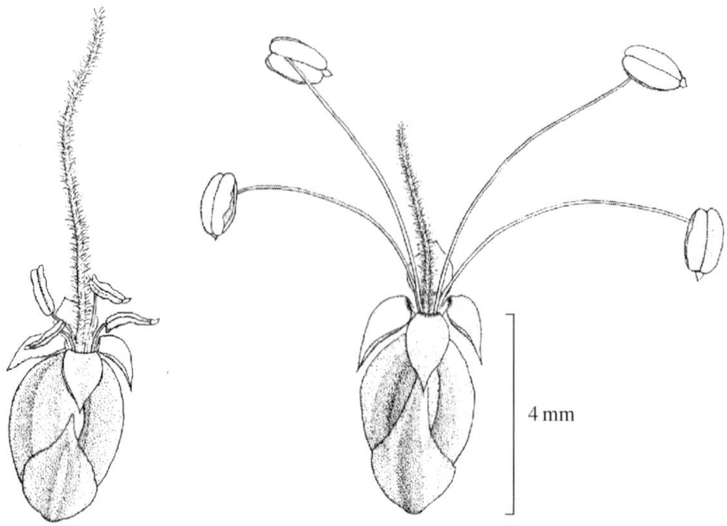

Fig. 8.8 Flower of a male-sterile and hermaphrodite individual of the gynodioecious *Plantago coronopus*. Drawing by J. van der Toorn. Reproduced with permission from Koelewijn and van Damme (1995), *Genetics*, Genetics Society of America.

advantage for the female (Lloyd 1975). The situation is different for a male in an otherwise selfing population; the total number of seeds the male can sire is restricted to the fraction $(1 - S)$ of the ovules on the cosexuals. The fitness of the common type is $W = (2 - 2S\delta)(1 - r)$ (substitute r for r_m in Eq. 8.2) and at the ESS: $r^* = (1 - S)/(2 - 2S\delta)$ (Eq. 8.4). Combining this gives the fitness of the cosexual at the ESS:

$$W = (2 - 2S\delta)(1 - r^*) = 1 - 2S\delta + S \tag{8.10}$$

Now suppose hermaphrodite plants can freely reallocate resources to the two sex functions (full compensation). This is almost certainly an overestimate of what they can do. Male-sterile plants often have the anthers aborted in a rather late stage. The filament is then produced and the anther contains infertile pollen grains. These useless remnants of the hermaphrodite flower may incur some cost (Fig. 8.8), so that compensation is only partial. Now by substituting $r_m = 0$ in Eq. (8.2), we find that the absolute fitness of a rare female mutant is $W_m = 1$. A rare female can invade if her fitness exceeds that of the common cosexual type (Eq. 8.10): $1 > 1 - 2S\delta + S$ or $\delta > 0.5$. A rare male ($r_m = 1$) has absolute fitness $W_m = 1/r^*(1 - S)(1 - r^*)$, which can be rewritten as $1 - 2S\delta + S$. The fitness of the male equals that of the cosexual at the ESS (Eq. 8.10): the male cannot invade. His fitness is at best, with full compensation, equal to that of the cosexual. The result is summarized in Fig. 8.9. The

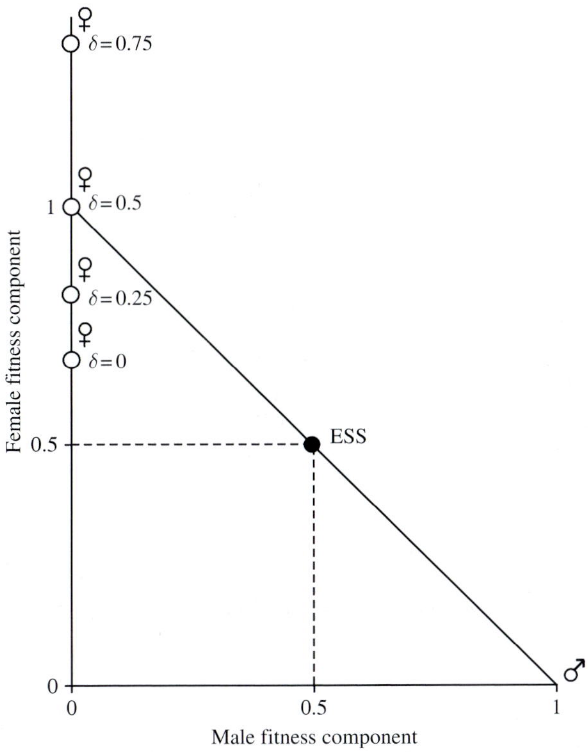

Fig. 8.9 So-called fitness set for a selfing species with inbreeding depression and full compensation. Graphs of this type were frequently used by Charnov (1982a). Fitness is scaled in such a way that at the ESS the male and female fitness contribution are both equal to 0.5. Next relative fitness of mutants with different sex allocations, ranging from full male to female, is computed using Eq. 8.2 and is plotted. In the figure the fitness set is a straight line: all mutants have fitness 1, except the female. This corresponds to the weak ESS of Fig. 4.1. Females are denoted as points on the y-axis. Females can invade if inbreeding depression (δ) is larger than 0.5. Pure males lie on the fitness set; they cannot invade, regardless of the value of δ.

females escape from selfing and enjoy an outbreeding advantage. If $\delta < 0.5$ females are selected against, if $\delta > 0.5$ they can invade. This analysis gives a possible explanation of why gynodioecy is so much more common than androdioecy.

We started with the highly optimistic assumption that females have full compensation. It is commonly assumed that the female boosts her seed production to $(1 + k)(1 - r)$, as compared to the seed production of the cosexual $(1 - r)$. The rare female can then invade if her fitness is larger than that of the cosexual:

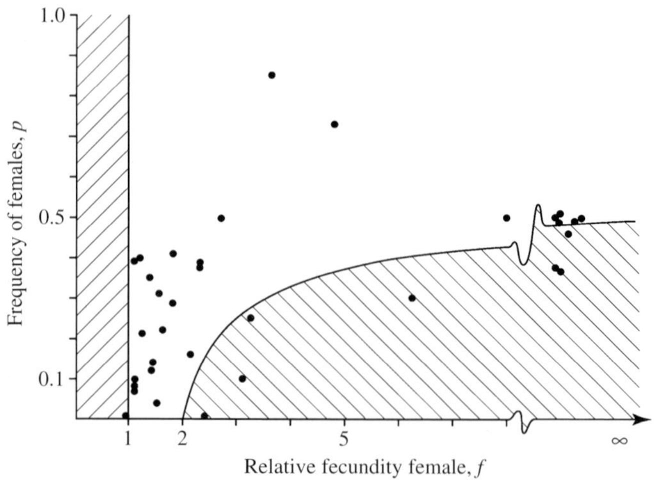

Fig. 8.10 Frequency of females p, as a function of the relative fecundity of females (f = seed production female/seed production hermaphrodite) in populations of different gynodioecious plants species. The drawn line is the theoretical relation, assuming either no selfing or no inbreeding depression in the hermaphrodite (Eq. 8.15). Redrawn with permission from Gouyon and Couvet (1987), Birkhäuser Verlag.

$$(1+k)(1-r^*) > (2-2S\delta)(1-r^*) \text{ or } k > 1-2S\delta \tag{8.11}$$

We adopt the $1+k$ notation of Charlesworth and Charlesworth (1978), one of the first papers on this subject, but some authors find it easier to define f as the relative fecundity of females which is seed production female/seed production hermaphrodite ($f = 1+k$). In that case females invade if:

$$\text{relative fecundity of female } (f) > 2-2S\delta \tag{8.12}$$

With Eq. (8.12) in mind it is often stated that the female should more than double the seed production of the cosexual to maintain herself in the population. Note that S and δ are multiplied: if one of these factors equals zero the 'factor two rule' holds. Also if both are small the 'factor two rule' is approximately correct. All the points to the right of $f = 2$ in Fig. 8.10 meet the 'factor two rule'. In many gynodioecious species f was found to be quite large, so that hermaphrodites are almost, but not fully, male (Fig. 8.10). The points between $f = 1$ and $f = 2$ denote species in which the females produce less than twice the number of seeds as hermaphrodites. Still, as Eq. (8.12) indicates, if selfing rate and inbreeding depression are high enough it would be possible for the female to maintain herself through an outbreeding advantage.

8.6 Are gynodioecious species typically self-compatible, with a high selfing rate and high inbreeding depression?

The female outbreeding advantage is based on self-compatibility (SC), a high selfing rate and a high level of inbreeding depression in hermaphrodites. Since selfing lowers the threshold for the females to invade, one expects that gynodioecy is much rarer in self-incompatible (SI) than in SC species. Hermaphrodite individuals in various gynodioecious species indeed appear to be SC in most cases. Jos van Damme (unpublished data, 2004) compiled a list of 120 gynodioecious plant species. From this list only three species were SI: *Plantago lanceolata*, *P. maritima* and *Hirschfeldia incana*. Collin and Shykoff (2003) reviewed selfing rates of cosexual plants in 20 species and found that selfing occurred frequently (Fig. 8.11). The majority of gynodioecious species are SC with considerable selfing.

Do gynodioecious species have high inbreeding depression? In hermaphrodite plants of *Cucurbita foetidissima* (buffalo gourd) $S = 0.71$ and $\delta = 0.73$ (Kohn and Biardi 1995), allowing maintenance of females even if they do not make more seeds than hermaphrodites (see Eq. 8.12). In *Thymus vulgaris* (thyme) the selfing rate was between 0% and 49% (Thompson *et al.* 1998). Inbreeding depression was measured in different life stages (Table 8.3). Because such effects are multiplicative, lifetime inbreeding depression was considerable: $\delta = 1 - (402/582) \times (29.9/45.2) \times (145.6/365.2) = 0.819$ (Table 8.3). Females produced between 1.5 and 3.5 times as many seeds as hermaphrodites, so average $f = 2.5$ ($k = 1.5$). At these values females can easily be maintained in the population.

In *Schiedea adamantis* (Caryophyllaceae) growing on Hawaii, females produced 2.32 times as many seeds as hermaphrodites (Sakai *et al.* 1997). Selfing

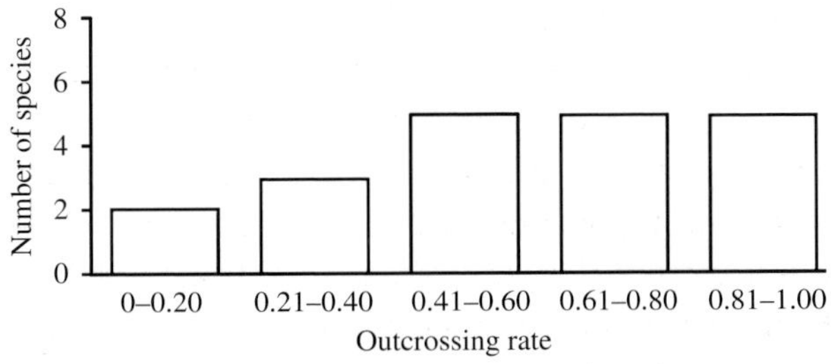

Fig. 8.11 Frequency distribution of the selfing rate of cosexual plants in field studies of 20 gynodioecious species. Redrawn with permission from Collin and Shykoff (2003), *American Journal of Botany*.

Table 8.3 *Inbreeding depression* δ *in the gynodioecious* Thymus vulgaris*;* δ *per life stage was calculated with Eq. (8.1)*

	Self	Outcross	δ
Seed production	402	582	0.31
Germination %	29.9	45.2	0.34
Seedling size	145.6	365.2	0.60

Source: Thompson *et al.* (1998).

was estimated at 0.68 and inbreeding depression at 0.60. Norman *et al.* (1995) compared the gynodioecious *Schiedea salicaria* and its hermaphrodite sister species *S. lydgatei*. Both species showed high inbreeding depression (*S. salicaria* $\delta = 0.79$, *S. lydgatei* $\delta = 0.62$) but differed in selfing rate ($S = 0.53$ for *S. salicaria* and $S = 0.14$ for *S. lydgatei*). This is consistent with the theory in this chapter. Schultz and Ganders (1996) studied another Hawaiian species, *Bidens sandvicensis*, in detail. They estimated $\delta = 0.69$, $k = 0.2$–0.5 and $S = 0.25$–0.50 in four populations (Schultz and Ganders 1996). Again inbreeding depression and selfing are high enough to favour maintenance of females (Eq. 8.12).

Data on some other species suggested that the outbreeding advantage was absent or insufficient for the maintenance of females. For instance, in *Echium vulgare* with $\delta = 0.5$ and $S = 0.1$, the relative seed fecundity of females should be higher than 1.9. This is much higher than the value $f = 1.2$, which has been observed in the field (Klinkhamer *et al.* 1991), so that the model predicts that females are rapidly eliminated from the population. The same conclusion holds for the SI *Plantago lanceolata*. With $S = 0$, females should double the seed production of the hermaphrodite to maintain themselves. Females produced 1.1 to 1.6 times as many seeds as hermaphrodites (Kuiper and Bos 1992), so they cannot maintain themselves in this model based on nuclear inheritance. For *P. lanceolata* cytoplasmic factors have been shown to be involved in male sterility, which facilitates the maintenance of females, a point to which we return in Chapter 16.

8.7 Female frequency increases with relative seed production of females

If the conditions for females to invade were satisfied, at what frequency would they be maintained in the population? Suppose at equilibrium a fraction p females and $(1 - p)$ cosexuals coexist. The $1 - p$ cosexual plants can then sire the seeds on other cosexuals and all seeds on females and their fitness is:

$$W_{\text{cos ex}} = (1 - S)(1 - r) + 2S(1 - \delta)(1 - r) + \frac{1}{1 - p}[p(1 - r)(1 + k) + (1 - p)(1 - S)(1 - r)] \quad (8.13a)$$

$$W_{\text{female}} = (1 - r)(1 + k) \quad (8.13b)$$

In equilibrium the fitness of the female equals the fitness of the cosexual. With some algebra this yields (Lloyd 1975; Charlesworth and Charlesworth 1978):

$$p = \frac{k + 2S\delta - 1}{2(k + S\delta)} \quad (8.14)$$

If either inbreeding depression or selfing rate is zero Eq. (8.14) reduces to:

$$p = \frac{k - 1}{2k} \quad (8.15)$$

If the relative fecundity of females (k) increases, their frequency increases until a maximum of 50% (Fig. 8.10). Selfing rate and inbreeding depression come as a combined parameter $S\delta$, so their effect on the frequency of females is symmetrical: more selfing and higher inbreeding depression both lead to a higher frequency of females in the population.

The predicted frequency of females can be compared with field data. For the *Schiedea adamantis* data from Sakai *et al.* (1997), the frequency of females according to Eq. (8.14) (with $k = 1.32$, $S = 0.68$, $\delta = 0.6$) should be 0.33 which is close to the observed frequency of 0.39. For *Thymus vulgaris* ($k = 1.5$, average $S = 0.25$, $\delta = 0.819$) the equation predicts that females maintain themselves at a frequency of 0.267. The average frequency of females was usually higher, *c.*60% (Thompson *et al.* 1998).

According to Eq. (8.14) female frequency should increase with selfing rate, the level of inbreeding depression and the relative fecundity of the females.

Selfing rates were measured by Sun and Ganders (1986) in populations of different *Bidens* species on Hawaii. They found an increase, from 10% to 40%, of the fraction of females with increasing levels of selfing in the cosexual (Fig. 8.12). A similar intraspecific trend was found in *Bidens sandvicensis* (Schultz and Ganders 1996) and in *Thymus vulgaris* (Thompson *et al.* 1998). We have found no studies that related female frequency to the level of inbreeding depression in different populations.

Relative fecundity of females was positively correlated with female frequency when comparing many species (Fig. 8.10) (Gouyon and Couvet 1987). Delph (1990a, b) found the same trend between and within species of *Hebe* (Scrophulariaceae) and Webb (1979) reported it for umbellifers.

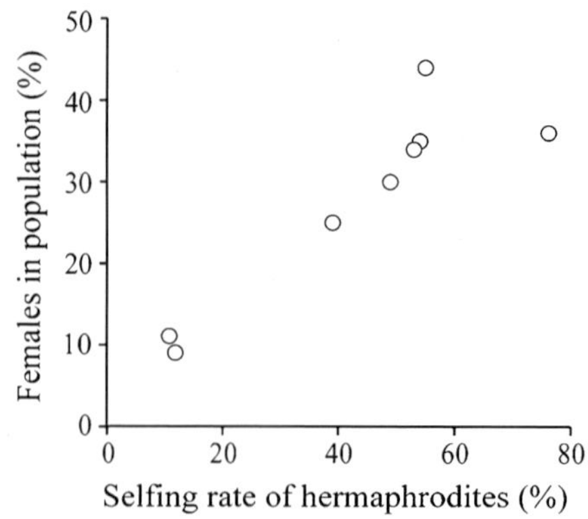

Fig. 8.12 The frequency of females increased with the selfing rate in populations of different species of *Bidens* in Hawaii. This indicates that females are maintained in populations, at least partly, because of an outbreeding advantage. Redrawn with permission from Sun and Ganders (1986), *American Journal of Botany*.

8.8 The effect of females on the sex allocation of the cosexual

How will the sex allocation of the hermaphrodite change if a fraction p females is present which each produce V seeds? Is there an ESS r^* value for the cosexual in such a mixed population? Absolute fitness of a rare cosexual mutant with allocation r_{m}, arriving in a mixed population of p females and $(1-p)$ cosexuals with allocation strategy r, is (see also Eq. 8.2):

$$W_{\mathrm{m}} = (1-S)(1-r_{\mathrm{m}}) + 2S(1-\delta)(1-r_{\mathrm{m}}) + \frac{r_{\mathrm{m}}}{r}\frac{1}{1-p}[pV + (1-p)(1-S)(1-r)] \qquad (8.16)$$

To simplify matters we set S equal to zero:

$$W_{\mathrm{m}} = (1-r_{\mathrm{m}}) + \frac{r_{\mathrm{m}}}{r}\frac{1}{1-p}[pV + (1-p)(1-r)] \qquad (8.17)$$

With Parker's method of differentiating the fitness function of the mutant to calculate the candidate ESS, we obtain:

$$\frac{\mathrm{d}W_{\mathrm{m}}}{\mathrm{d}r_{\mathrm{m}}} = -1 + \frac{1}{r}\frac{1}{(1-p)}[pV + (1-p)(1-r)] = 0 \qquad (8.18)$$

so that

$$r^* = \frac{1-p+pV}{2(1-p)} \qquad (8.19)$$

The candidate ESS for male allocation always increases with both p and V. When V is very high, so that females make up 50% of the population, r^* should be at its maximum value of 1, i.e. the hermaphrodite should become fully male. Note however that this not very likely because even in the most favourable case, with full compensation and $r = 0.5$, the female can at best double its seed production compared to the hermaphrodite. At lower, more realistic values of V, the hermaphrodite retains female function.

8.9 Sex allocation of the cosexual does not change with female frequency in *Thymus vulgaris*

Based on Eq. (8.19) one would expect that the presence of females favours more emphasis on male function in cosexuals. Manicacci *et al.* (1998) found this trend by comparing three different thyme species. However, within *Thymus vulgaris* male allocation did not seem to increase with the fraction of females in the population, contrary to the model predictions (Thompson and Tarayre 2000). Since both pollen and seed production have been shown to be genetically variable in thyme, this is a puzzling result. In Chapter 16 we will argue that with cytoplasmic inheritance of male sterility (as was shown to be the case in thyme) such a result can be explained.

8.10 The evolution of dioecy

Darwin (1878) stated that:

> a tree bearing numerous hermaphrodite flowers would rarely intercross with another tree ... Now the separation of the sexes, whether the plant were anemophilous or entomophilous, would most effectively bar self-fertilization.

Darwin was thus aware that geitonogamous selfing should increase with flower number and suggested that geitonogamy facilitated the transition towards dioecy. Can we quantify his suggestion?

We have already used Eq. (8.6) to study optimal sex allocation of a cosexual population when selfing depended on male allocation r_{m}. This equation was:

$$\frac{\mathrm{d}W_{\mathrm{m}}}{\mathrm{d}r_{\mathrm{m}}} = -(1 - S_{\mathrm{m}}) + (1 - r_{\mathrm{m}})(-S'_{\mathrm{m}}) - 2S_{\mathrm{m}}(1 - \delta) + 2(1 - \delta)S'_{\mathrm{m}}(1 - r_{\mathrm{m}}) + \frac{1}{r}(1 - S)(1 - r) = 0 \tag{8.20}$$

As stated above, S'_{m} is the derivative of the selfing rate of the mutant with respect to r_{m}. Equation (8.20) still contains r_{m} and S_{m} so now the second derivative of absolute fitness with respect to r_{m} can take on different values, both positive or negative. If the second derivative of fitness is smaller than zero this implies a fitness maximum at an intermediate value of r^*. Cosexuality is then stable. If the second derivative is positive, there is branching point. In that case cosexuality is not an ESS; instead a system with two separate sexes may evolve (de Jong and Geritz 2002). The second derivative of fitness with respect to r_m is:

$$\begin{aligned}\frac{\mathrm{d}^2 W_\mathrm{m}}{\mathrm{d}{r_\mathrm{m}}^2} &= S'_\mathrm{m} - S''_\mathrm{m} + S'_\mathrm{m} + r_\mathrm{m} S''_\mathrm{m} - 2S'_\mathrm{m}(1-\delta) \\ &\quad + 2(1-\delta)S''_\mathrm{m} - 2(1-\delta)(S'_\mathrm{m} + S''_\mathrm{m}) \\ &= 2S'_\mathrm{m} - S''_\mathrm{m} + r_\mathrm{m} S''_\mathrm{m} - 4(1-\delta)S'_\mathrm{m} \end{aligned} \tag{8.21}$$

In this equation S''_m is the second derivative of S_m with respect to r_m. Equation (8.21) can be rewritten as:

$$\frac{\mathrm{d}^2 W_\mathrm{m}}{\mathrm{d}{r_\mathrm{m}}^2} = (4\delta - 2)S'_\mathrm{m} + (1 - r_\mathrm{m})S''_\mathrm{m} \tag{8.22}$$

To interpret this equation it is good to reflect on what the first (S'_m) and second (S''_m) derivative of the selfing rate with respect to r_m mean. In the simplest case, case 1 in Fig. 8.13, the selfing rate is constant, independent of male allocation; both S'_m and S''_m are then zero. Equation (8.4) then gives r^*, which is a weak ESS. In case 2 the selfing rate increases linearly with r_m. The first derivative S'_m represents the constant slope of the curve for the selfing rate. The second derivative indicates if this slope changes; for case 2 the slope does not change and $S''_\mathrm{m} = 0$. Substituting this in Eq. (8.22) shows that the breaking point is at 50% inbreeding depression (Fig. 8.4). In case 2 with inbreeding depression less than 0.5, the second derivative of fitness is always negative, corresponding to an intermediate allocation r^* which constitutes a strong ESS. With inbreeding depression larger than 0.5 the second derivative of fitness is always greater than zero, corresponding to a branching point at r^*. Now cosexuality is no longer an ESS and at the branching point the population splits up into different types. Now consider case 3 when the selfing rate is a decelerating function of male allocation. In this case S''_m is negative. Because the last term, on the right-hand side in Eq. (8.22), is now negative it becomes more difficult for the sum of the two terms to be positive. A level of inbreeding higher than 0.5 is required before the second derivative of fitness becomes positive. A deceleration of the selfing rate with male allocation, as in case 3, stabilizes cosexuality.

So in the end we have quite simple conditions for stable cosexuality versus evolution towards a system with two sex types. The underlying mechanism is that with $\delta > 0.5$ and selfing as in case 2 (Fig. 8.13), the female gain curve is

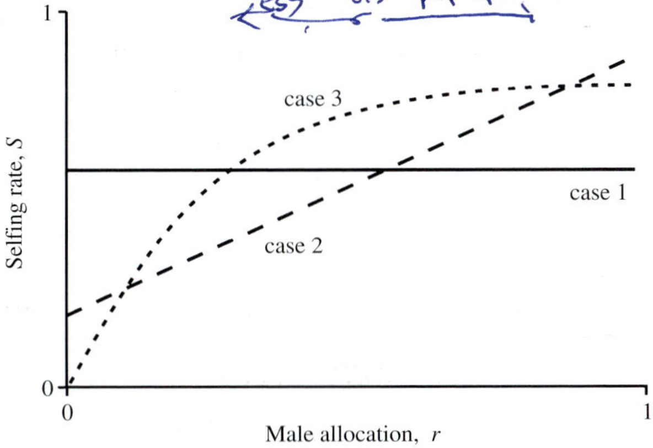

Fig. 8.13 Selfing rate (S) is a function of male allocation (r). In case 1 the selfing rate is constant and both the first and second derivative of the selfing rate with respect to male allocation are zero. In case 2 the selfing rate increases linearly with male allocation: the first derivative, the slope of the function, is a positive constant; the second derivative is zero. In case 3 a decelerating relation exists: the first derivative is positive and the second derivative negative.

accelerating. If a mutant allocates more to female function, direct fitness gain is proportional to the new seeds produced, i.e. proportional to $1 - r_m$. At the same time however the mutant allocates less to male function, so that the extra seeds produced are less likely to be selfed. With high inbreeding depression this 'extra' works out positive for fitness and the fitness gains from extra seeds produced increase more than proportionally with investment. The extra advantage for more female-biased plants on a gradual scale of sex allocation is similar to Lloyd's (1975) outbreeding advantage for females in a partially selfing cosexual population.

8.11 The route towards dioecy

A discussion on the major selective factors in the evolution of dioecy is still continuing (Freeman *et al.* 1997). Given that this was not a unique event but something that happened many times over in different plant families, both from monoecious and hermaphrodite ancestors, it should be possible to test a general theory in a number of specific situations. In the discussion authors have placed different emphases on the importance of outbreeding versus ecological factors. The old idea, dating back to Darwin (1877, 1878), was that dioecy evolved to overcome the negative effects of inbreeding, as an alternative to self-incompatibility. This is essentially what we have modelled here. Later several authors (Bawa 1980; Givnish 1982) focused on how ecological factors (power exponents of the male and female gain curves greater than 1; see Chapters 3 and 4) could lead to accelerating gain curves, even in self-incompatible species (Morgan 1994).

While both scenarios are theoretically possible and do not exclude each other in any way, we would argue that selection for outcrossing is the driving force behind the evolution of dioecy within the conditions set by ecology (the exponents of the gain curves). There are few convincing examples of accelerating gain curves due only to ecological factors (Chapter 3), while the combination of severe inbreeding depression and an increase in selfing with male allocation does readily produce such an accelerating relation (Eq. 8.22).

If the ecological hypothesis were true, this mechanism would work in both SI and SC species, while the outbreeding hypothesis only applies to SC species (Thomson and Barrett 1981). By examining the association of dioecy with the breeding system we can compare the value of the two different hypotheses. Charlesworth (1985) meticulously compiled data on the presence of SI in genera with dioecious plants. She found that out of 38 genera with dioecious species 16 genera contained both SC and SI species, 20 contained only SC species and only 2 genera contained only SI species. In 18 cases attempts were made to self the sporadically occurring cosexuals in (sub)dioecious species. All these cosexuals were found to be SC. In 22 cases cosexual species were compared which were a close relative of a dioecious species. All related cosexual species were SC. Nevertheless some SI genera with only SI members contained a few dioecious species. Charlesworth (2001) argued that these few exceptions may prove the rule that dioecy evolved as an outbreeding strategy in SC cosexuals. If an SI species becomes tetraploid, a doubling of the genome size, the self-incompatibility system is sometimes abolished. In such a case a previously SI species reverts back to SC and dioecy may evolve if the conditions outlined in this chapter are met. Data on North American *Lycium* species were consistent with this scenario (Miller and Venable 2002).

The outbreeding hypothesis can be tested further and, in principle, be rejected on its assumptions. If dioecious species can be traced back to an SI ancestor with the same ploidy level this refutes the outbreeding hypothesis. Similarly, if we can self female plants in some way or make brother–sister crossings and it turns out that inbreeding depression in dioecious species is below 50%, we must reject the outbreeding hypothesis. Finally, we can ask whether it is plausible that in the cosexual ancestor of a dioecious plant, geitonogamous selfing increases with male allocation as in Figs. 8.2 and 8.3.

Delesalle and Muenchow (1992) explicitly tested the outbreeding hypothesis by comparing closely related *Sagittaria* species with different sex systems. *Sagittaria latifolia* (broadleaf arrowhead) was predominantly dioecious,

Table 8.4 *Inbreeding depression δ in the dioecious* Sagittaria latifolia *and monoecious* S. australis; *δ for each life stage was calculated with Eq. (8.1), averaged over different habitats from data in Delesalle and Muenchow (1992)*

	S. latifolia			*S. australis*		
	Self	Outcross	δ	Self	Outcross	δ
Seed set %	43.0	49.5	0.131	72.5	81.0	0.105
Germination %	59.0	75.0	0.213	45.0	46.0	0.021
Survival %	55.5	62.5	0.112	45.5	47.5	0.042
Size (number of leaves) at six weeks	7.5	8.0	0.056	4.8	5.4	0.112

whereas *S. australis* (longbeak arrowhead) was monoecious. The dioecious species had more simultaneously open flowers and therefore more opportunities for geitonogamous selfing. Measurements of inbreeding depression included seed set, germination in incubators, seedling survival and the size of six-week-old seedlings (Table 8.4). When plants were then transferred to the field disaster struck and all plants were eaten by herbivores, terminating the experiment. Inbreeding depression over this six-week period can be estimated as in the thyme example (Table 8.3). In *S. latifolia* total inbreeding depression $= 1 - (43/49.5) \times (59/75) \times (55.5/62.5) \times (7.5/8) = 0.43$. In *S. australis* total inbreeding depression $= 1 - (72.5/81) \times (45/46) \times (45.5/47.5) \times (4.8/5.4) = 0.26$. Inbreeding depression was therefore higher in the dioecious species. It was not higher than 50%, contrary to predictions. However, 0.43 is close to 0.5 and life stages past the sixth week have not been included. Any inbreeding depression later in life, especially during reproduction, would increase the level of inbreeding depression. It could also be argued that measurements of germination and early survival were done in the lab. More stressful field conditions could result in larger differences in performance between self and outcross seeds. In a follow-up study on *S. latifolia*, Dorken *et al.* (2002) focused on populations in which the species was mostly dioecious, as in the study of Delesalle and Muenchow (1992), and on populations in which it was monoecious. Selfing rate as determined by isozymes was 41% in monoecious and 11% in subdioecious populations. Inbreeding depression was high ($\delta = 0.83$–0.84) in both populations. Since this species is monoecious, all selfing must be due to geitonogamous transfer of pollen between different flowers and inflorescences of this clonal, aquatic species. Rarely occurring hermaphrodite individuals had even higher (70%) selfing rates (Barrett 2003).

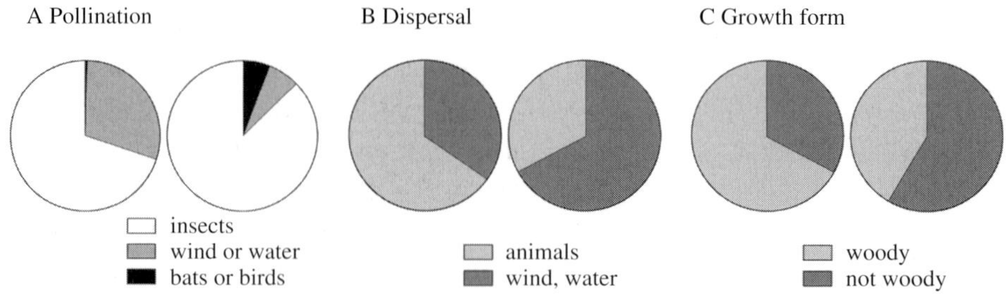

Fig. 8.14 Percentage of genera with various (A) pollen vectors, (B) types of seed dispersal and (C) growth forms when comparing *c*.800 genera that include one or more dioecious species (left) with all (*c*.13 500) angiosperm genera (right). Estimated from Renner and Ricklefs (1995).

As Barrett (2003) emphasized, the *Sagittaria* data are fully consistent with the outbreeding hypothesis for the evolution of dioecy.

What are the correlates of dioecy and do they give clues about its evolutionary history? Although the dioecious species are a mixed bag and occur in different genera, they are certainly not a random sample from all plants. Several studies of local floras have suggested that, at the species level, dioecy is associated with abiotic pollination, fleshy fruits, woody growth form, small inconspicuous flowers, inflorescences with many flowers and a tropical distribution (Fig. 8.14) (Bawa 1980; Givnish 1980). Vamosi *et al.* (2003) used the database from Fig. 8.14 in combination with a molecular phylogeny of the plant kingdom (Soltis *et al.* 1999), which allowed a phylogenetic correction of the data. Vamosi *et al.* (2003) found the same associations as reported before, but no indication that dioecious species had more numerous flowers.

As argued above, selection for outbreeding seems to be the driving force behind the evolution of dioecy. Nevertheless ecological factors can have a modifying effect. Strongly decelerating gain curves, due to efficiency factors or local competition, can stabilize cosexuality. The negative association between dioecy and biotic pollination could then result from a decelerating male gain curve in animal-pollinated plants. The association between dioecy and fleshy fruits could result from a release from local resource competition by animal dispersal.

To sum up: the theoretical prediction that selfing selects for female-biased allocation is strongly supported by the data. Females in gynodioecious plant species enjoy an outbreeding advantage. If more male allocation leads to a higher selfing rate and if inbreeding is high, this destabilizes cosexuality. We argued that selection for outbreeding in SC species drives evolution towards dioecy and is modified by ecological factors. Not everyone may agree with this last statement but all elements of the outbreeding hypothesis are testable and so far the hypothesis fared well.

Questions

(1) What is the outbreeding advantage for females in gynodioecious species?
(2) Do you expect that gynodioecious species are mostly SC or SI?
(3) *Plantago lanceolata* is an SI gynodioecious species. Females produce about 10–60% more seeds than hermaphrodites. Is this consistent with the models developed here?
(4) Why does the female gain curve accelerate when selfing increases with male allocation and when inbreeding depression is high?
(5) What is the ecological and what is the outbreeding hypothesis for the evolution of dioecy? Do these hypotheses exclude each other?
(6) How would you collect data to support or reject the two hypotheses for the evolution of dioecy?
(7) In the outbreeding hypothesis, does dioecy evolve gradually or with a big step, through gynodioecy?

9

Size-dependent sex allocation

In some animal species (for instance, in some reef fishes) that form groups consisting of a dominant male surrounded by a number of smaller individuals of the opposite sex, complete sex reversal can occur during an individual's life (Fricke and Fricke 1977). Individuals start their reproductive life as a female and, when the dominant male is taken out, grow rapidly to become the new dominant male. In most plant species, changes in sex allocation are more gradual. Nevertheless many plant species are capable of adjusting the relative allocation to male and female reproduction to size or circumstances. Sex allocation theory may explain these phenomena. Ghiselin (1969), Charnov and Bull (1977), Horovitz (1977), Charnov (1982a), Lloyd and Bawa (1984), Goldman and Willson (1986) and Brunet (1992) amongst others recognized that hermaphrodite individuals are not necessarily equally effective male and female parents. Obviously this should have consequences for the evolution of the division of resources over both functions.

In Chapters 4 and 8 on sex allocation strategies, we assumed that all individuals in a population were equal in size and had the same budget for investing in reproduction. In reality individual plants differ enormously in size. In contrast to most animals, 100-fold differences in size are not uncommon in plants. Whenever the opportunities for male and female reproduction depend differently on size, we can expect the ESS for sex allocation to depend on size as well.

Very few cosexual plants, fewer than 0.1% of all plant species (Freeman *et al.* 1980), show complete gender change and such species are called sequential hermaphrodites. Well-known examples are found in the genus *Arisaema*. Hotta (1971) studied gender expression in several *Arisaema* species and distinguished three categories. In all three categories small individuals are male. Larger individuals may shift towards monoecy (group 1), they may shift to monoecy and then become pure females (group 2) or shift directly from male to female (group 3). North American *Arisaema* species include *A. triphyllum*

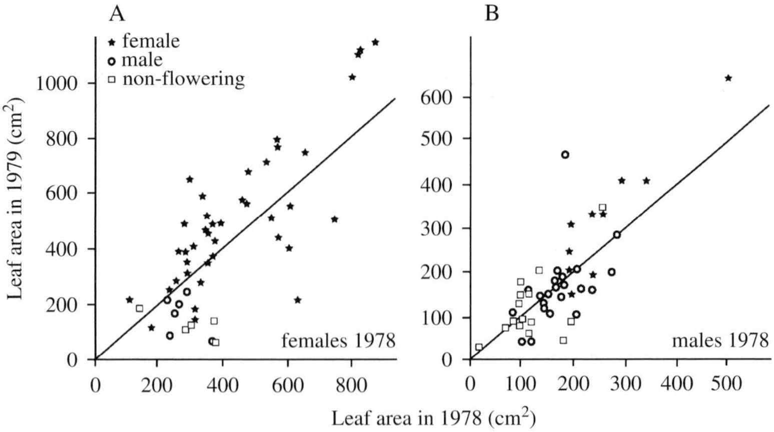

Fig. 9.1 Pattern of size and sex change displayed by (A) female and (B) male plants of *Arisaema triphyllum*. The chance of staying female or of changing from male to female increased with plant size. Redrawn with permission from Figs. 2 and 3 in Bierzychudek (1984b), *Oecologia*, Springer-Verlag.

(jack-in-the-pulpit, group 2) although monoecious plants are not always found (Fig. 9.1) (Bierzychudek 1984a, b) and *A. dracontium* (green dragon, group 1) (Clay 1993). In most cosexual plants, however, changes in sex allocation are subtler and have, for a long time, gone largely unnoticed. This is especially true for trees, for which there is scattered evidence for gradual or even sudden sex change. However, both the large size and the long timescale make experimental studies difficult. To express individual gender variation within populations Lloyd and Bawa (1984) used standardized phenotypic gender. Phenotypic femaleness of plant i is calculated as: $f_i/(f_i+r_i(\Sigma f/\Sigma r))$, in which f_i is the investment of plant i in female reproduction, r_i is its investment in male reproduction and Σf and Σr are the total investments in female and male function for the whole population. Phenotypic maleness can be calculated likewise. Phenotypic femaleness indicates, assuming fitness gains are proportional to investment, what fraction female function contributes to total fitness. A plant that produced fruits only, and no pollen, would score 1 for its femaleness, a plant that produced no fruits but did make flowers with pollen would score 0 and a plant with exactly the population average of investment in male and female function would score 0.5 for its phenotypic femaleness. Lloyd and Bawa (1984) showed large gender variation in populations of many species (Fig. 9.2) and their models predicted this variation to be linked to size and circumstances. They concluded, however, on the basis of the limited data that were available at the time that gender variation was rarely linked to size. Nevertheless, over the years, empirical evidence has accumulated showing

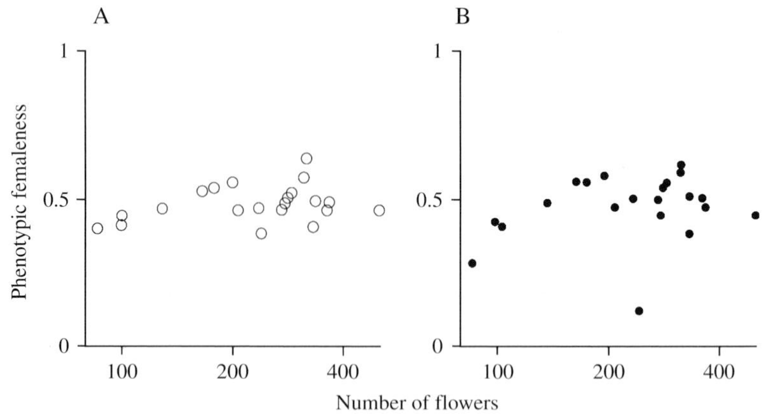

Fig. 9.2 Standardized phenotypic femaleness was independent of the size of the plant (measured by flower number) in *Conopodium majus* (pignut). Standardized phenotypic femaleness was estimated from (A) the proportion of hermaphrodite flowers and (B) the proportion of flowers that produced fruit. Redrawn with permission from Fig. 3 in Lloyd and Bawa (1984), *Evolutionary Biology*, Springer-Verlag.

unequivocally that some cosexual plants do change their sex allocation according to size and circumstances (Freeman *et al.* 1980, 1981; Brunet 1992; Bickel and Freeman 1993; Klinkhamer *et al.* 1997a). Several hypotheses have been put forward to predict the direction of size-dependent sex allocation (SDS) in these species. Unfortunately, as we will see, in most cases empirical data that would enable us to check the model assumptions are lacking. Despite these shortcomings some interesting generalizations can be made.

9.1 Six, mutually non-exclusive, hypotheses for SDS

Many hypotheses exist about SDS and Murray's (2001) critique of one species–one new hypothesis (Chapter 1) might apply here. However, as we will see it is possible to collect data that discriminate between the different hypotheses, even though this has not yet been done in a systematic manner. We start with discussing in detail what we consider to be the three most important hypotheses and three less important ones.

9.1.1 Hypothesis 1: Direct effects of plant size on the shape of the gain curves

Plant size can affect fitness returns directly. For instance, in wind-pollinated species distances over which pollen is dispersed are likely to increase with plant

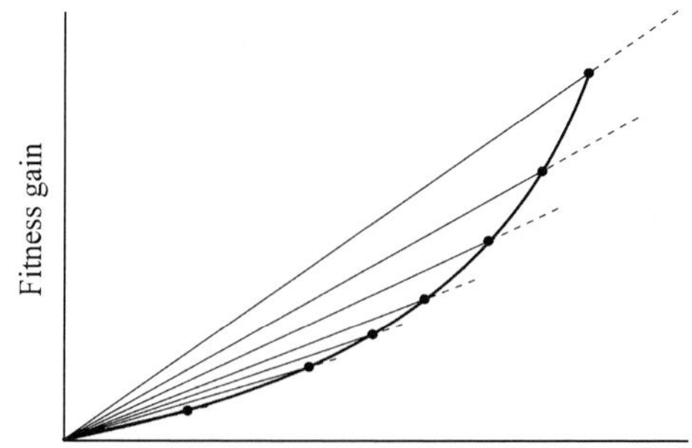

Fig. 9.3 Linear gain curves may appear non-linear if plants of different size are used. This is due to not controlling for direct effect of plant size on the shape of fitness gain curves. In this example the gain curves for larger plants have a steeper slope. Straight lines indicate the gain curves for plants of different sizes. Dots indicate sampling points. The curved thick line gives the resulting apparent 'gain curve' if fitness gains of differently sized plants are connected.

height (Burd and Allen 1988). Similarly, with ballistic dispersal or dispersal by wind, fruits on tall plants may be dispersed over greater distances than fruits on short plants. This effect of plant size on male or female gain curve is sometimes confused with the shape of the gain curves (Emms 1993; Klinkhamer *et al.* 1997a). Gain curves describe the hypothetical relationship between the investment in male or female reproduction and the fitness gained through that function (Chapter 3). 'How would the male reproductive success of a plant increase if it made 150 flowers, instead of 100?' is a typical question about gain curves (Chapter 3). The easiest way to test this would be to search for a plant with 100 flowers and one with 150 flowers and estimate their success. The underlying assumption in this procedure is that the two plants only differ in absolute investment and that everything else is equal. Data for drawing gain curves are frequently collected in this way, by selecting plants of different sizes in the field. When gain curves are determined in this way small and large investments become associated with small and large plants. Perhaps large plants grow on better sites with higher resource levels and better opportunities for their offspring. The plant with 150 flowers is perhaps taller than the one with 100 flowers. Short and tall plants may differ in the efficiency of pollen and seed dispersal. Therefore, small and large plants are expected to have different gain curves and using plants of different sizes to estimate gain curves may lead to apparently non-linear gain curves while in fact they are linear (Fig. 9.3). In some cases we can test how important these direct effects of plant size on the shape of the gain curve. For instance in our example we may want to keep the flower number of the first plant at 100, but reduce the flower number of the second plant by pruning also to 100. When reproductive success of the two types is equal after this manipulation, apparently no direct effects of plant size on fitness existed.

We will calculate how a rare plant, which is smaller or larger than the average size in a population, should change its sex allocation. This indicates the direction of SDS and is therefore a useful first step. We do not address the much more complicated problem how, given a size distribution in the population, a plant should adjust its sex allocation exactly to size. Mathematically we can express the effect of plant size on the gain curves as a simple power function of plant size. We consider an outcrossing plant that varies its allocation to male function depending on its size. To separate the direct effect of plant size from that of increasing budget with increasing size (see hypothesis 2), we assume that the fitness gains through male and female function are proportional to the investment. The population plays the ESS with respect to allocation to male function (r) and all individuals have the same size (X). With size we mean a parameter of the plant, like height, that affects the gain curves. The assumption that all plants have the same size is for the sake of simplicity; the argument would not change qualitatively if we would assume a specific size distribution. The total budget for reproduction is denoted as T. We can write the absolute fitness of a rare plant with different size X_{m} and budget T_{m} in a population of plants with size X, budget T and allocation r as:

$$W_{\mathrm{m}} = a{X_{\mathrm{m}}}^{c}(1 - r_{\mathrm{m}})T_{\mathrm{m}} + \frac{{X_{\mathrm{m}}}^{d} r_{\mathrm{m}} T_{\mathrm{m}}}{X^{d} r T} a X^{c}(1 - r)T \tag{9.1}$$

The first term in Eq. (9.1) represents the fitness contribution through female reproduction with a being a scaling factor and ${X_{\mathrm{m}}}^{c}$ giving the effect of size on female success of the rare plant. The second term represents the rare plant's share in siring the seeds produced in the population with ${X_{\mathrm{m}}}^{d}$ being the effect of plant size on its male success. Differentiating with respect to r_{m} gives:

$$\frac{\mathrm{d}W_{\mathrm{m}}}{\mathrm{d}r_{\mathrm{m}}} = aT_{\mathrm{m}}\left[-{X_{\mathrm{m}}}^{c} + {X_{\mathrm{m}}}^{d}\frac{X^{c}(1 - r)}{X^{d} r}\right] \tag{9.2}$$

The derivative does not depend on r_{m}. Whether the derivative is positive or negative depends on the size of the plant (X_{m}). If the derivative is positive, fitness always increases with investment in male function and is optimized when the plant is completely male. When the derivative is negative the reverse is true. In other words we would expect a complete change of gender with increasing size. With linear gain curves, ESS allocation to male function

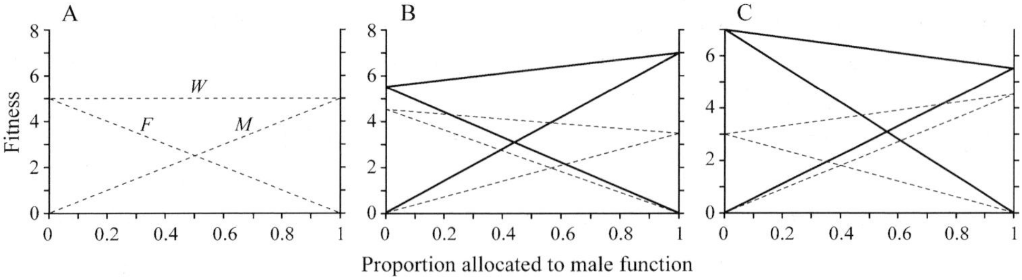

Fig. 9.4 (A) In a population with equally sized individuals and linear gain curves, the ESS is an equal proportional investment in male and female reproduction. If the population is at the ESS, fitness returns from male (M) and female function (F) are equal and proportional to investment. (B) With size differences in the population, a large individual (solid line) should be completely male if increasing plant size increases the slope of the male gain curve more than that of the female gain curve ($d > c$). Similarly, a small individual (broken line) should be completely female. (C) The reverse holds if with increasing plant size the slope of the female gain curve increases more than that of the male one ($d < c$). Redrawn from Box 1 in Klinkhamer *et al.* (1997a), *Trends in Ecology and Evolution*, with permission from Elsevier.

equals: $r^* = 0.5$. Substituting this as the average value for the resident population and dividing by X^c (which is positive) yields that the derivative of fitness of the rare plant is proportional to:

$$aT_{\mathrm{m}}[(X_{\mathrm{m}}/X)^d - (X_{\mathrm{m}}/X)^c] \tag{9.3}$$

if $d > c$ the derivative is positive for all $X_{\mathrm{m}} > X$. In other words, for plants larger than the average size the investment in male function should be maximal ($r = 1$) (Fig. 9.4A). If $d < c$ the derivative is negative for all $X_{\mathrm{m}} > X$. For plants larger than the average size the investment in male function should be as low as possible ($r = 0$) (Fig. 9.4B). The reverse is true for plants smaller than the average size so that we expect complete sex change.

Few data are available to test whether plant size affects the male gain curve more than the female curve, i.e. whether d is smaller or larger than c. The direct effect of size on the shape of the female gain curve will depend on the mode of seed dispersal. With wind-dispersed seeds especially, fitness returns may increase with plant size because seeds on larger plants may be carried over greater distances by wind. For plants with fruits that are dispersed by animals direct effects of plants size are, in all likelihood, small. Likewise, the direct effect of size on the shape of the male gain curve will depend on the mode of pollen dispersal. Especially with wind-dispersed pollen we may expect that fitness returns increase with plant size because pollen is released from a greater height and dispersed over a greater distance. This positive effect can, depending on plant geometry, be partly offset if larger plants have a greater canopy that prevents pollen from being lifted up by the wind (Bickel and Freeman 1993). Direct effects on the shape of the male gain curve in animal-pollinated plants are most probably small, because pollinators are more likely to respond to factors such as the number of flowers than to plant size itself.

9.1.2 *Hypothesis 2: Size only affects fitness gains indirectly through budget effects – the case of outcrossing plants*

In the previous section we discussed direct effects of plant size on fitness, assuming that the shape of the relationship between (absolute) investment in male or female function and plant fitness was different for small and large plants. With no direct effect of plant size on the shape of the gain curves, fitness returns for a given absolute investment are equal for small and large plants. However, the budget T that can be invested will be greater for larger plants. As a consequence, when gain curves are plotted as a function of the proportion of resources invested in male function (r) or female function ($1-r$) they are qualitatively different for small and large plants except when they are linear (Fig. 9.5). This qualitative difference also leads to SDS. Assume that plants are outcrossing and that direct effects of plant size are absent. Gain curves are no longer assumed linear but instead they may decelerate. The population is at the ESS for allocation to male function (r) and all individuals have a total budget

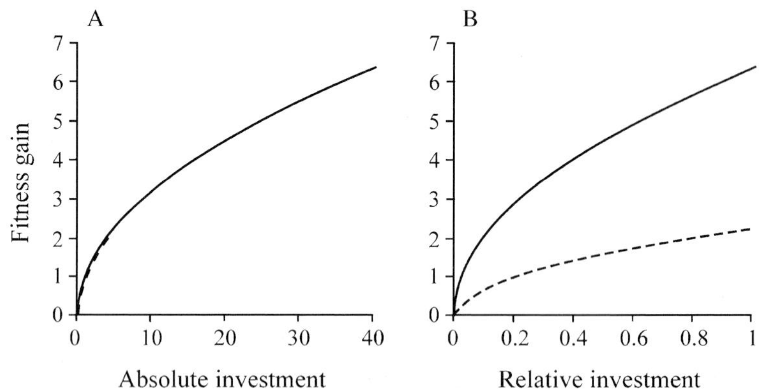

Fig. 9.5 Non-linear gain curves that are equal for small and large plants when absolute investment is plotted become quantitatively different when relative investment in either male or female reproduction is plotted. (A) The relationship between absolute investment and fitness gain. (B) The relationship between relative investment and fitness gain for a plant with in total 5 units of resources (broken line), and a plant with 40 units of resources (solid line).

for reproduction (T). As in the previous section, the latter assumption is for the sake of simplicity but the argument would not change qualitatively if we would assume another budget distribution. We can write absolute fitness of a rare plant with different budget T_{m} as:

$$W_{\mathrm{m}} = a[(1-r_{\mathrm{m}})T_{\mathrm{m}}]^{\beta} + \frac{(r_{\mathrm{m}}T_{\mathrm{m}})^{\alpha}}{(rT)^{\alpha}}a[(1-r)T]^{\beta} \tag{9.4}$$

As before, the first term gives the female component of fitness and the second term the individual's share in siring (outcrossed) seeds in the population. As before (Chapter 3), the exponents of the male gain and female gain curve are α and β, respectively. For this individual with different size, we want to find the optimal allocation to male function. We do this by taking the derivative with respect to r_{m} and setting it equal to zero. This yields:

$$\frac{\mathrm{d}W_{\mathrm{m}}}{\mathrm{d}r_{\mathrm{m}}} = -a\beta {T_{\mathrm{m}}}^{\beta}(1-r_{\mathrm{m}})^{\beta-1} + \alpha\frac{{r_{\mathrm{m}}}^{\alpha-1}{T_{\mathrm{m}}}^{\alpha}}{(rT)^{\alpha}}aT^{\beta}(1-r)^{\beta} = 0 \tag{9.5}$$

This can be rewritten as:

$$\frac{\beta}{\alpha}{T_{\mathrm{m}}}^{(\beta-\alpha)} = \frac{{r_{\mathrm{m}}}^{\alpha-1}}{(1-r_{\mathrm{m}})^{\beta-1}}\,\frac{T^{\beta}(1-r)^{\beta}}{T^{\alpha}r^{\alpha}} \tag{9.6}$$

The term $T^{\beta}(1-r)^{\beta}/(T^{\alpha}r^{\alpha})$ only depends on the characteristics of the resident population and can be considered constant (C). We can rewrite Eq. (9.6) as:

$$\frac{\beta}{\alpha}{T_{\mathrm{m}}}^{(\beta-\alpha)} = \frac{(1-r_{\mathrm{m}})^{1-\beta}}{{r_{\mathrm{m}}}^{1-\alpha}}C \tag{9.7}$$

Both α and β are smaller than 1. When $\beta > \alpha$, the term on the left-hand side of this equation increases with T_{m}. For the term on the right-hand side to increase as well, r_{m} should become smaller. In other words, if the female gain curve is less decelerating than the male gain curve ($\beta > \alpha$), the optimal proportional allocation to male function decreases with increasing budget for reproduction. If the male gain curve is less decelerating than the female one ($\beta < \alpha$), proportional allocation to male function should increase with increasing budget. A graphical analysis of this budget effect is given in Fig. 9.6. The left-hand side represents the situation where the male gain curve is more strongly decelerating than the female curve ($\beta > \alpha$), leading to increased femaleness with a larger budget, and the right-hand side represents the opposite situation ($\beta < \alpha$), leading to increased maleness with a larger budget. Note that in contrast with the direct effects of size, in these cases a gradual shift in allocation pattern with size is selected for.

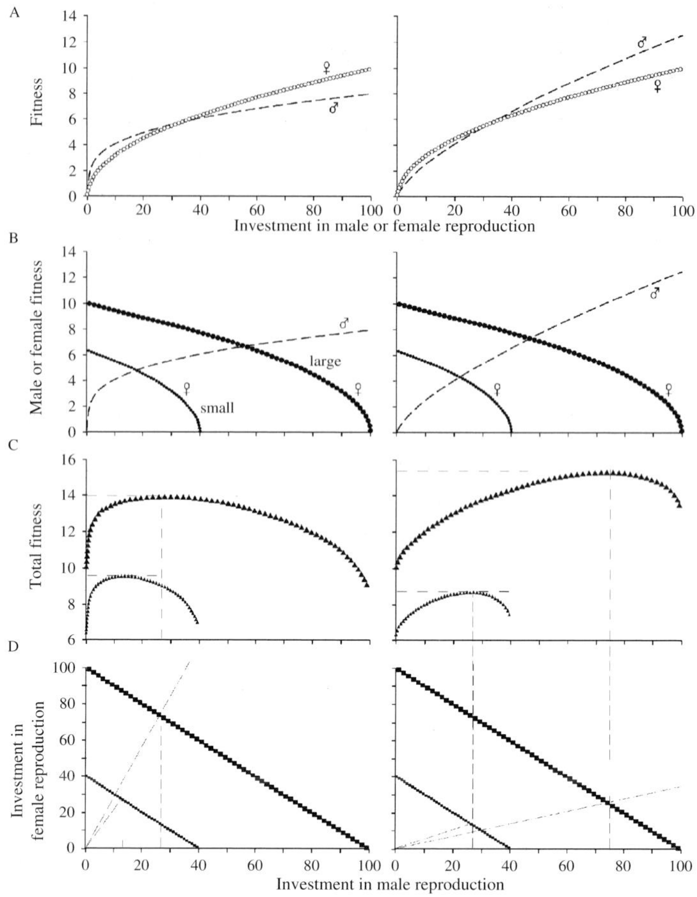

Fig. 9.6 Graphical representation of the optimal sex allocation of two rare individuals with different levels of resources in a population at the ESS with regards to sex allocation. For a large individual the level of resources is arbitrarily set at 100 units, for the small individual it is set at 40 units. (A) Male and female gain curves. On the left-hand side the male curve is more strongly decelerating than the female curve ($\alpha < \beta$). The right-hand side represents the opposite situation ($\alpha > \beta$). (B) Male and female fitness as a function of the absolute investment in male reproduction. Female fitness is calculated using the gain curves and the trade-off between investment in male and female reproduction given in (D). (C) Total fitness, the sum of male and female fitness. The vertical broken line indicates highest total fitness. (D) Trade-off between investment in male and female reproduction. The vertical lines indicate the optimal investments in male and female reproduction for a small and a large plant. At the intersection of the vertical line and the trade-off line we find the point that gives the investment in female reproduction, which corresponds to the optimal investment in male reproduction. The slope of the line that connects the origin with this point has a slope that equals the ratio of female to male component of fitness. The higher this slope the higher the femaleness. On the left-hand side, the slope is higher for the large plant; when $\beta > \alpha$, femaleness should increase with plant size. On the right-hand $\beta < \alpha$, so that maleness should increase with plant size.

If the rare plant has the same size as all other plants in the population ($T_m = T$) we can find the candidate ESS for allocation by applying Parker's method to Eq. (9.5), which yields the familiar result $r^*/(1 - r^*) = \alpha/\beta$ (Eq. 4.8). In Chapter 3 we discussed the fact that both male and female curve are likely to level off. Here it is relevant whether the male curve decelerates more strongly than the female curve. We have already discussed in Chapter 4 that this is likely to be the case in animal-pollinated plants where more male allocation induces longer bout lengths and less efficient pollen export, an effect which is non-existent in wind-pollinated plants.

9.1.3 *Hypothesis 3: A larger budget increases the selfing rate*

So far we have considered self-incompatible (SI) species without selfing. In many self-compatible (SC) species the selfing rate depends on the total investment in male reproduction (Fig. 9.7). Large plants with a high absolute investment in male reproduction are likely to have a higher selfing rate and selfing selects for a lower relative investment in male reproduction (Chapter 8). Here, we will prove this by comparing the ESS for male allocation in populations that differ in the average size of the individuals, using a number of simplifying assumptions. We assume that both pollen export to other plants and selfing rate (S) are linearly related to absolute investment in male reproduction (Tr, with T being the total investment in reproduction and r the fraction allocated to male reproduction). We consider a mutant in a resident population with r allocation to male fraction. In such a population the fitness of the mutant was already given in the previous chapter by Eq. (8.2), which was scaled to $T = 1$. Because now we want to consider size effects explicitly, we rewrite the equation taking the average size of individuals in the population as T and their selfing rate as S. The selfing rate is given by $S = bTr$ for the resident

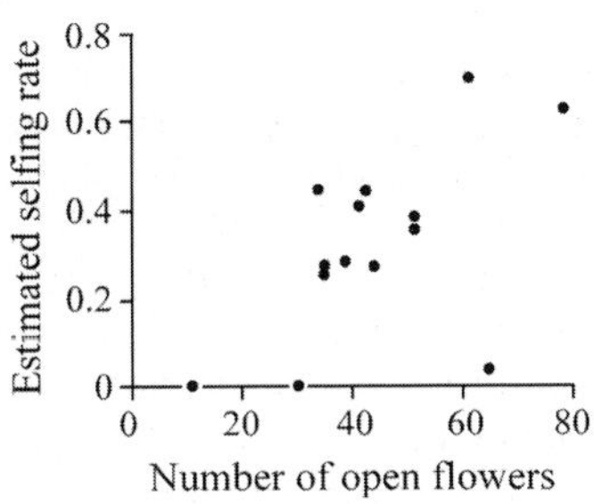

Fig. 9.7 The fraction selfing increased with the number of open flowers in a field experiment with potted plants of *Cynoglossum officinale*. Redrawn with permission from Vrieling *et al.* (1999), *Ecology Letters*, Blackwell.

and $S_m = bTr_m$ for the mutant. For simplicity we assume that there is no inbreeding depression. The mutant's absolute fitness W_m is then given by:

$$W_m = (1 - S_m)(1 - r_m)T + 2S_m(1 - r_m)T + \frac{r_m T}{rT}(1 - S)(1 - r)T \qquad (9.8)$$

The first term of the equation gives the fitness contribution through outcrossed seeds, the second term gives the fitness contribution through selfed seeds, and the last term gives the fitness contribution through seeds sired on other plants. We find, as usual, the candidate ESS for allocation to male function by Parker's method of differentiation of the fitness of the mutant (W_m) with respect to r_m and setting the derivative equal to zero. We write S'_m for dS_m/dr_m.

$$\begin{aligned} dW_m/dr_m &= -(1 - S_m) - S'_m(1 - r_m) + 2S'_m(1 - r_m) - 2S_m + (1 - r)(1 - S)/r \\ &= 0 \end{aligned} \qquad (9.9)$$

Taking $r_m = r$ and $S_m = S$ gives: $-(1 + S) + S'_m(1 - r) + (1 - S)(1 - r)/r = 0$. Because $S_m = bTr_m$ and consequently $S'_m = bT$, we obtain:

$$-(1 + bTr) + bT(1 - r) + (1 - bTr)(1 - r)/r = 0 \qquad (9.10a)$$

We rewrite this as:

$$bTr^2 + 2r - 1 = 0 \qquad (9.10b)$$

and because we can replace bTr by S, this is equivalent to $Sr + 2r - 1 = 0$ so that the candidate ESS for male allocation is:

$$r^* = 1/(2 + S) \qquad (9.11)$$

We can see in Eq. (9.10b) that b and T have the same effect; high values increase selfing, which selects for less male allocation.

We may also ask how the ESS for male allocation is related to size differences within population if all individuals can adjust their allocation to male function according to their individual sizes. To do this we have to assume a size distribution of the individuals within the population. De Jong *et al.* (1999) assumed that a population consisted of two size classes, and that each parent contributed equal numbers of offspring to both classes, i.e. half of the offspring lands on a favourable site and grows up to be large and half off the offspring lands on an unfavourable site and grows up to be small. The fitness of a genotype is thus the sum of the fitnesses through selfed and outcrossed seed production on a small and a large plant and the share for a small and a large plant of siring the outcrossed seeds produced in the population. Although somewhat more elaborate the principle is the same as in Eq. (9.8). De Jong *et al.* (1999) calculated the

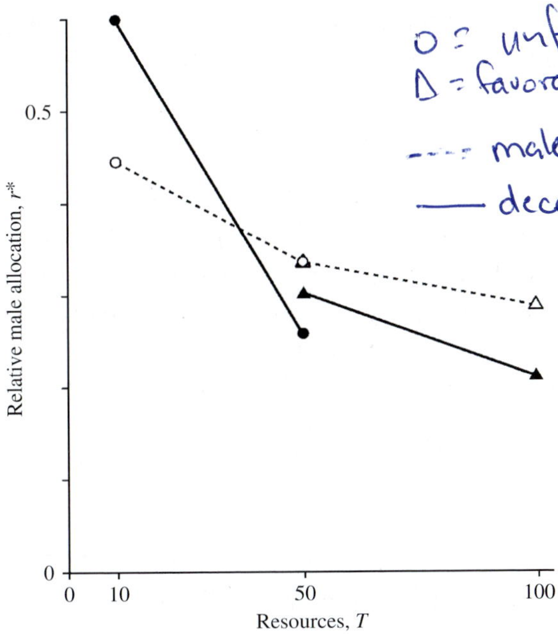

Fig. 9.8 ESS for relative allocation to male function (r^*) for plants growing in populations consisting of 50% large and 50% small plants, which differ in resource availability (T). Under unfavourable conditions (circles) T equals 10 and 50, respectively for small and large plants. Under favourable conditions (triangles) T equals 50 and 100. Selfing rate increases with male allocation. Optimal sex allocation was calculated for conditions under which the male gain curve is linear (broken line) or decelerating (solid line) (de Jong *et al.* 1999). Only with a decelerating male gain curve we find support for Charnov's (1982a) suggestion that an individual of a given size ($T = 50$) should behave differently when growing together with small ($T = 10$) or large plants ($T = 100$). Redrawn with permission from de Jong *et al.* (1999), *Journal of Evolutionary Biology*, Blackwell.

ESS allocation to both the small and the large plant. Their results showed that it is optimal for genotypes to adjust allocation plastically to plant size with, as expected, less male allocation in large plants (Fig. 9.8). This effect was stronger for the animal-pollinated species (decelerating male gain curve) compared to wind-pollinated species (linear gain curve).

9.1.4 Hypothesis 4: Male and female reproduction have different effects on mortality

This hypothesis states that plants emphasizing female function invest more in reproduction (Delph 1999), remain small and therefore have a higher chance of dying. This would select for a strategy of investing more into male function when small and into female function when large (Iwasa 1991). The hypothesis applies especially to perennial plants.

9.1.5 *Hypothesis 5: Female reproduction takes longer and has a higher chance of failure due to mortality*

Day and Aarssen (1997) focused on mortality during the reproductive season. In some plant species, mortality rates per unit of time diminish strongly with plant size. Female reproduction includes seed ripening and lasts longer. Therefore especially small female plants may run a high risk of not surviving before reproduction is completed. The hypothesis applies especially to short-lived annuals that live in an unpredictable environment with a high death rate, for instance through disturbance.

9.1.6 *Hypothesis 6: Genes dispersed through pollen have a reduced chance of ending up in the environment of the mother*

Genes dispersed through male function travel further than genes dispersed through seeds because they are first dispersed within pollen grains and subsequently within seeds. Genes transmitted through male function are therefore more likely to sample over all patch qualities within the population as a seedling than genes transmitted through female function. The latter are more likely to sample over patch qualities that are biased towards that of the mother plant. If plant size is correlated with patch quality this means that genes of large plants that are transmitted through seeds are more likely to end up again in high-quality patches than the genes transmitted through pollen. For small plants it is the other way around. We would expect selection for an emphasis on female reproduction in large plants and on male reproduction in small plants. Both theoretically and experimentally this hypotheses deserves further development. As for hypotheses 4 and 5, few data exist to test them.

9.2 Predictions of the direction of SDS

We have discussed three major and three minor hypotheses on SDS and stressed that their relative importance still needs to be established. This does not mean, however, that we cannot make any predictions on the direction of SDS. We can at least make six testable predictions. We will discuss the support for each prediction afterwards.

Prediction 1 Given the hypotheses above we expect SDS to be a common phenomenon in plants.

Prediction 2 The graphical analysis in Fig. 9.6 showed that budget effects lead to a gradual change in gender with size. Only in the absence of budget effects and with strong direct effects of size may we get a complete sex change. A more detailed analysis for outcrossing organisms (Cadet *et al.* 2004) indeed showed that conditions that select for a complete sex change are very restricted. In addition we may expect complete sex change in selfing organisms if a small investment in male reproduction would strongly compromise female reproduction through inbreeding depression. Because of these restricted conditions we expect complete sex change in plants to be rare.

Prediction 3 Hypotheses 3, 4, 5 and 6 all predict increasing femaleness with plant size. The other two hypotheses can work out both ways, but there are in fact only two factors that would lead to selection for increasing maleness with plant size: strong local resource competition and strong direct effects of plant size on male gains, as could be the case in wind-pollinated species. Because most factors select for increasing femaleness with plant size we expect this to be the commonest form of SDS, at least in animal-pollinated plants.

Prediction 4 What are large or small plants depends on the size distribution of the breeding population. Charnov (1982a) considered that this offers a good possibility for a critical test for SDS. Patterns of size-dependent gender change should vary among populations with different size distributions (but see Fig. 9.8). Data on the sequential hermaphrodite *Arisaema triphyllum* (Bierzychudek 1984a, b) seem to confirm this idea. The size at which individuals changed sex increased with the mean size of the individuals in the population. It would be worthwhile to apply this test to more populations.

Prediction 5 In contrast to animal-pollinated plants we expect positive direct effects of plant size on the shape of the male gain curve in wind-pollinated plants. In addition we have argued in Chapter 4 that the male gain curve should be more decelerating in animal-pollinated plants as compared to wind-pollinated plants. If this were true we could predict that increasing maleness with plant size would occur more often in wind-pollinated than in animal-pollinated plants.

Prediction 6 Because hypothesis 3 only applies to SC species we would expect that in SC species increasing femaleness with size is more pronounced than in SI species. This prediction can be tested by comparing the patterns of SDS in pairs of closely related SC and SI species.

9.3 Plant size and gender in *Cynoglossum officinale*

One of the species for which data on SDS are available is the monocarpic perennial *Cynoglossum officinale*. In the sand dunes of Meijendel (near the

Hague, the Netherlands) where we studied this species, flowering is restricted to a period of three to four weeks in May and June while usually seeds are ripe near the end of July. Cymes carry usually one to three open flowers simultaneously. With ten cymes, plants have 20–30 simultaneously open flowers. The most common pollinators are *Bombus* spp. Each flower has four ovules and can potentially produce up to four seeds. Mostly however, flowers produce far fewer seeds. The species is fully SC.

The first hypothesis about the direction of SDS (on direct effects of plant size) probably plays no role. Differences in plant height among flowering individuals are small and unimportant for foraging bees and are also unimportant for seed dispersal. Hypothesis 2 (on budget effects) was tested by Rademaker and de Jong (2000). They found no significant deviations from linearity for the female gain curve while the male gain curve, measured as dye grains exported to stigmas of other plants, was slightly decelerating (exponent of the gain curve: 0.96). Budget effects are therefore expected to have a relatively small effect on selection for SDS. Hypothesis 3 (on increased selfing rates) does play a role for *C. officinale*. Vrieling *et al.* (1999) found that the selfing rate increased from 10% in a plant with 10 simultaneously open flowers to 45% in a plant with 80 open flowers (Fig. 9.7). Because *C. officinale* is a monocarpic species, hypothesis 4 (on differential survival) does not apply. Most probably hypothesis 5 (on different times required for male and female reproduction) is not important either, because the period of seed production is rather short while the mortality rate during seed set is very low. Hypothesis 6 (on differential gene dispersal through male and female function) can potentially play a role. In the field, however, there is no strong relationship between plant size and patch quality. Taking these arguments together, we predict, mainly on the basis of hypothesis 3 that femaleness should increase with plant size in *C. officinale*. Plants may change their sex allocation by varying the number of pollen grains or seeds produced per flower. Seed and flower production were studied in a large number of populations and in different years. In addition data were collected on the number of pollen grains per flower for small and large plants.

Number of seeds was typically proportional to plant mass (Fig. 9.9A) (Chapter 2), so that per gram plant mass small and large plants produce equal amounts of seeds. Mass per seed is similar for small and large plants.

Flower number was a decelerating function of plant mass: in the regression of log (flower number) versus log (plant mass) the regression coefficient was significantly smaller than 1. In other words, per gram plant mass, small plants produced more flowers than large plants. From the regression it was estimated that plants weighing 1 g produced 90 flowers per gram plant mass

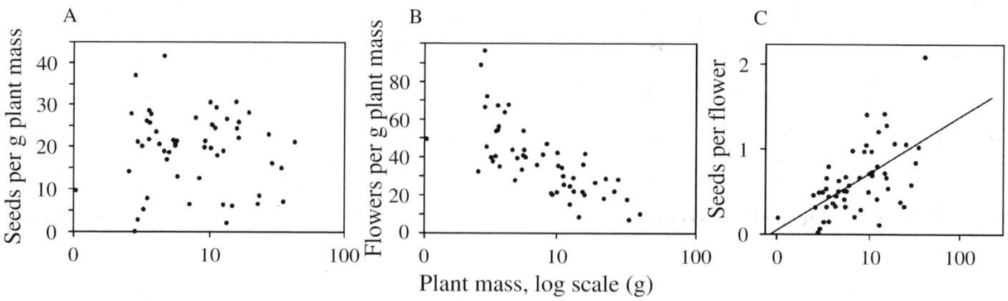

Fig. 9.9 (A) Number of seeds per gram plant mass, (B) number of flowers per gram plant mass and (C) and number of seeds per flower (as a measure of ratio of allocation to female and male) in relation to plant mass in a natural population of *Cynoglossum officinale*. Redrawn from Klinkhamer and de Jong (1997) with permission from Elsevier.

whereas plants weighing 60 g produced only 12 flowers per gram plant mass (Fig. 9.9B).

Seed-to-flower ratio was low. Typically 60% of all flowers did not produce seeds at all, while only about 5% produced the maximum number of four seeds. Mean seed-to-flower ratio per plant was in the range 0–2. Because small plants produced, per gram plant mass, more flowers and about equal numbers of seeds as compared to large plants, the seed-to-flower ratio increased with plant mass (Fig. 9.9C).

Mean pollen production per flower ranged from 30 000 to 155 000. Variation in number of pollen grains per flower was unrelated to plant mass. Both small and large plants produce on average 54 000 pollen grains per flower.

The direction of SDS is according to our prediction: femaleness increases with size. Small plants emphasize male reproduction (producing relatively many flowers with pollen, but with few seeds) while large plants emphasize female reproduction (producing relatively few flowers with more seeds per flower).

What causes SDS in *C. officinale*? Although the pattern of gender change is according to our prediction, it is still an open question how the pattern arises. Perhaps this is a plastic allocation response of the plant to its size. Alternatively, environmental factors correlated with plant size may cause the pattern. First, one could think that small plants receive fewer visits by pollinators and are perhaps poorly fertilized. However, additional hand pollinations showed that in *C. officinale* seed production was not pollen limited (de Jong and Klinkhamer 1989). Second, one could argue that large plants grow on sites with higher soil nutrient or water status and have higher internal nutrient concentrations and water potential and these in turn affect the seed-to-flower ratio. For *C. officinale* this explanation appeared not to be true. Leaf-nutrient concentrations were not related to plant size at the beginning of stem elongation (Klinkhamer and de Jong 1993). Furthermore, if this explanation were true we would expect the pattern of size-related gender to disappear if environmental conditions were made equal for small and large plants. Two experiments in which plants were grown under uniform experimental conditions showed, however, that the pattern of SDS was maintained (Klinkhamer and de Jong 1993; Rademaker and Klinkhamer 1999). This suggests that SDS in *C. officinale* is a plastic response of the plant and should be considered adaptive.

9.4 How common is SDS?

In addition to *C. officinale*, Klinkhamer and de Jong (1997) studied another seven monocarpic species in the same habitat. For five out of seven species, the regression coefficient for the relationship between log flower number and log plant mass was smaller than 1. In two cases this was significant, while in one case the regression coefficient was significantly larger than 1. The coefficient for the regression between log number of seeds and log plant mass was larger than 1 in five out of seven species (in two cases this was significant). As a result, the number of seeds per flower increased with plant mass in all but one species (in five out of seven cases this was significant). The exception was *Echium vulgare*. Also three other endemic *Echium* species, *E. leucophaeum*, *E. virescens* and *E. wildpretii*, studied at Tenerife (Spain) showed no SDS (T. de Jong and P. G. L. Klinkhamer, unpublished data). It is unclear why *Echium* species do not behave as predicted. Perhaps genetic variation is lacking or perhaps there is no mechanism by which these plants can 'measure' their size.

Unfortunately, complete data on flower, seed and pollen production in relation to plant mass are not available for most species. Taking also data on number of seeds per fruit or fruits per flower as measures for sex allocation, Klinkhamer *et al.* (1997a) found, based on literature data, significant SDS in 26 of 44 monocarpic species. In 25 from the 26 significant cases femaleness increased with plant size. When only studies with at least a five-fold difference in plant size were included, 79% of all species showed a significant SDS. Although in one of the best-studied perennial species, *Asclepias syriaca*, large plants strongly emphasized male reproduction (Willson and Rathcke 1974) the general pattern in perennials resembles that of monocarpic species: most species showed SDS with increasing femaleness when large (Wolfe 1983; Charlesworth 1984; Stocklin and Favre 1994; Fritz and Nilsson 1996; but see Pickering and Ash 1993). The analysis above is largely based on data on seeds per fruit, fruits per flower or seeds per flower. This weakens the conclusions because hardly any data on number of pollen grains per flower in relation with plant mass are available.

9.5 Comparing animal- and wind-pollinated species

In the discussion of the hypotheses on SDS we suggested differences in SDS between wind- and animal-pollinated species (Table 9.1). For the direct effects of plant size (hypothesis 1) on the male gain curve we expect a minor effect for animal-pollinated plants and a major positive effect (a steeper slope with increasing size as in Fig. 9.3) for wind-pollinated species. For the budget effect, we argued in Chapter 3 that the male curve decelerates more strongly in animal-pollinated species than in wind-pollinated species. Both effects together suggest that increasing femaleness with plant size is more common in animal-pollinated than in wind-pollinated species. If the direct effects of plant size on female gains are small we might even find increasing maleness with plant size for SI wind-pollinated species. For SC wind-pollinated species the effects of increased selfing with size may counteract the direct effects and it is less clear what to expect.

The data on hermaphrodite plants that we discussed above included only two wind-pollinated species which both showed increasing femaleness with size. A better chance to test the prediction comes from data on monoecious plants. Usually monoecious species produce more male flowers than female ones (Ganeshaiah and Shaanker 1991 but see Ågren and Schemske 1995) and the male bias is stronger in trees than in herbs (Ganeshaiah and Shaanker 1991). Out of a total of 22 monoecious species studied by Bickel and Freeman (1993), 8 species were animal-pollinated and all these species

Table 9.1 *The predicted direction of SDS for animal- and wind-pollinated plant species that are either self-compatible (SC) or self-incompatible (SI)*

Hypothesis	Compatibility	Animal pollination		Wind pollination	
		Condition[a]	SDS[b]	Condition[a]	SDS[b]
1. Direct effect of size	SC + SI	$c>d$	M → F	$c<d$	F → M
2. Budget effect of size	SC + SI	$\alpha<\beta$	M → F	$\alpha>\beta$	F → M
3. Selfing depends on size	SC		M → F		M → F

[a] Condition: c and d are the exponents of the direct effect of plant size on female and male fitness, respectively; α is the exponent of the male gain curve and β of the female gain curve.

[b] M → F indicates an increase of femaleness with size and F → M a decrease.

showed increasing femaleness with size. From the 14 wind-pollinated species, 8 species (all members of the Gramineae or the Asteraceae) showed increasing maleness with size. For this data set it is not known in which species self-pollination occurs. Other scattered data suggest similar trends. In the wind-pollinated *Quercus gambelii* (gambel oak) tall plants in the canopy had proportionally more male inflorescences. Flowers on short plants, with presumably small opportunities to disperse pollen were almost all female (Freeman *et al.* 1981). In Spanish populations of *Mercurialis annua*, which is androdioecious in part of its range, maleness also increased with plant size in monoecious individuals (Pannell 1997).

To sum up: there is now a coherent body of theory that predicts how plants should optimally adjust sex allocation to size. In animal-pollinated species, increasing femaleness with size is the general pattern. Limited evidence in wind-pollinated species suggests that increasing maleness with size is more common. This is in line with model predictions, based on the assumption that male gains in wind-pollinated species are not saturating and are to some extent, directly affected by plant size (height). However, especially for wind-pollinated species more data on the shape of the gain curves would be most useful. Theory also predicts different patterns of SDS for SC and SI species. Collecting systematic data on SDS taking phylogeny, SI/SC, seed dispersal and pollination system into account is clearly the way forward.

Questions

(1) Name the three major hypotheses on SDS.
(2) Why do we expect to find increasing maleness with size in some wind-pollinated plants, but not in animal-pollinated plants?
(3) Under what conditions is complete sex change expected?

10

Sex ratios in dioecious plants

Dioecious plants with separate male and female individuals are quite rare; less than 4% of all plant species belong to this category. Examples are *Humulus lupulus* (hop), *Carica papaya* (papaya), *Salix* spp. (willow), *Asparagus officinalis* (asparagus) and *Actinidia* spp. (kiwi fruit). Sex determination mechanisms differ between species (summarized recently by Ainsworth *et al.* 1998). Most often the male is the heterogametic sex (XY or *Aa*) and the female the homogametic sex (XX or *aa*), as in humans. In few cases, the best-studied example being *Silene latifolia* (white campion), sex chromosomes can be morphologically distinguished under a microscope. In other cases, such morphological differences are not obvious, but sex determination works in a similar way. This has been demonstrated by selfing plants, which can be achieved by application of hormones that make the male or female individual cosexual (Westergaard 1958). When the offspring segregates into males and females the parent is heterozygous at the sex determination locus. In a number of cases there exists environmental sex determination (ESD): the sex of a plant changes with environmental conditions (Korpelainen 1998). In other cases the sex of the plant is quite stable and does not change with environment within the range of conditions plants encounter in the field (genetic sex determination, GSD) (Westergaard 1958).

The evolutionary problem for a dioecious plant with GSD is to optimize the sex ratio in the seeds. For maximal seed production of the species it would be best if just enough males were produced to fertilize all females in the population (Chapter 4). This is indeed what a farmer would do if he wished to maximize his yield of, for instance, papaya. However, in such a female-biased population, plants producing many sons would enjoy high male reproductive success and transfer more copies of their genes to the next generations than the common type with a female bias. Optimizing sex ratio in the seeds is a game between different players and the relevant question is: 'Which sex ratio in the seeds is an ESS?' There could be constraints to this problem. If the male is the

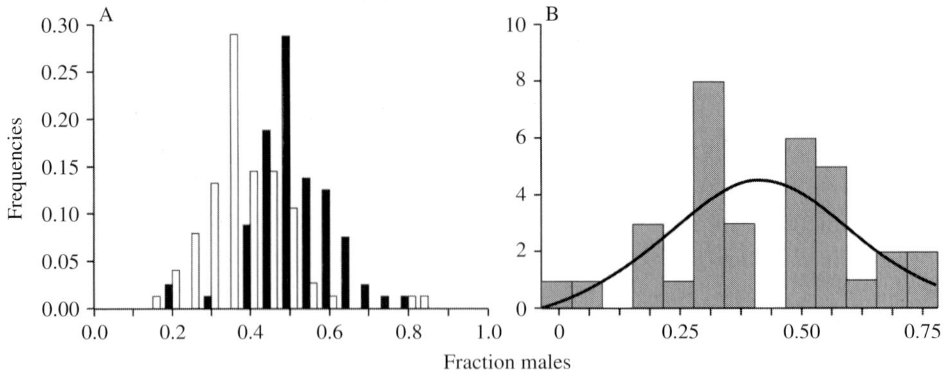

Fig. 10.1 (A) Variation in seed sex ratio of *Silene latifolia* between seed batches from different mother plants. Two male plants (open and closed bars) were crossed with many female plants (Taylor 1999). (B) Variation in seed sex ratio in *Urtica dioica* (stinging nettle) between seed batches from different mother plants after open pollination in the field (de Jong *et al.* (2005)). Redrawn with permission from Taylor (1999), *Evolution* and de Jong *et al.* (2005), Thieme Verlag.

heterogametic sex with genotype *Aa* and the female is *aa*, then Mendelian inheritance would always produce a sex ratio of 0.5 sons. Nevertheless it is quite likely that other genes can modify this sex ratio in the seeds in a number of ways. First, genes in the father could, in some way, change the 1 : 1 ratio of pollen carrying the male *A* allele and that carrying the female *a* allele. Second, genes in the mother could differently affect germination or pollen tube growth of pollen carrying the *A* or *a* allele. Both points involve gametophytic selection. Third, the mother could abort seeds selectively, depending on their gender. Fourth, other genes in the offspring could affect sex determination, overruling the major sex determination genes. Species with multilocus sex determination are also known (for instance *Mercurialis annua*) and in such cases biased seed sex ratios can be produced even with Mendelian inheritance (Durand and Durand 1991). The two species in Fig. 10.1 show abundant variation in seed sex ratio, allowing genetic change in this character under natural selection. We now first calculate the ESS sex ratio in the seeds without genetic constraints and then examine how predictions hold up against the data.

10.1 ESS for seed sex ratio

We can be brief here because the problem of the sex ratio in seeds of dioecious plants is mostly identical to the problem of sex allocation in cosexual plants (Chapter 8). It is only slightly more complex to write down the equations.

Maynard Smith (1978) did this for animals and we supplement his model for dioecious plants. Suppose an annual dioecious plant has 1 unit of resource available, male and female seeds are equally costly to produce and males and females differ in their probability C of surviving until reproduction. The common genotype in the population makes a fraction r male seeds and a fraction $1-r$ female seeds, so that the ratio of surviving adult females to males is $C_{♀}(1-r)/C_{♂}r$. We assume that there are always enough males to fertilize all females. Success of sons depends on the number of females per male in the population. If this ratio is high, a male can fertilize many females; if it is low, males are relatively unsuccessful. Note that by taking success of the sons into account we calculate fitness as the number of alleles passed on to the grandchildren; each daughter produces a constant number of grandchildren but the number of grandchildren sired by a son varies.

Suppose also, as Maynard Smith (1978) did, that there is some structure in the population so that a fraction S of the matings is between brothers and sisters. For plants it would seem reasonable to make the extra assumption that the offspring of these matings can be a fraction $(1-\delta)$ times less viable than offspring from matings between unrelated individuals. The fraction sib mating is analogous to the selfing rate and δ corresponds to inbreeding depression in the model of the cosexual plant. Absolute fitness of the rare mutant with a fraction r_{m} male and $1-r_{\mathrm{m}}$ female seeds in a population in which all other plants have the strategy of producing in proportion to r male and $1-r$ female seeds, is then:

$$W_{\mathrm{m}} = (1-S)(1-r_{\mathrm{m}})C_{♀} + 2S(1-\delta)(1-r_{\mathrm{m}})C_{♀} + \frac{C_{♂}r_{\mathrm{m}}}{C_{♂}r}(1-S)(1-r)C_{♀} \tag{10.1}$$

This is the Shaw–Mohler equation with selfing from Chapter 8. The first term denotes the success through female seeds (daughters) that cross with unrelated males. The second term denotes the success through female seeds (daughters) that mate with their brothers. The probability that these daughters pass on a gene (allele) present in the mother is twice as high as it would be if they mated with unrelated males. This is analogous to the effect of selfing in previous chapters. The third term denotes the success through male seeds (sons), which depends on the number of females per male at the population level. Daughters are a sure thing, whereas sons are a gamble; they contribute to fitness in proportion to their ratio of females to males in the population. We find the ESS again by applying Parker's method of differentiation of the fitness function of the mutant with respect to the evolutionary variable, in this case the fraction of males in the seeds (r_{m}).

This gives $\mathrm{d}W_{\mathrm{m}}/\mathrm{d}r_{\mathrm{m}} = (-1+S)C_{♀} - 2S(1-\delta)C_{♀} + (1/r)(1-S)(1-r)C_{♀} = 0$. The parameter $C_{♀}$ cancels and we end up with the following equation for the ESS fraction of sons in the offspring:

$$r^* = (1-S)/(2-2S\delta) \qquad (10.2)$$

Sib mating results in a female bias in the seeds, similar to the effect of selfing on sex allocation (Eq. 8.4). The rationale for this result is clear if we consider the extreme situation when all females are mated by their brothers, $S=1$. A plant with an equal proportion of sons and daughters in 100 seeds yields 50 fertilized daughters that pass on their genes to the grandchildren. A plant that produces 99 daughters and only one son that fertilizes all his sisters yields 99 fertilized daughters. If selfing is guaranteed, the latter is the superior strategy.

Differences in mortality between males and females do not matter for r^*. This result emerges because if males have, for instance, 50% higher mortality than females, the surviving males will enjoy a doubling of their reproductive success. This exactly balances the higher mortality risk for males. The sex ratio in the seeds is the evolutionary variable that the plant can optimize and this is not necessarily the same as the sex ratio of the flowering plants. We must be careful in making this distinction. We may well find that in a plant species with 50% higher mortality for females, the natural populations contain only 33.3% female plants. Natural selection would still favour 50% sons in the seeds of this species. The above model was for annuals with non-overlapping generations, but the results are more general, and apply to long-lived perennials as well (Pen and Weissing 2002).

10.2 Effects of local competition on seed sex ratio

Bulmer and Taylor (1980) wrote: ‘it is clearly possible to obtain either a male-biased or a female-biased sex ratio, depending on the relative dispersal distance of seed and pollen’. This is quite a different prediction than that from Eq. (10.2), which suggested only unbiased or female-biased sex ratios. Another effect on the seed sex ratio is not in the Shaw–Mohler equation and first became apparent in the models of Bulmer and Taylor (1980) and P. D. Taylor (1994). This effect is that sons always encounter less local resource competition (LRC), selecting for a male bias. In general, the models of Bulmer and Taylor showed that you should bias allocation towards the sex that is dispersed best. Why should this be the case? When a dioecious plant produces a female seed, then the seeds that this daughter produces are dispersed in her seed shadow.

If the plant produces a male seed, the pollen of this son is first transferred to a female plant, after which the seeds are dispersed. The dispersal of genes through sons involves an extra step. In higher plants, the male gametophyte (the pollen grain) disperses, whereas the female gametophyte (the egg cell or embryo sac) does not. If a plant produces sons, its grandchildren are more evenly spread out over the whole population than if it makes daughters. Producing more sons always reduces LRC. Since pollen dispersal distance have been shown to be up to hundreds of times the seed dispersal distance (Hu and Ennos 1997), the effect could be considerable, especially in tall wind-dispersed plants such as pine trees.

The question is how the two effects, sib mating and LRC, sum up in realistic models of populations of dioecious plants. Lifelike models have not been developed yet, but the chessboard model of de Jong *et al.* (2002) may, at the moment, come closest. In this model pollen and seeds were allowed to disperse one or more squares, like squares covered by the king in one or more moves on a chessboard (Fig. 10.2). Thus sib mating can take place and one can also

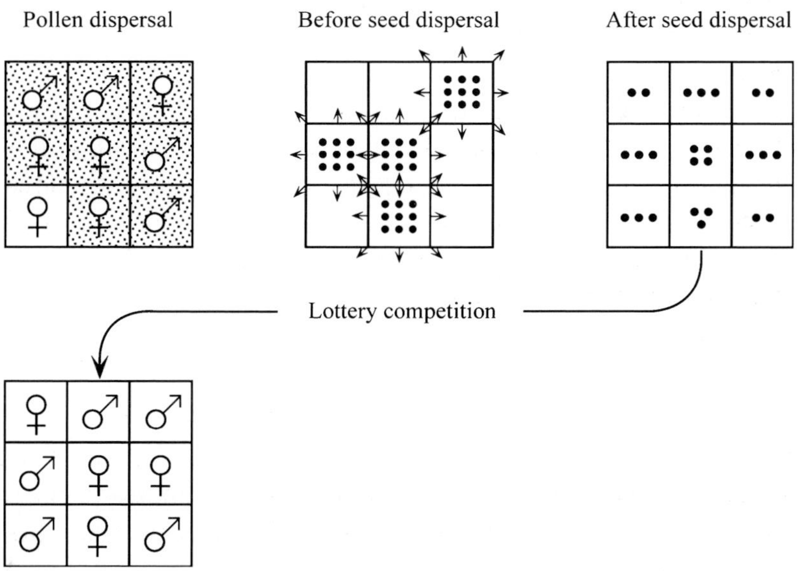

Fig. 10.2 Schematic outline of the chessboard model of de Jong *et al.* (2002). In the example pollen is dispersed one square, like the moves of the king on the chessboard. Next the seeds are dispersed one square, after which lottery competition takes place and only one of the plants, male or female, can occupy a square. By varying the number of squares that pollen and seeds migrate one can examine, by means of simulation, the effect of dispersal distance on the seed sex ratio. For convenience only a small fraction of a large population is shown. Redrawn with permission from de Jong *et al.* (2002), *Journal of Evolutionary Biology*, Blackwell.

Table 10.1. *Predicted shift in the ESS sex ratio in seeds according to the chessboard model of de Jong* et al. *(2002)*

	High pollen dispersal	Low pollen dispersal
High seed dispersal	(1) No sib mating, no LRC *Unbiased sex ratio*	(2) More sib mating (–), LRC similar for sons/daughters *Female bias*
Low seed dispersal	(3) More sib mating (–), sons less LRC than daughters (+) *Male or female bias*	(4) Highest sib mating (–), sons less LRC than daughters (+) *Female bias*

Notes: All comparisons are with case 1, the panmictic population with high pollen and seed dispersal. A plus sign indicates a factor that selects for a higher sex ratio (more males), a minus sign indicates selection for a lower, more female-biased sex ratio.

adjust the distances travelled by pollen and seeds. By simulation it can then be examined which seed sex ratio is selected in different scenarios. The results of this model were as follows (Table 10.1):

(1) When pollen and seeds are both dispersed over great distances the ESS sex ratio is 0.5; this is a logical result since we approach the panmictic situation.
(2) When, compared to the panmictic case, pollen dispersal is decreased, this results in more sib mating, while the difference in LRC encountered by male and female offspring remains small. This selects for female bias in the seeds.
(3) When, compared to panmixis, seed dispersal is reduced, this results in more sib mating. It also results in less LRC for male offspring than for female offspring. The first factor selects for a female bias, while the second selects for a male bias. The seed sex ratio could thus become smaller or larger than 0.5, depending on the exact parameter values.
(4) When both pollen and seed dispersal are reduced the effect is complex. In this case sib mating is high and in the simulations this effect was strongest, selecting for a female bias in the seeds.

10.3 The Waterloo of sex allocation theory?

How well do dioecious plants conform to these theoretical predictions? Data on seed sex ratios are not abundant in the literature, especially not for long-lived dioecious plants. The reason for this publication bias is obvious. Most dioecious plants do not have sex chromosomes that are visible under a microscope and it is a lifetime job to record the flowering of a dioecious tree, keeping mortality of vegetative plants as low as possible to exclude any possible effects of gender-dependent mortality. More abundant are data on sex ratios of flowering plants in field populations (Kay and Stevens 1986). Such data are certainly of interest, but will include differences in germination, mortality and age of first reproduction between the sexes. Such data do not provide the most appropriate test of model predictions. We therefore have only compiled data for species in which the seed sex ratio was known with reasonable certainty. The grouping was according to seed and pollen dispersal. Comparative studies (Hamrick and Godt 1990) showed that wind pollination, as compared to insect pollination, and dispersal features of seeds (fleshy fruits or plumed seeds) promote gene flow, i.e. reduce genetic differentiation between populations. This review also showed that there is much variation within a group so that such generalizations should be handled with care.

Table 10.2 shows that there is quite some variation in seed sex ratio, with both female bias and male bias occurring, consistent with the Bulmer and Taylor (1980) model and inconsistent with the Maynard Smith (1978) model. In nine cases a biased sex ratio was found and in eight cases a balanced ratio. This may represent a publication bias; species with a conspicuous sex bias in the flowering plants may have attracted more attention than an unbiased species. Nevertheless the inescapable conclusion from Table 10.2 is that it does not support the predictions from Table 10.1. Also there exists zero support for the simpler prediction from Eq. (10.2), that species with seed dispersal over short distances have more sib mating and therefore a more female-biased sex ratio. Of course, the data in Table 10.2 are fragmentary and many objections could be raised. Nevertheless, the present conclusion should be that, based on the available data, one cannot predict the seed sex ratio from life-history parameters of the species. This is a remarkable conclusion as for the animal kingdom people have claimed that the optimal sex ratio of the offspring under LMC and LRC is well understood (Hardy 2002) and our models of cosexual plants fared quite well in predicting effects of selfing on allocation (Chapter 8). What is the explanation for this discrepancy?

Most important, in our opinion, is the realization that our phenotypic optimization model assumes selection on autosomal sex ratio modifiers. In cosexual plants each nuclear gene is represented in the seeds and the pollen and the proportion of seeds and pollen in the ESS favours the transmission of all nuclear genes. However, as soon as two sexes are created, as in dioecious plants, this divides the interests of nuclear genes (Cosmides and Tooby 1981). The chromosomes that carry the sex determining factors will develop separately and crossing-over will become impossible for the non-homologous part of these chromosomes. Driving factors on the X and Y chromosomes may

Table 10.2. *Seed sex ratios (SSR) of dioecious plants with and without features for seed dispersal and with either insect or wind pollination. Provided wind transfers pollen over greater distances than insects and fleshy fruits and plumes promote seed dispersal, the scheme can be compared with the predictions of Table 10.1*

	Wind pollination	Bias of SSR	Reference	Insect pollination	Bias of SSR	Reference
High seed dispersal (fleshy fruit/ plume)	*Hippophae rhamnoides*	Male	de Jong and van der Meijden (2004)	*Asparagus officinalis*	Equal	de Jong and van der Meijden (2004)
	Populus tremula[a]	Male	Kay and Stevens (1986)	*Bryonia dioica*	Equal	Kay and Stevens (1986), de Jong and van der Meijden (2004)
				Potentilla fruticosa	Equal	Kay and Stevens (1986)
				Salix repens	Female	de Jong and van der Meijden (2004)
				Salix viminalis	Female	Alström-Rapaport *et al.* (1997)
				Viscum album	Female	Kay and Stevens (1986)
Low seed dispersal	*Distichlis spicata*	Female	Eppley *et al.* (1998)	*Dioscorea* spp.	Male	Martin (1966)
	Mercurialis annua	Equal	Kay and Stevens (1986)	*Silene dioica*	Female	Kay and Stevens (1986)
	Rumex acetosella	Equal	Kay and Stevens (1986)	*Silene latifolia*	Female	Taylor (1999)
	Rumex acetosa	Equal	Kay and Stevens (1986)	*Silene otites*	Equal	Soldaat *et al.* (1997)
	Urtica dioica	Equal	de Jong *et al.* (2005)			

[a] Based on flowering plants; all other data refer to seed sex ratios.

be selected and the interests of these genes are in sharp conflict (Hamilton 1967; D. R. Taylor 1994). This is known as meiotic drive. In addition, cytoplasmic genes are usually transmitted through the seeds and not through the pollen. If cytotypes can affect seed sex ratio, a type that results in 100% female seeds will be selected. We return to intragenomic conflict over sex allocation in Chapter 16.

To sum up: data that are now available on seed sex ratios in dioecious plants are not consistent with sex allocation theory. In a sense the classical sex allocation theory, based on nuclear genes, has met its Waterloo here. The theory would need to be modified to take intragenomic conflict into account and especially the ways in which this conflict is possibly mediated. New insight is most likely to come from precise studies of the genetics of seed sex ratio in model species rather than from broad scale species comparisons.

Questions

(1) Do differences in mortality between males and females select for a different seed sex ratio in the Shaw–Mohler equation?
(2) Why is intragenomic conflict stronger in dioecious plants as compared to cosexual plants?
(3) What is the primary and secondary sex ratio and which does evolution optimize?
(4) Why should sex allocation become biased towards the sex that disperses best?
(5) What did Bulmer and Taylor say with regard to dispersal of gametophyte and sporophyte in plants?

11

Outcrossing, selfing or no sex at all?

We will cover a number of related topics in this chapter. We first examine when plants should be selfing or outcrossing. The conditions for either strategy to be optimal depend in a simple way on the level of inbreeding depression. It is more difficult to explain the conditions that favour a combination of selfing and outcrossing. Nevertheless many plant species with such a mixed mating system exist. One group of species that we highlight here produces a mix of cleistogamous and chasmogamous flowers. The cleistogamous flowers usually do not even open and are always selfed. The chasmogamous flowers can outcross. We list the various hypotheses about the evolution of such a mixed strategy. Plants can also abandon sex altogether. Plants of *Taraxacum officinale* (dandelion) produce seeds on their own and some clonal plants have stopped flowering altogether, and rely on vegetative reproduction for their persistence. We briefly sketch advantages and disadvantages of asexual reproduction.

11.1 Fisher's automatic advantage of selfing

From all advantages of selfing, the transmission advantage or the automatic advantage (Fisher 1941) is probably both the strongest one and the one most easily overlooked. To illustrate this let us consider our familiar example (Chapters 4, 8) of a monoecious plant that optimizes the fraction of male flowers. Suppose the population is fully outcrossing, so that 50% allocation to male and 50% to female is an ESS. At the ESS, the fitness gain through female flowers equals that through male flowers. Suppose now a rare mutant with an allele for 100% selfing is introduced into the population. In this mutant, for instance, the anthers briefly touch the stigma just before the flower opens and from the 10 000 pollen grains produced per flower 10 pollen adhere to the stigma. This low number is sufficient to fertilize all ovules before any outcross pollen arrives, but hardly reduces the pollen export of the mutant. The mutation also has no effect on the sex allocation; the mutant allocates r to male function, just as the common type in the population. The fully selfing mutant would transfer two allele copies to the selfed seed, while it still transmits one allele copy through pollen. The rare mutant transfers 1.5 times as many copies of the 'selfing' allele to the next generation as the 'outcrossing' allele in the common type. The mutant can invade and increase in frequency, first rapidly and then more slowly as the automatic advantage erodes with more selfers in the population. This is called Fisher's automatic advantage of selfing. We now generalize Fisher's argument to a partially selfing population, using the models from Chapter 8. At the ESS the fitness of the common cosexual with relative male allocation r in a population with selfing rate S (we write S for selfing rate of the common cosexual and S_m for the selfing rate of the mutant) and level of inbreeding depression δ was (Eq. 8.2):

$$W = (2 - 2S\delta)(1 - r)^{\beta} \tag{11.1a}$$

As before β is the exponent of the female gain curve and the exponent of the male gain curve is α. A mutant has a different selfing rate (S_m) but the same male allocation (r) as the common type.

Absolute fitness of such a mutant is:

$$W_m = (1 - S_m)(1 - r)^{\beta} + 2S_m(1 - \delta)(1 - r)^{\beta} + \frac{r^{\alpha}}{r^{\alpha}}(1 - S)(1 - r)^{\beta} \tag{11.1b}$$

This is the same as in Chapter 8, except that now the mutant has a different selfing rate, but not a different male allocation; therefore Lloyd's competitive share (r^{α}/r^{α} in Eq. 11.1b) equals 1 and can be crossed out in Eq. (11.1b). A rare mutant with full selfing ($S_m = 1$) still exports r^{α} pollen grains, the same amount as the common type. This fully selfing mutant produces 100% selfed seeds and sires seeds on cosexual plants with selfing rate S and has absolute fitness:

$$100\%\ \text{selfing:}\ W_m = 2(1 - \delta)(1 - r)^{\beta} + (1 - S)(1 - r)^{\beta} \tag{11.2}$$

Can this type invade in an outcrossing population? The fitness of a common outcrossing type ($S = 0$) is $2(1 - r)^{\beta}$ (Eq. 11.1a). Without inbreeding depression the fitness of the rare selfer in the outcrossing population is $3(1 - r)^{\beta}$ (Eq. 11.2). This shows again Fisher's 1.5-fold automatic advantage of selfing, mentioned above; the selfer can easily invade. Full selfers also invade, and take over, a population of partially selfing plants if $W_m > W$, i.e. if $2(1 - \delta) + 1 - S > 2 - 2S\delta$ or $-2\delta + 2S\delta > S - 1$ or $\delta < (S - 1)/2(S - 1)$ or $\delta < 0.5$.

When can a rare outcrossing mutant ($S_m = 0$) invade into a population with an intermediate selfing rate? Its absolute fitness consists of the outcross seeds it makes plus the seeds it sires on common cosexual plants with S selfing:

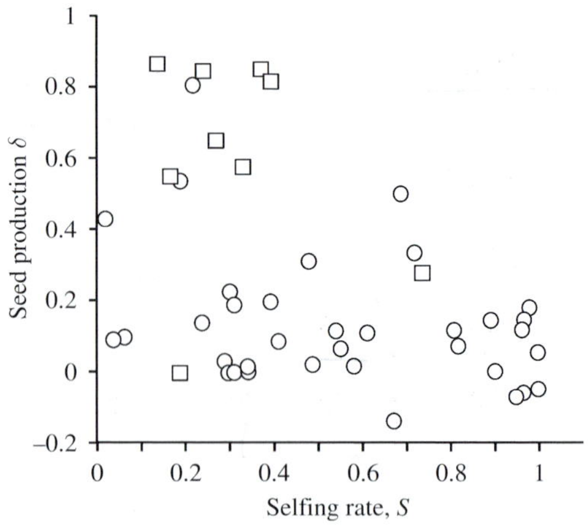

Fig. 11.1 Early inbreeding depression in seed set declines significantly with the selfing rate. The nine squares are gymnosperms; the 35 circles refer to angiosperm species. Redrawn with permission from Husband and Schemske (1996), *Evolution*.

$$\text{100\% outcrossing: } W_\text{m} = (1-r)^\beta + (1-S)(1-r)^\beta = (2-S)(1-r)^\beta \tag{11.3}$$

Now compare Eq. (11.1a) and Eq. (11.3). The outcrosser invades in the population of a partial selfer if $W_\text{m} > W$ or $2 - S > 2 - 2S\delta$ or $\delta > 0.5$. So if inbreeding depression is smaller than 50% full selfers can invade and drive the partially selfing type out of the population; if inbreeding is larger than 50% outcrossers invade and win from the selfer.

Theory then predicts that selfers are typically species with low inbreeding depression, whereas inbreeding depression should be high in outbreeders. This pattern can be reinforced if lethal alleles are expressed in the offspring of selfers. Selfing removes these alleles from the population; this is called purging of lethal alleles (Crnokrak and Barrett 2002). As a result the level of inbreeding depression in the next generation is reduced and selfing becomes even more favourable. A lower level of inbreeding depression in selfers was indeed found by Husband and Schemske (1996) who reviewed data from different species. On average they found $\delta = 0.23$ in predominantly selfers and $\delta = 0.53$ in outcrossers (Fig. 11.1). Theory also predicts that intermediate selfing rates are not stable and should be rare in plants.

11.2 Bimodal distribution of the selfing rate

Barrett and Harder (1996) compiled data on selfing rate involving 129 species (Fig. 11.2). Full selfers and outcrossers are distinct peaks in the distribution, as

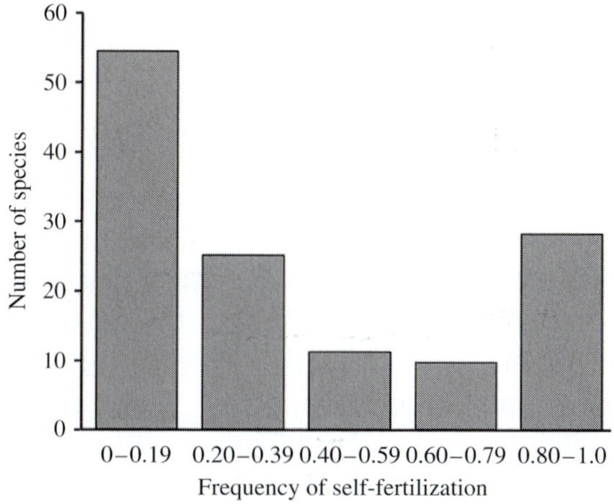

Fig. 11.2 Frequency distribution of measured selfing rate in 129 plant species. Redrawn from Box 1 in Barrett and Harder (1996), *Trends in Ecology and Evolution*, with permission from Elsevier.

theory predicts. However, approximately one-third of the species has an intermediate selfing rate, contrary to the model predictions.

11.3 Can intermediate selfing rates be an ESS?

Can we explain the rather large group of plants with an intermediate selfing rate? There is no simple general answer to this question and perhaps the answer depends on the details of the species in question. First, it has been argued that even with high inbreeding depression, it may be favourable to retain some selfing, since this provides reproductive assurance. Second, if inbreeding depression is greater than 50%, self-incompatibility (SI) may not always evolve. The species may lack genetic variation for an SI system or other obstacles exist (see below). In this situation species will be selected to reduce selfing as much as possible, for instance by evolving floral traits that prevent selfing within the flower. Without SI, it may be difficult for a species to avoid selfing completely. Third, it has been argued that selfing costs some pollen and therefore reduces pollen export. While this is unlikely for prior selfing, for other forms of selfing in which outcross and self pollen grains compete, this is a distinct possibility. We detail each of these three points below.

11.3.1 Reproductive assurance

Usually reproductive assurance is named first in listing the advantages of selfing. One could indeed easily imagine that after a seed has made the highly improbable journey to an island or an isolated patch, it is unlikely to find a mate there and obligate outcrossing would be a most unfortunate handicap. Self-compatibility (SC) would be best for species with long-distance colonization; this is known as Baker's law. However, this advantage of retaining some selfing has hardly been put at a test. If selfed offspring were of inferior quality, perennial plants in established populations might do better by not producing selfed seeds in years with few pollinators and just wait and reserve resources for better pollination conditions. Herlihy and Eckert (2002) showed for the perennial *Aquilegia canadensis* (wild columbine) that selfed seeds are of poor quality and producing these seeds is just a waste of resources. Herrera *et al.* (2001) studied the perennial herb *Helleborus foetidus* (stinking hellebore). Genotypes from different regions in Spain differed in selfing rate when pollinators were excluded. However, genotypes from regions with the fewest pollinators did not have the highest rate of self-pollination, as one would expect if the main function of self-pollination were reproductive assurance. The reproductive assurance argument is pervasive, especially for annuals, but should still be tested critically.

11.3.2 Self-incompatibility does not automatically evolve

The genetics of SI is described in detail in Richards (1997). In gametophytic SI the stigma recognizes the pollen so that germination of pollen with the same allele at the SI locus as the mother is prevented. For instance, a mother plant with genotype S_1S_3 rejects all S_1- and S_3-bearing pollen, whether it comes from itself or from another plant, but accepts pollen with S_2, S_4 etc. Suppose all plants in the population are SC and have the S_0S_0 genotype. When the first allele for SI (S_1) arises in the population most mothers will have genotype S_0S_1. When these mothers reject S_1 pollen this does not mean that they completely avoid selfing. Half the self pollen grains, those with genotype S_0, will still germinate on the stigma. However, S_1 pollen is rejected on all S_0S_1 and S_1S_1 plants in the population. Models that investigated how SI systems come into existence, indicated that, for the spread of the first SI allele, higher levels of inbreeding depression are required than 0.5 (for instance, Charlesworth and Charlesworth 1979a).

There are more reasons why SI does not automatically evolve *de novo*, even if inbreeding depression is greater than 0.5. In Chapter 15 we will show that there is conflict between genes expressed in the parent plant (the sporophyte) and genes expressed in the pollen (the male gametophyte). Furthermore many plant families lack genetic variation for SI (Richards 1997) and SI systems are conserved over species (Richman and Kohn 1996), i.e. species inherit the SI from a common ancestor. These are serious concerns that the phenotypic model is too simple: it should not be taken for granted that with high inbreeding depression SI evolves automatically. If the transition to SI is not made and inbreeding depression is higher than 50%, flower traits that reduce selfing in a less rigorous way will be selected. If some self-pollination then still occurs, for instance through geitonogamy, this should probably be regarded as non-adaptive (Barrett 2003).

11.3.3 Pollen discounting selects for intermediate selfing rates

Holsinger (1992) developed models in which selfing is not prior to the arrival of outcross pollen. In his models a pollinator delivers self and outcross pollen at the stigma at the same time and the two pollen types compete for access to ovules. In such a model more selfing may be accompanied by a reduced export of pollen to other plants in the population and this negative relation between pollen export and selfing is known as pollen discounting. Geitonogamous pollination by bees is a clear case of pollen discounting which we discussed in Chapter 3. In another case, you could imagine that selfing rate and pollen export become negatively correlated, while male allocation stays the same.

What is the ESS with respect to the selfing rate in this case? Consider a rare mutant in a monoecious species with alternative selfing rate S_{m} in a population in which all plants have selfing rate S. All plants, mutant included, allocate r to male and $1-r$ to female. The mutant exports E_{m} pollen grains and the common type export E pollen grains to the outcross pollen pool. Pollen export is a decreasing function of selfing rate. We drop, for simplicity, the exponents of the gain curves. The absolute fitness of a mutant with selfing rate S_{m} and pollen export E_{m}, in a population of plants with selfing rate S and pollen export E, is:

$$W_{\mathrm{m}} = (1-S_{\mathrm{m}})(1-r) + 2S_{\mathrm{m}}(1-\delta)(1-r) + \frac{E_{\mathrm{m}}}{E}(1-S)(1-r) \qquad (11.4)$$

We find the ESS selfing rate by using Parker's method of differentiating mutant fitness with respect to the evolutionary variable (S_{m} in this case) and setting the derivative equal to 0. This yields:

$$\frac{\mathrm{d}W_{\mathrm{m}}}{\mathrm{d}S_{\mathrm{m}}} = -(1-r) + 2(1-\delta)(1-r) + \frac{1}{E}\frac{\mathrm{d}E_{\mathrm{m}}}{\mathrm{d}S_{\mathrm{m}}}(1-S)(1-r) = 0 \qquad (11.5)$$

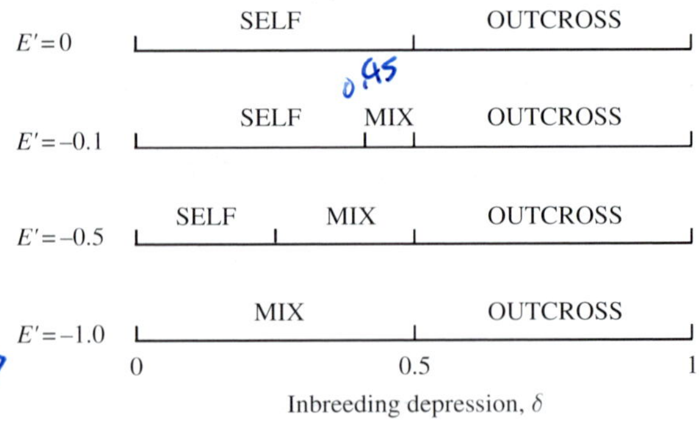

Fig. 11.3 When the efficiency of pollen export decreases with the selfing rate (pollen discounting), then at some levels of inbreeding depression an intermediate selfing rate becomes an ESS. To calculate this example we simply chose $E_m = a + bS_m$ so that $dE_m/dS_m = E' = b$.

With some algebra, scaling pollen export relative to $E = 1$ and writing E' for dE_m/dS_m we obtain the candidate ESS for the selfing rate S^*:

$$S^* = 1 - \frac{2\delta - 1}{E'} \tag{11.6}$$

With $\delta \geq 0.5$ it is still always best to be fully outcrossing ($S^* = 0$). With $\delta < 0.5$ results are however quite different from the model without pollen discounting. It is convenient to describe the pollen discounting by a simple linear function $E_m = a + bS_m$, so that $dE_m/dS_m = b$. For $b = -1$ this describes pollen discounting in the exponential model for geitonogamous selfing by bees that we examined in Chapter 3. With $b = 0$ no pollen discounting exists. Values of the slope b between 0 and –1 correspond to partial pollen discounting. By solving Eq. (11.6) for different values of slope b (Fig. 11.3) we find that without pollen discounting, when b approaches 0, we are back to the old situation of Eq. (11.2) and Eq. (11.3): full selfing is an ESS when $\delta < 0.5$ and full outcrossing is an ESS when $\delta > 0.5$. With some pollen discounting, $b = -0.10$, there is a small range, δ between 0.45 and 0.5, where an intermediate selfing rate is an ESS. With more pollen discounting ($b = -0.5$) this range increases, so that now δ needs to be between 0.25 and 0.5. Finally, with complete pollen discounting ($b = -1.0$) in the whole range of δ between 0 and 0.5 intermediate selfing is an ESS. The, rather surprising, result of Holsinger's models is that mixed mating is often stable if some pollen discounting exists.

Holsinger's (1992) model does not explain why species do not evolve prior selfing, without reduced pollen export, when $\delta < 0.5$. Constraints seem unlikely in this case; any mutation that brings the stigma closer to the anthers or that synchronizes ripening of pollen and stigma would lead to more selfing and would be selected for. As argued in the previous section, a combination of mixed mating and high inbreeding depression may be maintained when genetic variation for SI is lacking and when conditions for *de novo* evolution of such a SI system are restricted.

11.4 Plants play a mixed strategy: cleistogamous and chasmogamous flowers

In some species individual plants produce a mix of cleistogamous (CL) flowers, flowers that do not even open and which are 100% selfed, and chasmogamous (CH), flowers that can be outcrossed. These plants have an intermediate selfing rate, but the system is more complex with a number of associated features. The CL flowers are usually smaller with reduced anthers. Many violets (*Viola* spp.) display this strategy. *Impatiens capensis* (jewelweed) is another example. In some species the CL flowers are produced underground. This is called geocarpy (Fig. 11.4). The two types of fruits produced can be of different size or form; this is called amphicarpy. In grasses the CL seed is usually heavier (Campbell *et al.* 1983). The seeds can also differ greatly in dispersal capability, especially if the CL seed is formed underground. If both types of seeds are formed on the flowering stem, like in the ant-dispersed tropical herb *Calathea micans*, the difference in dispersal may be negligible (Lecorff and Horvitz 1995).

When is such a mixed strategy evolutionarily stable? Why don't plants produce one type of flower, the type with the highest contribution to fitness? Constraints, as discussed in paragraph 11.3, are not likely here; one can easily imagine a mutant that produces only one seed type. A number of hypotheses have been advanced, and we will list the three most important ones.

11.4.1 Environmental change

The first hypothesis states that in one type of environment either CL or CH is best, but the plant should change its strategy when conditions change. This denies that plants play mixed strategies in constant environments. Some plants seem to confirm to this idea. For instance, *Ceratocapnos heterocarpa*, a Spanish annual, makes CL flowers in winter and switches in spring to CH flowers when pollinators have become more abundant (Declavijo and Jimenez 1993). Many others species produce the two flower types simultaneously, so here this hypothesis does not apply.

11.4.2 Bet-hedging

In this hypothesis plants play a game against nature in which they are hedging their bets in an unpredictable environment (Venable 1985), as in Cohen's (1966) model of optimal germination fraction in annuals (Chapter 7). If, for instance, a dry year is best for seeds of type A and a wet year for seeds of type B, then the best strategy is a conditional one; produce the seed type appropriate for the conditions. However, if the plant has no information as to whether it will be a dry or wet year, the best strategy in such an unpredictable environment is to make a mix of the two seed types. CL flowers seem a safe, low-risk option that provides reproductive assurance. The seeds in CL flowers are produced independent of pollinators, economizing on attraction and male function of flowers. CH flowers seem a high-risk option; seed set is dependent on pollinators and the availability of mating partners, but if pollination is successful the seeds are of higher quality. For grasses with geocarpy it has been argued that in some years, grazers eat the flowering stem with the entire seed crop, while the seeds produced by underground CL flowers in the soil escape (Campbell *et al.* 1983). Fire could have similar effects (Fig. 11.4) (Cheplick 1994). Many plants with CL/CH seeds are ruderals that live in disturbed, and therefore unpredictable, environments. Berg and Redbo-Torstensson (1998)

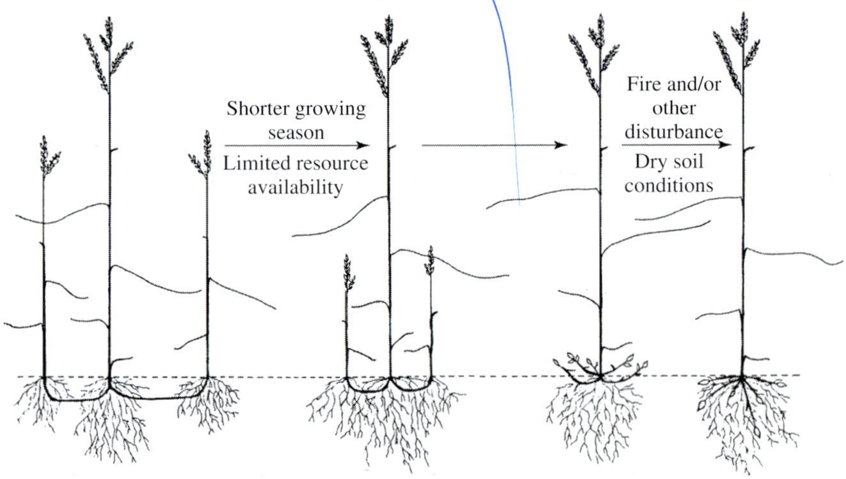

Fig. 11.4 Hypothetical scheme for the evolution of two flower types in *Amphicarpum purshii* (blue maidencane). The reduced panicles with few spikelets evolve into subterranean tillers. The plant on the right produces outcrossed chasmogamous flowers aboveground and underground it produces selfed cleistogamous flowers. Redrawn with permission from Cheplick (1994), *Plant Species Biology*, Blackwell.

found differences in seeds per fruit of CL and CH flowers in different years and between different populations of *Oxalis acetosella* (wood sorrel), consistent with the bet-hedging hypothesis.

11.4.3 Decelerating fitness gain curves

Schoen and Lloyd (1984) modelled this by assuming near–far dispersal for the two types of seeds and increasing LRC if increasing numbers of seeds are produced. In their model the CL seed are dispersed in the first ring around the plant and are probably the first to be affected by LRC. The CH seeds can disperse further into a second ring where they escape from LRC. That LRC can be intense for seeds from CL flowers is clear for the grass *Sporobolus vaginiflorus*: in the first square decimetre near the parent plant on average 145.8 seedlings were found, while in the adjacent square on average 5.1 seedlings emerged (Cheplick 1996).

Schoen and Lloyd's (1984) near–far model is similar to the problem of the car manufacturer in Chapter 2 who needed to decide on making small or large cars. First, the plant should produce the seed with the highest fitness gain per unit of investment. When fitness gains decrease, then at some point it should also start producing the alternative. It would seem logical to begin with the safe option of making CL flowers and to switch later towards CH flowers. Masuda *et al.* (2001) considered the opposite situation where it is initially favourable to produce CH flowers. When many flowers are simultaneously open on the same plant this may result in more geitonogamy. When there are so many CH flowers that all are selfed through geitonogamy, production of less costly CL flowers becomes more economical. Under both scenarios one would expect a strong dependence of the ratio of CL to CH flowers on the size of the plant.

11.5 Size dependency of the ratio CH : CL flowers

Porras and Muñoz (2000) presented relevant data for evaluating the near–far hypothesis for *Centaurea melitensis* (Maltese star thistle). This annual produces two types of flowers, CL and CH. The initial CL seeds are heavy (2.31 mg) with short plumes (Fig. 11.5). The CH seeds weigh less (1.33 mg) and the plume is longer, allowing long-range dispersal. Porras and Muñoz (2000) sowed seeds at different densities in the greenhouse and found that total yield was a decelerating function of density. At the lowest density the fitness return, measured as biomass of the seedling divided by biomass per seed, was highest for CH seeds. At the two higher densities, however, the fitness return of

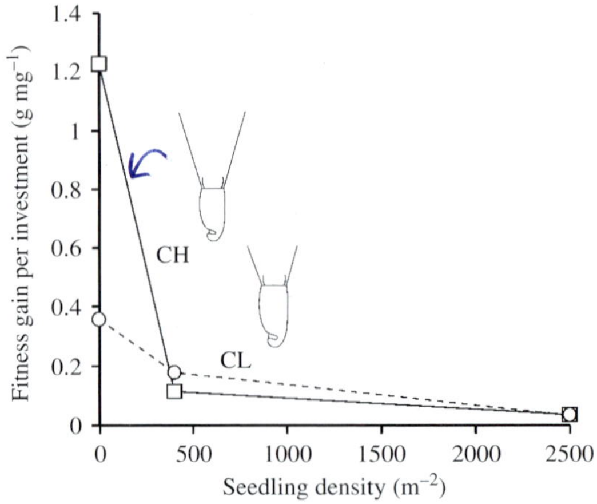

Fig. 11.5 Fitness gain (plant biomass per mg of seed biomass) in *Centaurea melitensis* for seeds from initial cleistogamous (CL) and chasmogamous (CH) flowers at three different sowing densities. Fitness gain is the final biomass per seedling divided by initial seed weight. Seeds from CL flowers are heavier and have a shorter plume. From the figure one would predict that plants begin by making CH flowers and when the fitness gains per unit of investment drop below the gains of making the first CL flower, plants should switch. After Porras and Muñoz (2000).

the seeds produced by CL was higher. One would then expect a small plant to produce CH flowers and a large plant perhaps to switch to producing CL flowers later, when seedling densities are high. This expectation is contrary to the observed strategy of *Centaurea melitensis*. Small plants, grown at the highest density, tended to produce CL flowers only. The reasons for this are unclear to us.

In other species size dependency of the fraction of CH flowers is also frequently observed. Examples in which the fraction of CH flowers increases with plant size seem to be most numerous. In addition to *Centaurea melitensis* (Porras and Muñoz 2000), these examples include *Impatiens* spp. (Waller 1980), *Collomia* spp. (Wilken 1982), *Amphicarpaea bracteata* (hog-peanut) (Schnee and Waller 1986), *Impatiens capensis* (Schmitt and Gamble 1990) and *Ceratocapnos heterocarpa* (Ruiz de Clavijo and Jimenez 1993). Species in which the opposite trend holds include *Danthonia spicata* (poverty oat grass) (Clay 1982), *Oxalis montana* (mountain wood sorrel) (Jasieniuk and Lechowicz 1987) and *Viola mirabilis* (Mattila and Salonen 1995).

Clearly we are just in the stage of listing hypotheses. We still have to await critical experiments to demonstrate the possible advantages of producing both CL and CH flowers.

11.6 Why bother about sex?

The questions why sex evolved and why it is maintained are still far from being solved, despite the many articles and books written about the subject. We will only briefly address this issue here, comparing sexual and asexual genotypes within the same plant species.

It is important to realize that sex has strong disadvantages. First, the costs of sex are quite high; any investment in male allocation is inefficient. If a rare mutant discards sex completely and is still able to produce seeds, its fitness will be double, compared to an outcrossing cosexual plant with equal allocation to both sexes. This is referred to in the literature as the two-fold cost of sex (Maynard Smith 1978). The cost of sex may be even higher when we consider clonal plants that grow only vegetatively and do not waste time or energy on flowering and seed production. Second, sex breaks up favourable gene complexes. If two parents are well adapted to their environment, the different new combinations of chromosomes in the offspring almost certainly are not.

Sex generates genetic variation and the positive effects of this variation feature prominently in the textbooks. First, if the environment changes rapidly, in time or in space, and the best genotype in one set of conditions is worst in another set of conditions, there is a premium on producing variable offspring. Second, if the environment is constant but genotypes degrade rapidly because of high mutation rates, sex will result in many bad genotypes with numerous deleterious mutations but also, by chance, in some very good genotypes that lack the mutations. If selection is then so strong that only the best genotypes survive and not the average ones, sex can be of selective advantage (Kondrashov 1984), because it increases the variance in offspring quality. Whether this effect can offset the disadvantages of sex depends critically on the mutation rate. Third, a possible advantage of sex is that all offspring are different from their parents. If parasites (for plants one might think of pathogens or herbivores) develop a matching genotype for a common plant genotype, this allows them to exploit the plant. A plant with a rare genotype then enjoys an advantage: not because its genotype is inherently better, but because it is different from the common genotype that is targeted by the parasite. This advantage of being rare is known as the 'Red Queen hypothesis', named after the Red Queen from Lewis Carroll's story *Through the Looking-Glass*. In the book the Red Queen tells Alice: 'it takes all the running you can do, to stay in the same place.'

These and many other theories about the evolution of sex have been discussed at length – see for instance the April 2002 issue of *Nature Genetics* – without a final conclusion being reached. Some people are now searching in different directions. For instance, Meselson (Anonymous 2003) recently asked: 'Why does meiosis exist?' He pointed out that during the first stage of meiosis the homologous chromosomes are aligned before being pulled towards opposite parts of the cell. During this alignment they are in close physical contact and scanning occurs. If the homologous chromosomes are sufficiently different, meiosis stops. Meselson suggested that this could be an extraordinary way to prevent transmission of selfish genetic elements such as retrotransposons. Retrotransposons replicate by incorporation at various places within the DNA of a cell. To test the idea that sex is a protection against retrotransposons, Welch and Meselson (2000) chose bdelloid rotifers as a study system. These animals of 0.5 mm or less are filter-feeders that live in water; they can withstand frost and long periods without water and resume their activities when conditions have become suitable again. No males have ever been observed in bdelloid rotifers and molecular evidence suggested that they had no sex for at least 40 million years (Arkhipova and Meselson 2000). Interestingly, the bdelloids contained no retrotransposons. This is consistent with the idea that asexual reproduction is superior without retrotransposons in the host DNA, but a dead end when infected by retrotransposons. In Arkhipova and Meselson's (2005) view meiosis helps to limit the deleterious load of selfish mobile elements. It remains to be seen if such ideas can be substantiated but it does show there is sufficient room for fresh new ideas and that the problem of the evolution of sex is by no means solved.

11.7 Apomicts

We now examine some examples of plant species with both asexual and sexual genotypes. We distinguish between apomicts, plants that produce seeds without sex, and clonal plants that propagate through vegetative growth. Apomixis occurs in some plant families, but is completely lacking in others, probably because genetic variation for this complex trait does not exist (Mogie 1992). *Taraxacum officinale* is a well-known apomict. When you remove all pollen from a dandelion flower before it opens so that pollination is prevented, the flower will still set seed. In most apomicts the seeds are genetically identical to the mother. In *Taraxacum*, however, early meiosis does take place and in this stage some recombination can apparently take place (van Baarlen *et al.* 2000); seeds are still very similar to their parent but need not be 100% identical.

When comparing sexuals and apomicts, the apomicts usually have a higher ploidy level. Most asexual *Taraxacums* have 3*n* chromosomes and sexuals have 2*n*. In a literature review Bierzychudek (1987) reported that asexuals usually occurred under more extreme conditions, over a larger geographical range, at the highest elevations and in once-glaciated areas, and were sometimes associated with disturbance. As clear as this pattern is, as unclear is the correct explanation for it. It has been argued that asexuals do worse in competition but have a superior colonizing ability. This hypothesis could be tested by placing seeds of the sexual plant into the environment of the apomict; after a little help with their dispersal, you would expect the sexuals to outcompete the apomicts. There is however a more fundamental problem in using species of *Taraxacum* to understand the benefits of sexual or asexual reproduction. The apomicts have a different ploidy level and perhaps this allows plants to grow in such harsh conditions, irrespective of the sexuality (Bierzychudek 1987). Studies of sexuals and apomicts with the same ploidy level will give a clearer answer to the question about the pros and cons of sex. Such a study was carried out with sexual and asexual populations of the perennial *Antennaria parlinii* (Parlin's pussytoes) in the USA (Michaels and Bazzaz 1986, 1989; O'Connell and Eckert 2001). Asexuals produced twice as many seeds, in field and greenhouse, and their higher reproductive effort was associated with reduced survival and reduced clonal propagation. Seeds of asexuals were smaller and more highly barbed, allowing dispersal over greater distances. Sexual populations occurred in wooded habitats, while asexuals were apparently excluded from this habitat. The asexuals were common in disturbed roadsides, ditches and fallow fields. Instead of either the sexual or asexual always outcompeting the other form, the forms have differentiated and have each moved into their own habitat.

11.8 Clonal plants

Clonal plants use all their resources for efficient growth. Some indication of the sort of problem that clonal plants face come from Eckert's (2002) study on the aquatic scrub *Decodon verticillatus* (swamp loosestrife) in New England, USA. An investigation in 28 populations across a 500-km latitudinal gradient showed a marked reduction in several components of sexual reproduction in the northern part of the range of the species. On average, plants in the north produced only 15 seeds compared to 1139 in the south. Fertility in the field populations corresponded with fertility in a common greenhouse, showing the genetic basis of reduced fertility. Eckert (2002) argued that seedling recruitment was severely limited in northern populations, selecting for more clonal growth (see also the ESS model of Olejniczak (2001) which clearly illustrates this). When the northern plants of *D. verticillatus* flowered, it became evident that sexual traits such as percentage viable pollen and synchrony of different flower parts had become degraded (Eckert 2002).

Mable and Otto (1998) emphasized that the haploid stage is important for selection. In higher plants the haploid stage is restricted to the pollen grain and the embryo sac, which depend strongly on the flowering plant, i.e. the diploid sporophyte. One of the genetic consequences of being a diploid is that deleterious mutations within the genome are masked. There is always the other homologous chromosome that can take over if mutation renders a gene inactive. Masking is, however, a two-edged sword: it is beneficial for the individual who survives, but disadvantageous to the offspring who inherit more mutations. In the haploid phase the deleterious mutations are expressed in the gametophyte and this is an easy, and probably cheap, way of selecting the best quality from the many pollen grains formed. If selection occurs mainly in the gametophyte stage, the less sex plants have the more deleterious mutations are accumulated in the diploid sporophyte. This would spell the decline of clonal plants in the long run, consistent with Eckert's (2002) observations on northern plants of *D. verticillatus*.

To sum up: depending on the level of inbreeding depression 100% selfing or 100% outcrossing are alternative steady states. While many plants fit into this dichotomy, one-third of all species has an intermediate selfing rate. Reproductive assurance may be important in this respect. Alternatively, cost-free selfing or self-incompatibility may not be so easily achieved as the phenotypic models assume. With constraints on SI, partial selfing may be non-adaptive and just difficult to avoid. Plants that produce both CH and CL flowers play a mixed strategy, which requires an explanation. We just listed the most important hypotheses, which stated that (1) the environment is changing in time, (2) the environment is unpredictable and (3) diminishing gains exist from producing each flower type.

Questions

(1) What is the difference between apomixis as in *Taraxacum* and clonal growth?
(2) Is selection more effective in the haploid or in the diploid stage?
(3) What is pollen discounting and how does it affect the optimal selfing rate?
(4) Which two advantages of selfing were mentioned?

12

Heterostyly

Heterostyly is the coexistence of two or three, genetically controlled, hermaphrodite plant types (morphs) in a population, in which the morphs display reciprocal differences in style and anther height. Least rare is distyly with two morphs and most rare is tristyly with three coexisting morphs (Barrett 1993). Historically heterostyly has attracted much attention. It was one of the main subjects of Darwin's (1877) book *The Different Forms of Flowers on Plants of the Same Species*. Ernst worked out the genetics of the system in the 1930s and 1940s (reviewed in Richards 1997). In this context we can only address a few issues that are close to the themes of this book. For a broader perspective see the books by Barrett (1992) and Richards (1997). First, we outline the basic biology of distyly. Second, we consider optimal sex allocation in populations of heterostylous species. Third, although heterostylous species are typically outcrossing, sometimes selfing mutants invade heterostylous populations. We look at the example of *Primula vulgaris* (primrose) to examine the fate of such mutants. Fourth, we examine the most likely evolutionary route from hermaphroditism to distyly, using the model of Lloyd and Webb (1992a, b).

12.1 Introduction

Tristyly occurs only in some members of the Oxalidaceae, Lythraceae, Pontederiaceae, Amaryllidaceae, Connaraceae and Linaceae (Barrett 1993; Barrett *et al.* 1996, 1997; Thompson *et al.* 1996). In this chapter we restrict ourselves to distyly, which occurs in at least 25 families and appears to have evolved independently at least 23 times (Lloyd and Webb 1992a). Well known examples include members of the genera *Primula*, *Pulmonaria*, *Limonium* and *Linum*. The morphs typically have reciprocal herkogamy (RH). Herkogamy means that the stigma and anthers are separated in space. Reciprocal refers to the fact that the two morphs are each other's opposites. In one morph a high

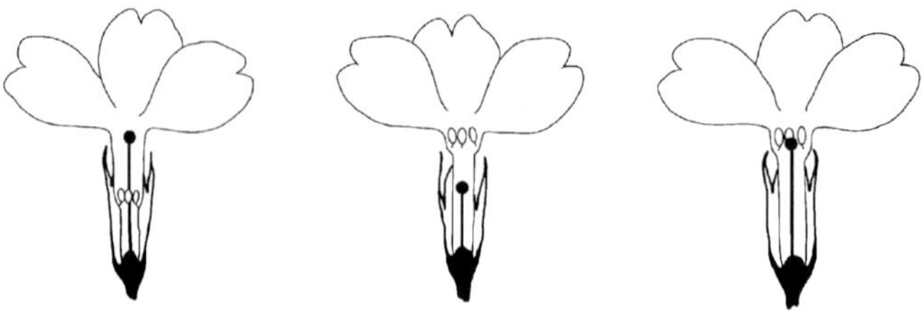

Fig. 12.1 The two flower morphs in distylous *Primula vulgaris* and the long homostyle. (A) Flowers with anthers in the low and stigma in the high position are called pin. (B) Flowers with the reciprocal position, anthers high and stigma low, are called thrum. (C) Through crossing-over, both long and short homostyles can arise. In *Primula vulgaris* the long homostyle is found in some English populations. A supergene controls the differences between the morphs, with thrum characters dominant over pin characters. Redrawn with permission from Crosby (1949), *Evolution*.

position of the stigma is combined with a low, more hidden, position of the pollen. This long-styled morph is called the pin. The other short-styled morph, the thrum, combines a low position of the stigma with a high position of the pollen (Fig. 12.1). Sometimes morphs are encountered in which pollen is positioned close to the stigma; these morphs self-pollinate and are called homostyles. All species with RH have flowers in which the petals form a bell, a tube, or at least a funnel, with concealed nectar presentation at the bottom and are pollinated primarily by bees and diverse Lepidoptera. The spatial separation of pollen and stigma prevents self-pollination, both within the flower and between flowers on the same plant and promotes mating between pins and thrums (Fig. 12.2). Later in this chapter we examine data on how well this system works.

In *Primula vulgaris*, two genes determine whether a plant is a thrum or a pin. Thrum characters are dominant; *G/g* controls style length (*G* for short styles dominant) and *A/a* controls anther height (*A* for long anthers dominant). The thrum has genotype *Gg/Aa* and the pin has genotype *gg/aa*. In *Primula* the two genes are very close together on the S chromosome and also other genes, for instance for pollen shape and size, lie inside this cluster. For this reason the term 'supergene' is used (Kurian and Richards 1997). We can denote the thrum as *Ss* and the pin as *ss* for all characters at the supergene. Because the genes are so close together on the chromosome, crossing-over during meiosis is unlikely. If this rare event occurs this could lead to, for instance, *Gg/aa* (homostyles with both short styles and anthers) or *gg/Aa* (homostyles with both long styles and anthers, as in Fig. 12.1C). Such long-styled homostyles have been found in

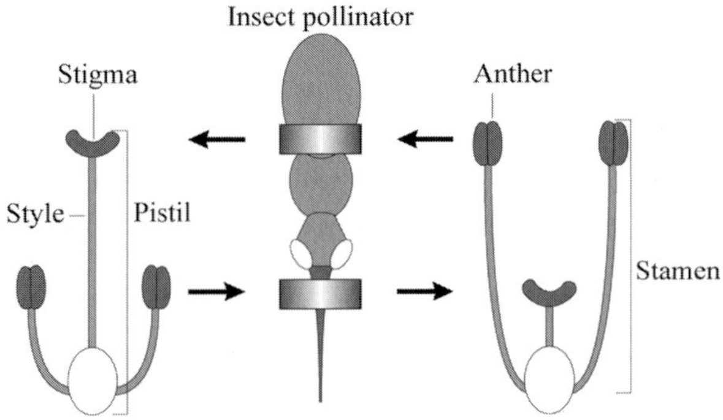

Fig. 12.2 Pollen and stigma in the low position mostly contact the head of the insect pollinator, while in the high position thorax and abdomen are contacted. Redrawn with permission from Barrett (2002), *Nature Genetics*, Nature Publishing Group, http://www.nature.com.

some field populations, whereas the short-styled homostyles are rarely observed in crosses. The homostyles are usually fully self-compatible (SC). The genes for the self-incompatibility system are still functioning but if the recombination has resulted in, for instance, a flower with pin-type pollen and a thrum-type stigma, the germination of self-pollen is vigorous. In cases where distyly evolved into dioecy, the heterozygote thrum has typically become the male and the pin has become the female (Bawa 1980). Males are then the heterogametic sex, as is the rule for most dioecious plants.

A striking feature of heterostyly is that other genes have become associated with the genes controlling RH. Many species with RH, *Primula* being again the genus best studied, have sporophytic self-incompatibility (sSI). This means that pollen of a thrum plant cannot successfully fertilize another thrum plant, it can only be successful on a pin plant. Pollen of a pin plant is only successful on a thrum plant. The system is a bit leaky. For instance, Crosby (1949) estimated for *Primula vulgaris* that thrums were fully outcrossing but pins had 10% selfing. Because selfed pins produce 100% pin offspring (remember pins are homozygous), this leads to a small excess of pins in *Primula vulgaris* populations. Sporophytic SI prevents self-pollination, but also stops pollen from germinating on the stigma of another plant of the same morph. This greatly reduces the male reproductive success and, if pollen is limiting, also reduces seed set. Darwin (1877, p. 265) already wondered: 'How can it be an advantage to a plant to be sterile with half its brethren?'

In *Primula* sSI is typical, but recently other systems have been reported also. SI occurs in the genus *Narcissus*, but morph and mating type are not coupled (Arroyo *et al.* 2002) and some species with RH are self-compatible (Barrett and Cruzan 1994).

12.2 Reciprocal herkogamy promotes inter-morph mating

The different positions of the anthers in the thrum and pin morph will result in different placement of pollen on the bee (Fig. 12.2). Darwin (1877) suggested that this results in more legitimate pollen flow (pin–thrum) than would be expected from random pollen movement. This directionality in pollen flow is referred to as inter-morph or disassortative mating. Pin and thrum pollen on a stigma can be distinguished under a microscope. In the first studies of inter-morph mating, researchers collected stigmas in the field and compared the fraction of legitimate pollen with the frequencies at which the two pollen types are produced in the population. They frequently found high fractions of illegitimate pollen. The initial studies were therefore sceptical of whether RH established inter-morph mating at all. However, as Lloyd and Webb (1992a) pointed out, the method of just collecting stigmas in the field is not appropriate. Plants usually have many flowers open at the same time, all of the same morph, and a flower on a pin plant is likely to receive much pin-type pollen from anthers in the same flower and from all the neighbouring flowers through geitonogamy. Therefore the fraction of legitimate pollen on the stigma cannot simply be compared with the population average. Emasculating all flowers on a receptor plant eliminates self-pollination. Emasculation before the flowers open and counting the pin and thrum pollen on the stigma after natural pollination, already gives a better idea if RH promotes legitimate pollen transfer. In *Primula vulgaris* the ratio of the pollen types differed on the pin and thrum stigmas of emasculated flowers, indicating that legitimate pollination occurs frequently (Table 12.1). The system is not perfect; some of the pollen is deposited on the wrong stigma and these illegitimate pollen grains are unsuccessful.

Table 12.1 *Average number of legitimate (bold) and illegitimate pollen grains deposited on the stigmas of open pollinated flowers of* Primula vulgaris *on plants on which all flowers were emasculated*

	Pin stigma	Thrum stigma
Pin pollen	183	**305**
Thrum pollen	**267**	145

Source: Piper and Charlesworth (1986).

12.3 Sex allocation

How will sex allocation evolve in plant species with two flower morphs? This problem was addressed by Casper and Charnov (1982) for distylous and by Taylor (1984) for tristylous plants, using the methods we have become familiar with in the previous chapters. A plant is either pin or thrum and each crossing between a pin and thrum plant will produce again the two morphs. The genetic determination of the morphs means that, with Mendelian inheritance, they are produced in equal proportions. However, other genes could affect the ratio between pins and thrums, for instance by selective seed abortion. Once the pin and thrum offspring have been formed the sex allocation problem becomes how to divide resources over male and female function, and the optimal allocation may well be different in pins and thrums. At the ESS no mutant is expected to do better if it makes different decisions about allocation.

Assume that seeds of the common type contain a fraction f pins and $(1-f)$ thrums. In each offspring that these plants produce the allocation to male or female function may also vary. We denote the fraction of male allocation in pin plants as g, so that due to a trade-off $1-g$ is allocated to female. The fraction male allocation in thrum plants is denoted as h, corresponding with $1-h$ female allocation. At the ESS no mutant with a different allocation with respect to one of the three evolutionary variables f, g and h can do better than the common type (Fig. 12.3) (Casper and Charnov 1982). This requires that at the ESS, the derivative of mutant fitness with respect to each evolutionary variable is zero. Therefore at the ESS when the common type allocates f^*, g^* and h^*:

$$\frac{\mathrm{d}W_{\mathrm{m}}}{\mathrm{d}f_{\mathrm{m}}} = 0; \quad \frac{\mathrm{d}W_{\mathrm{m}}}{\mathrm{d}g_{\mathrm{m}}} = 0; \quad \frac{\mathrm{d}W_{\mathrm{m}}}{\mathrm{d}h_{\mathrm{m}}} = 0 \tag{12.1}$$

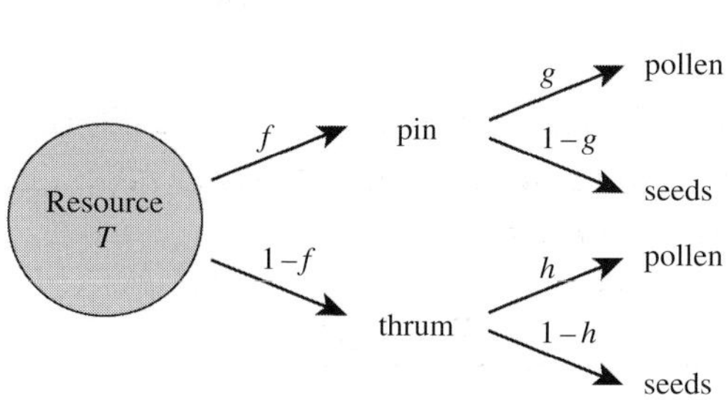

Fig. 12.3 The sex allocation problem in distylous plants. Plants can affect allocation by varying the fraction of pin or thrum offspring they produce and by varying the relative male allocation for each morph.

f = fraction pins
1−f = fraction thrums
g = fraction ♂ allocation in pin; (1−g) pin ♀
h = fraction ♂ allocation in thrum; (1−h) thrum ♀

The equation for plant fitness is complex, because it not only contains how much pin and thrum offspring the plant makes but also their success through both seed and pollen production. Each pin offspring can be successful through (1) the seeds it produces or (2) through its male reproductive success, siring seeds on thrums. The same holds for the thrum offspring; they can be successful through (3) seed production or (4) through male success on pins. We find the seed production by multiplying the fraction allocated to that morph (f to pins or $1-f$ to thrums) by the relative female allocation in the morph. Similarly the pollen produced is the product of the fraction allocated to the morph and the relative male allocation. The absolute fitness of a rare mutant with strategy f_m, g_m, h_m in a population playing f, g and h is then (Casper and Charnov 1982) (1) female success pin + (2) male success pin + (3) female success thrum + (4) male success thrum:

$$W_m = f_m(1-g_m) + \frac{f_m g_m}{fg}(1-f)(1-h) + (1-f_m)(1-h_m) + \frac{(1-f_m)h_m}{(1-f)h} f(1-g) \qquad (12.2)$$

First, what is the ESS fraction of pins? To find f^* we use Parker's method; first differentiate mutant fitness with respect to f_m, set the derivative equal to zero and set $g_m = g$ and $h_m = h$. It follows that $(1-f)(1-h)/f - f(1-g)/(1-f) = (1-h) - (1-g)$. This is only true if $(1-f)/f = f/(1-f)$ so that $f^* = 0.5$. With linear gain curves it is an ESS to produce pin and thrum offspring in equal proportions. This result is unaffected if pin and thrum offspring differ in growth or mortality before flowering, so that the ratio of pins to thrums in the flowering plants in the field differs from 1:1. Second, to calculate ESS sex allocation we differentiate mutant fitness with respect to g_m. This yields $g^* = (1-f)(1-h)/f$. Third, differentiating with respect to h_m yields $h^* = f(1-g)/(1-f)$. At $f^* = 0.5$ different combinations $g^* = (1-h^*)$ are an ESS. For instance, $g^* = h^* = 0.5$ is an ESS, but also $g^* = 0.8$ and $h^* = 0.2$ or even $g^* = 0$ and $h^* = 1$, no mutant with a different male allocation in pins or thrums could do better than the resident. In the last case ($g^* = 0$ and $h^* = 1$) pins are fully female and thrums are male. With linear gain curves this system could easily evolve towards dioecy. Of course, Eq. (12.2) can further be modified for decelerating returns from each allocation, in the same way as we modified our equations for sex allocation in earlier chapters. With decelerating fitness gain curves both types should remain cosexual. However, there is no reason for sex allocation in the two morphs to be the same, nor should we always expect equal proportions of pin and thrum offspring. If one morph is more suited for a particular sexual function, specialization towards that function is expected. There are still surprisingly few studies that have followed up on Casper and Charnov's (1982) idea to compare systematically sex allocation between morphs. Barrett (1980) found that the female fertility of the short-styled (thrum) morph was 56% higher than that of the long-styled (pin) morph in *Nymphoides indica* (water snowflake). This trend is opposite to gender specialization in dioecious species with distylous ancestors; as we have already noted, in such cases the pin morph typically became the female according to Bawa (1980). Two fairly recent studies have found no evidence of strong sexual specialization. Thompson and Dommée (1993) found that thrums produced more seeds per flower than pins in *Jasminum fruticans* (wild jasmine). However, after additional hand-pollination the seed set of the morphs was nearly identical, suggesting strong pollen limitation in pins in the field. In *Primula farinosa*, Mazer and Hultgard (1993) found lower pollen/ovule ratios in thrums, suggesting thrums are more female. Thrum pollen was however much larger than pin pollen so it is not clear how the morphs compare with respect to resource allocation. It would certainly be of interest for understanding gender specialization, if more systematic data on sex allocation in distylous plants were collected.

Box 12.1 Crosby's (1949) genetic model for *Primula vulgaris*

The genotypes of the morphs at the supergene controlling RH are known in *Primula vulgaris*. The thrum is Ss, the pin is ss and the homostyle is $s's$ with s' dominant over s but recessive when compared to the S allele. When the homostyle is selfed, a fourth genotype $s's'$ is produced, which is also a homostyle. The homostyle behaves as a thrum, so it crosses only with the pins. Following Crosby's notation, the frequencies of plants in the population are q for thrums, p for pins, r for heterozygous homostyles and s for homozygous homostyles; $p+q+r+s=1$. Crosby assumed that homozygous homostyles produce fewer seeds; in his model their seed production is reduced by a factor V (viability). Based on the field data, Crosby also assumed 10% selfing for the pins, but to simplify the model we have left this out. Pins sire seeds of thrums. Either thrums or both homostyle types sire seeds of pins. Seeds of homostyles are produced through selfing. The pins are mated by thrums with a probability of $q/(q+r+s)=q/(1-p)$, they are mated by heterozygous homostyles with a probability $r/(1-p)$ and by homozygous homostyles with probability $s/(1-p)$. This leads to the mating scheme shown in the table.

Genotype	Frequency	Mother	Mate	Mating frequency	Genotypes in next generation			
					Ss	ss	$s's$	$s's'$
Ss	q	Thrum	Pin	q	$q/2$	$q/2$	–	–
ss	p	Pin	Thrum	$\frac{pq}{1-p}$	$\frac{pq}{2-2p}$	$\frac{pq}{2-2p}$	–	–
			Homo1	$\frac{pr}{1-p}$	–	$\frac{pr}{2-2p}$	$\frac{pr}{2-2p}$	–
			Homo2	$\frac{ps}{1-p}$	–	–	$\frac{ps}{1-p}$	–
$s's$	r	Homo1	self	r	–	$r/4$	$r/2$	$(r/4)V$
$s's'$	s	Homo2	self	s	–	–	–	sV

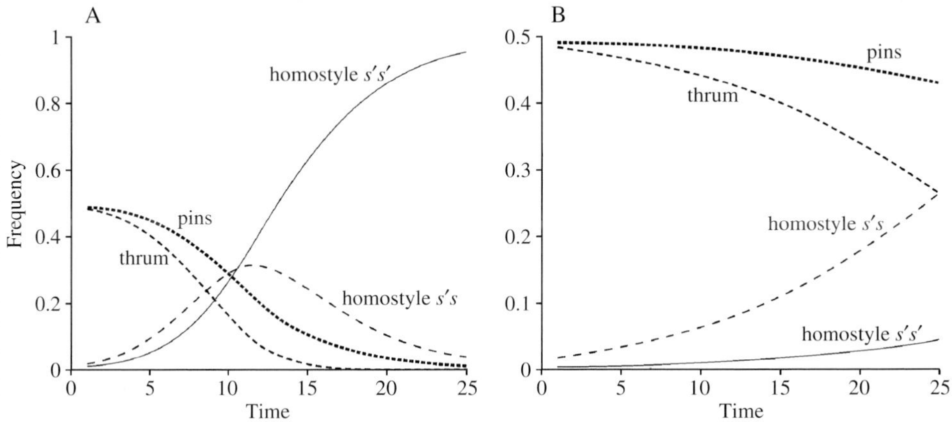

Fig. 12.4 Model calculations of the time course, in units of generation time, of genotype frequencies in *Primula vulgaris*. (A) Selfing homozygous homostyles produce the same number of seeds as the other types ($V = 1$). The thrums go down first, followed by the pins after which the populations evolves towards 100% homostyles of the $s's'$ genotype. (B) Homozygous homostyles produced 40% of seeds that the other types make ($V = 0.4$). Thrums are eliminated as before, but more slowly. After a large number of generations pins reach a frequency of 0.32 and coexist with homostyles.

The frequencies of the next generation genotypes on the right-hand side of the table sum up to $T = p + q + 0.75r + 0.25rV + sV$; only if $V = 1$ this adds up to 1. To calculate the genotype frequencies in the next generation we have to make sure all frequencies sum up to 1. This is the case if we divide all new the frequencies by total T.

The equations that relate the old frequencies of the four genotypes to the frequencies in the next generation (denoted by the prime) are then:

$$p' = \left(\frac{q}{2} + \frac{pq}{2-2p} + \frac{pr}{2-2p} + \frac{r}{4}\right)/T \qquad (12.3a)$$

$$q' = \left(\frac{q}{2} + \frac{pq}{2-2p}\right)/T \qquad (12.3b)$$

$$r' = \left(\frac{pr}{2-2p} + \frac{ps}{1-p} + \frac{r}{2}\right)/T \qquad (12.3c)$$

$$s' = ((r/4)V + sV)/T \qquad (12.3d)$$

These equations allow us to calculate the time course of allele frequencies (Fig. 12.4). A calculation of the equilibrium frequencies is possible, because at the equilibrium the thrums are always extinct ($q = 0$) and this simplifies Eq. 12.3. For the maths the reader is referred to Crosby (1949).

12.4 Wars of the primroses

In a study that was far ahead of its time, Crosby (1949) documented invasion of homostyles in populations of the distylous *Primula vulgaris* in Somerset, England. The geneticist R. A. Fisher was one of the volunteers who helped to document the morph frequencies in local populations. In the long-styled homostyles that occurred in these populations, anthers were at the thrum and stigmas at the pin position (Fig. 12.1C). The pollen of homostyles is of the thrum-type and therefore the homostyles are competing with thrum plants for pollination of pins. Crosby (1949) assumed that homostyles were selfers and this was later found to be correct using isozymes (Piper *et al.* 1984). Selfers pass on more allele copies to the next generation and, without inbreeding depression, should increase in frequency. This is Fisher's automatic advantage of selfing that we discussed in Chapter 11. What will happen with the homostyles in the Somerset populations? Crosby (1949) built a genetic model, using the detailed information about the genetics of the morphs that was already available at the time (Box 12.1). In his first simulations Crosby (1949) assumed that homostyles produced the same amount of seeds as the other morphs. Later studies confirmed this was the case if seed set of pins and thrum was not pollen limited (Piper *et al.* 1986). Does the s' allele that determines homostyly go readily to fixation in these populations, solely because of Fisher's transmission advantage? The first calculations by Crosby (1949), outlined in Box 12.1, showed that this is the case. Figure 12.4A shows that the thrum morph goes down first and, as this morph disappears from the population, breeding options for the pin morph decline and pins disappear also. In the end the population consists only of homozygous homostyles.

The prediction of fixation of the s' allele was however inconsistent with the observations on morph frequency in the Somerset populations; no *Primula vulgaris* population was found with 100% homostyles. The thrum morph was absent from the populations, consistent with the model prediction. However, in populations with a high fraction of homostyles still about 20% pins were present. Apparently pins were not completely eliminated as the model suggested. To account for this finding Crosby made a second series of calculations, now assuming that homozygous homostyles produce seeds of low quality (discussed in Richards 1997). If the viability of seeds produced by homozygous homostyles was reduced by 19% or less, simulations showed elimination of all pins. If the seed production of homozygous homostyles was further reduced, pins were maintained in the population. In Fig. 12.4B an example is given for $V = 0.4$ (60% reduction of seed production).

Twenty-seven years after Crosby made his observations, Curtis and Curtis (1985) revisited the area and re-counted populations that were still there. Their data supported the frequency dependence from Crosby's second model. Homostyles at low frequency in 1956 showed an increase, while the populations at high frequencies (80–100%) in 1956 showed no change (Fig. 12.5). The data suggest coexistence of homostyles and pins.

In other species or under other conditions homostyles may be eliminated from the population or may take over. For instance, in the western Himalayas, homostylous plants of *Primula floribunda* make up 100% of the population at the highest altitude but are not reported from sampling locations at low altitude (Richards 1997). Richards suggested that on high-altitude sites, reproductive assurance of the selfer makes it a superior colonizer. When homostyles remain rare, inbreeding depression probably prevents their invasion (Kery *et al.* 2000). A systematic study of inbreeding depression in distylous plants would certainly be worthwhile.

In *Primula vulgaris*, apparently the selfed offspring of the homozygous homostyle have reduced viability (Crosby 1949; Richards 1997). More generally, all selfed offspring may show some inbreeding depression. When pins coexist with homostyles, the model is very similar to the phenotypic model of gynodioecy from Chapter 8. Pins are maintained because they produce outcrossed seeds that do not suffer from inbreeding depression. In gynodioecy, females are also maintained because of an outbreeding advantage. In analogy to gynodioecy, we can use Eq. (8.14) to predict the equilibrium frequency of pins. If the selfing

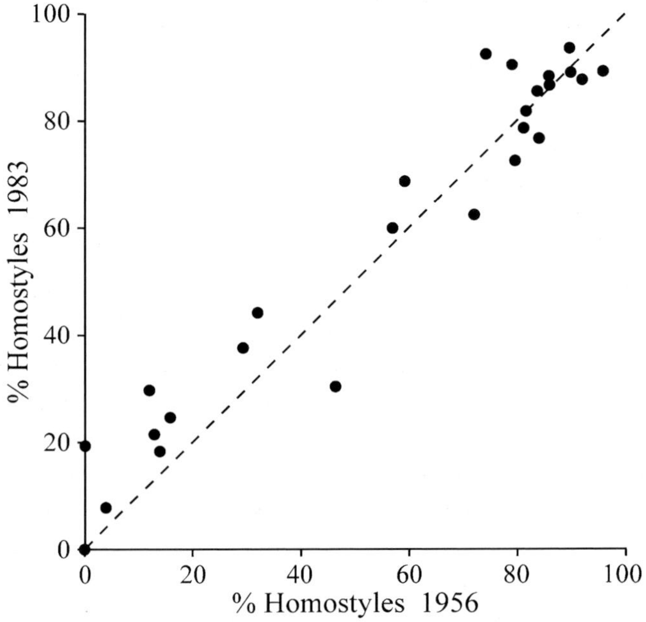

Fig. 12.5 Curtis and Curtis (1985) revisited Crosby's populations of *Primula vulgaris* after 27 years (approximately three generations). The dashed line indicates the situation without change. The points on the left-hand side were above the line, suggesting homostyles increased when rare. Redrawn with permission from Curtis and Curtis (1985), *Heredity*, Nature Publishing Group, http://www.nature.com.

rate of the homostyle is 100% and the seed production of pins and homostyles is equal (no compensation, $k = 0$), Eq. (8.14) becomes:

$$\text{fraction of pins at equilibrium} = \frac{2\delta - 1}{2\delta} = 1 - \frac{1}{2\delta} \qquad (12.3)$$

For the fraction of pins to be positive, more than 50% inbreeding depression is required. In Crosby's (1949) model more than 19% reduction in viability for the homozygous homostyle was required for the maintenance of pins. The two models differ in their assumption about which type has reduced viability, but both Eq. (12.3) and the Crosby model illustrate that reduced offspring vigour of the homostyles allows maintenance of plants that are functionally female.

12.5 The evolutionary path towards distyly

Considering the many studies on the genetics of distyly one would expect that the probable order of the evolutionary events is well known. This is not the case. Fisher inserted in his 1958 edition of *The Genetical Theory of Natural Selection* (p. 278 in the Variorum edition):

> No satisfactory selective model has been set up competent even to derive a distylic species like the primrose from a monostylic species of the same genus. Possibly, therefore, the course of evolutionary change has been complex and circuitous.

At present the model of Lloyd and Webb (1992b) gives the most satisfactory account and we will sketch their model in detail. Essential in their model is the transfer of pollen between morphs. They defined the proficiency of pollen transfer q between the morphs as the quantity of pollen deposited on the stigma after visiting one flower of another (or the same) morph. If we have two morphs A and B we have q_{AB}, q_{BA}, q_{AA} and q_{BB} as the four different possibilities. Lloyd and Webb started in as simple a way as possible with an outcrossing species with a single flower, visited only once by a pollinator depositing all of its pollen on the first flower visited. A logical extension of their model to species with many flowers would be to describe pollen transfer more exactly in terms of pollen removal from flowers, carry-over between different flowers on the same plant and pollen deposition (as in Fig. 3.5). Such an extension has not yet been published. In an ancient population in the Lloyd and Webb model, all plants can cross with each other. In order for two flower types to coexist each type must do better than the other when it is rare. If we consider a population with 100% plants of the A type, then each plant receives and donates q_{AA} pollen when a pollinator visits two A flowers in succession. If seed set is fully pollen limited, each receives and donates q_{AA} pollen so that both male and female reproductive success equals q_{AA}. The fitness of the common morph is therefore $2q_{AA}$. A rare new morph B would only receive pollen from plants of type A and will only donate pollen to plants of type A. Its fitness is then $q_{AB}+q_{BA}$ and the mutant can invade if: $q_{AB}+q_{BA}>2q_{AA}$. Similarly when the B type is common the A type can invade when $q_{BA}+q_{AB}>2q_{BB}$. If both conditions are fulfilled each type wins when rare, so that the two types will coexist. The model can easily be transferred to the situation where seed set is limited by resources rather than pollen. In such a case the fitness of type A when common, consists of a constant component, its seed set, and a variable component, the seeds sired on another plant (q_{AA}). The fitness of mutant B consists of the same constant seed set plus q_{BA}. The mutant can invade if $q_{BA}>q_{AA}$. Similarly an A mutant in a population in which all plants are of morph B, can invade if $q_{AB}>q_{BB}$. Equation (12.4) gives the conditions for coexistence of morphs A and B.

pollen limitation of seed set : $q_{AB}+q_{BA}>2q_{AA}$ and $q_{AB}+q_{BA}>2q_{BB}$ (12.4a)

Table 12.2. *The probabilities that a pollen grain picked up in one flower is deposited on the stigma of the next flower visited in* Jepsonia heterandra (*foothill jepsonia*)

Receptor:	Pin (A)	Thrum (B)
Donor:		
Pin (A)	2.98×10^{-3}	$\mathbf{2.71 \times 10^{-3}}$
Thrum (B)	$\mathbf{4.46 \times 10^{-3}}$	1.29×10^{-3}

Note: Thrum flowers contained 91 000 pollen grains and pin flowers contained 211 000 pollen grains. Bold type indicates legitimate or inter-morph pollen flow.
Source: Ganders (1974); Lloyd and Webb (1992b).

$$\text{resource limitation of seed set}: q_{BA} > q_{AA} \text{ and } q_{AB} > q_{BB} \qquad (12.4b)$$

For an ancestral species in which the two morphs produce the same amount of pollen, Lloyd and Webb's (1992b) proficiency of pollen transfer is clearly defined as the male component of fitness. It requires some more thought to put a number on the proficiency of pollen transfer for a real distylous species in which the morphs are already developed and produce different amounts of pollen per flower, and that may not fit the other model assumptions. Lloyd and Webb (1992b) argued that if pin and thrum flowers received the same number of similar visits, the probabilities of a pollen grain being deposited are proportional to the proficiency of pollen export and we follow this assumption. They then used data from Ganders (1974) to calculate this quantity (Table 12.2). These data are more detailed than in Table 12.1, as we now examine what happens to individual pollen grains when a bee moves once between flowers of different morphs. In the case of pollen limitation (Eq. 12.4a) $q_{AB} + q_{BA} = 2.71 + 4.46 = 7.17$, which is larger than both $2 \times q_{AA} = 2 \times 2.98 = 5.96$ or $2 \times q_{BB} = 2 \times 1.29 = 2.58$, so the morphs should coexist. Without pollen limitation (Eq. 12.4b) both the condition $q_{BA} > q_{AA}$ ($4.46 > 2.98$) and $q_{AB} > q_{BB}$ ($2.71 > 1.29$) is met. Inter-morph pollination occurs more frequently than expected under random mating and this helps to explain the coexistence of two morphs in the population before sSI evolves.

Next Lloyd and Webb (1992b) applied their concepts to the evolutionary events that led to distyly. They started from a species in which the style protrudes past the anthers so that an incoming bee touches the stigma first before alighting on a flower. They called this approach herkogamy (AH) and documented that many relatives of distylous plants have AH. The original A form does not change and is later called the pin form. The first mutation of A is

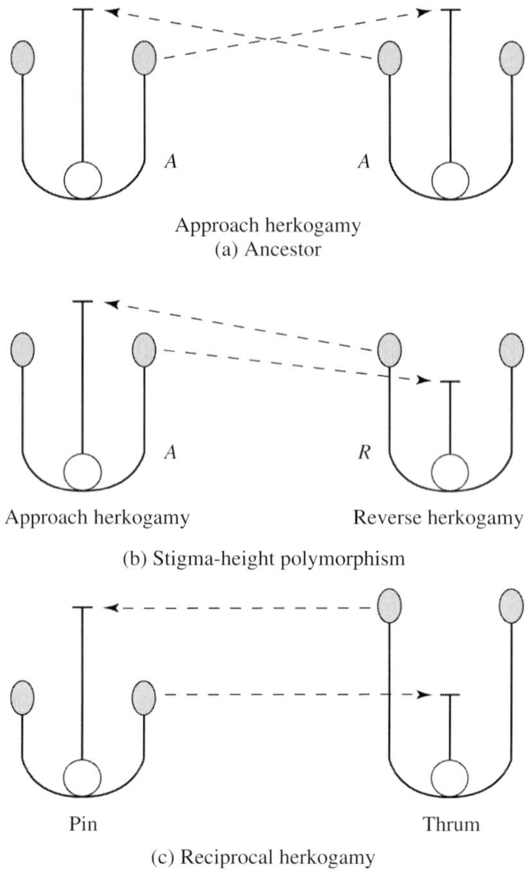

Fig. 12.6 Postulated changes in the evolution of reciprocal herkogamy. The ancestor has approach herkogamy (AH) with the style protruding past the anthers. This is the *A* morph. Next a short-styled mutant (*R* for reverse) invades. A second mutation in the *R* morph shifts the anthers to a high position so that the two morphs become reciprocal. The original *A* morph is the pin and the new morph is the thrum. Redrawn from Fig. 1 in Lloyd and Webb (1992b) with permission from Springer-Verlag.

then towards a short style (Fig. 12.6). This produces the so-called reverse morph *R*. The second mutation is towards increased anther height in the *R* morph, so that this becomes the thrum. Together the two mutations work fine in promoting inter-morph mating as we noticed in Table 12.2, but are there conditions under which the first mutation can succeed and *A* and *R* morphs coexist?

Stone and Thomson (1994) tested the first step in Lloyd and Webb's scenario, using artificial flowers (Fig. 12.7). The proficiencies of pollen export for the *A* and *R* morphs are given in Table 12.3. Conditions for coexistence in the Lloyd and Webb model were met, both with and without pollen limitation of

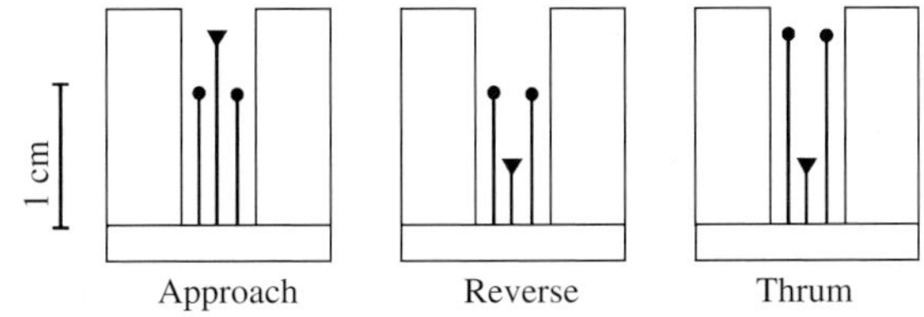

Fig. 12.7 Experimental flowers used by Stone and Thomson (1994). Redrawn with permission from Stone and Thomson (1994), *Evolution*.

Table 12.3 *Pollen counts on stigmas after a single visit to an artificial flower of type* A *(approach herkogamy) or* R *(reverse) after visiting an* A *or* R *flower*

Receptor:	*A*	*R*
Donor:		
A	$q_{AA} = 31.9$	$\mathbf{q_{AR} = 36.1}$
R	$\mathbf{q_{RA} = 60.6}$	$q_{RR} = 32.7$

Note: Bold type indicates legitimate or inter-morph pollen flow.
Source: Stone and Thomson (1994).

seed set. These results support the idea that a dimorphism in style length can arise before the dimorphism in anther height. The differences in pollen transfer are due to subtle changes in pollinator behaviour. Stone and Thomson (1994) video-taped bees and noted that presence of a long style changed the way the bee entered the flower. As a result the area on the bee on which pollen was placed was different for short- and long-styled morphs. The second step in the Lloyd and Webb model is a mutation towards increased anther height in the *R* morph to make it a thrum. Because the stigma in the original *A* morph is already in the high position, this second mutation is considered 'easy', an assumption that was confirmed by Stone and Thomson's (1994) experiments.

Lloyd and Webb (1992b) further argued that heterostyly evolves independently, preceding the establishment of sSI, contrary to a model by Charlesworth and Charlesworth (1979b) in which sSI evolves first. The Lloyd and Webb (1992b) model is now favoured for a number of reasons (see discussion in Richards 1997). First, it is the only model that offers an explanation for heterostylous species lacking sSI. Second, no examples exist of species with an sSI system with only two compatibility types in which the types have the same morphology. Third, sSI systems with only two types are difficult to establish because of the disadvantage Darwin (1877) mentioned: a plant cannot cross with 'half its brethren'. The pollen that is delivered on the stigma of a conspecific has, with random mating, a 50% chance that it is unsuccessful because this plant is of the same morph. Lloyd and Webb (1992b) pointed out that this disadvantage becomes smaller if inter-morph pollen flow already prevails so that only a small fraction of the pollinations is illegitimate. In their view the two morphs emerge first, directing pollen flow as Darwin argued, after which sSI evolves. Fourth, in the Charlesworth and Charlesworth (1979b) model the ancestral phenotype has the same anther and style length. This morphology is likely to promote considerable selfing and the inbreeding depression that is necessary for the establishment of diallelic SI is unlikely to be present in such ancestral populations (Ganders 1979).

To sum up: heterostyly promotes inter-morph mating and promotes outcrossing in plants pollinated primarily by bees. Nevertheless when inbreeding depression is sufficiently low, there is always a chance that a homostylous mutant invades, solely because of Fisher's automatic advantage of selfing. Models indicate that sex allocation could easily diverge between the morphs but until now this topic has received little attention. It is likely that in the ancestral species anthers and style differed in length. A mutation towards a short style, followed by a mutation towards long anthers resulted in the reciprocal morphs. The incompatibility between the morphs probably evolved later.

Questions

(1) Why are thrums eliminated and can pins coexist with long homostyles in *Primula vulgaris*?
(2) What stops the elimination of pins by homostyles in *Primula* populations?
(3) What is inter-morph pollination and why was it first doubted whether this occurred often in distylous species?
(4) What is the disadvantage of a sporophytic SI system with only two morphs?

13

Selective embryo abortion

In many cases, even with the best possible pollination some ovules do not develop into seeds. Seed/ovule ratios averaged 85% in developing fruits of annuals and only 50% in perennial plants (Wiens 1984). Including non-developing fruits would even lower these percentages. Burd (1994) found low fruit-to-flower ratios, even in species that were not pollen limited (Fig. 13.1). Apparently, embryos within fruits and whole fruits are aborted. Some of these embryos may not be viable because they carry lethal alleles, but as we will show later in this chapter, potentially viable embryos are aborted too. Studies in which ovaries were dissected (for instance, *Oxalis magnifica*: Guth and Weller 1986; *Epilobium angustifolium* (fireweed): Wiens *et al.* 1987) give most direct evidence for embryo abortion (see also Fig. 13.2). When using this method in *Cynoglossum officinale* the abortion level was estimated to be at least 39%, increasing from 16.4% in the beginning, to 82% towards the end of reproduction (Melser 2001). In gymnosperms the same pattern exists. For example, in *Pinus sylvestris* (Scots pine) on average 30% of the developing seeds was aborted (Karkkainen *et al.* 1999).

13.1 Three evolutionary hypotheses for low seed-to-ovule ratios

First, it has been suggested that the production of excess ovules is a bet-hedging strategy in variable and unpredictable environments. The overproduction of ovules gives an opportunity to rapidly adjust seed set when conditions improve (Lloyd 1980; Kozlowski and Stearns 1989). Under this hypothesis fruit- or seed-to-flower ratios should be close to their maximum under favourable conditions. There is little experimental support for this idea. For instance, data on *Cynoglossum officinale* collected in five different years and a total of 11 populations (Melser 2001) showed that the mean seed-to-flower ratio in populations ranged from 0.62 to 1.36, which is far below the maximum of 4 when all ovules in the flower would have produced a seed.

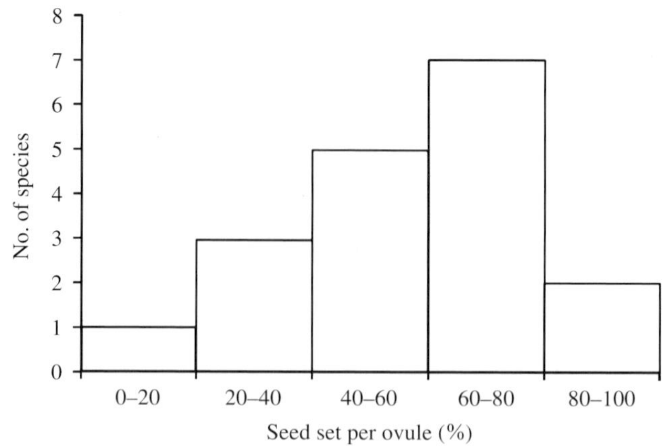

Fig. 13.1 Frequency distribution of percentage seed set per ovule. Only species that were not pollen limited were selected from a review by Burd (1994).

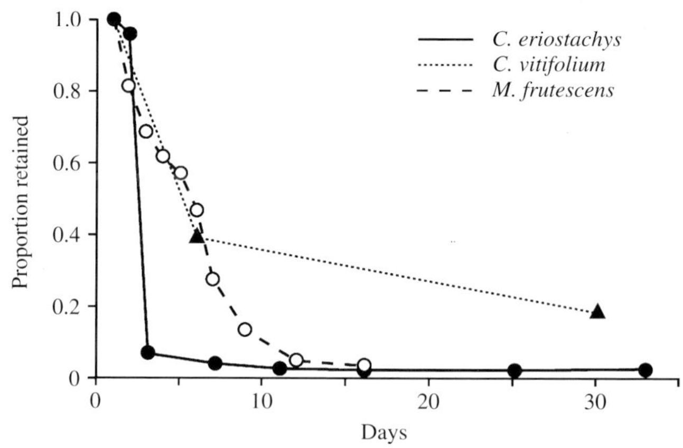

Fig. 13.2 Timing of flower and fruit abortion in *Caesalpinia eriostachys*, *Cochlospermum vitifolium* and *Myrospermum frutescens* as proportion of original number of flowers retained after day 1. In *C. eriostachys* abortions during the first three days and in *C. vitifolium* and *M. frutescens* during the first five days were considered flower abortions, while later abortions were considered fruit abortions. Redrawn with permission from Bawa and Webb (1984), *American Journal of Botany*.

Second, sex allocation theory can explain low fruit-to-flower ratios in hermaphrodite species (see Chapters 4 and 8). The 'excess' flowers may primarily serve male function (Willson 1979; Queller 1983; Sutherland and Delph 1984). Recall Eq. (4.8), which showed that, for a self-incompatible species, the ESS ratio of investment in male and female reproduction equals the ratio of the exponents of the male and female gain curve. This means that with equal exponents of the gain curves, plants should allocate as much resource to male as to female function. Fruits and seeds (female reproduction) are usually much more costly than pollen and flowers (male reproduction), even if we count costs of attraction as male. This means that more flowers than fruits should be produced. Because in female individuals in dioecious or gynodioecious species the flowers do not serve a male function, sex allocation theory would predict maximal fruit set in these females. In contrast to this prediction, a low number of seeds per flower (1.1 out of a potential maximum of 4) were found in male steriles of *Echium vulgare* (Klinkhamer *et al.* 1994a). Therefore in such cases sex allocation theory cannot explain low fruit-to-flower ratios of female plants.

Third, another, non-exclusive, explanation is given by the selective embryo abortion (SEA) hypothesis. This hypothesis states that embryos with a relatively low chance to survive and reproduce later in life are selectively aborted to save resources for higher-quality offspring, thereby increasing the fitness of the maternal plant. We will discuss several aspects of this hypothesis first and then return to models of fruit-to-flower ratios in plants.

13.2 Mechanisms of embryo abortion

Several mechanisms may lead to SEA (Lee 1988). First, there can be developmental regulation of abortion. Embryos are nourished by the endosperm, which gathers resources by dissolving surrounding cells. If the embryo is a strong sink for nutrients this may increase the flow towards the embryo. A similar process can regulate the abortion of multiple-seeded fruits, if more developing embryos together make a stronger sink for resources. When compared to surviving fruits, aborted fruits tended to have fewer developing seeds in *Cassia fasciculata* (partridge pea) (Lee and Bazzaz 1982) (Fig. 13.3), *Asclepias speciosa* (showy milkweed) (Bookman 1984) and *Cochlospermum vitifolium* (silk cottontree) (Bawa and Webb 1984). Especially with large, costly fruits containing few seeds, abortion may be the best option for the mother plant. Note, however, that abortion based on seed number per fruit does not result in higher-quality offspring.

Second, abortion can depend on the position of a seed within the fruit. In some fruits with linearly arranged ovules, a resource gradient exists from the basal to the distal end of the fruit, and seeds at the basal end are less likely to abort. For *Raphanus sativus*, Marshall and Ellstrand (1988) suggested that pollen donors sort themselves out along the gradient during fertilization. The combination of non-random fertilization and a gradient in abortion levels can

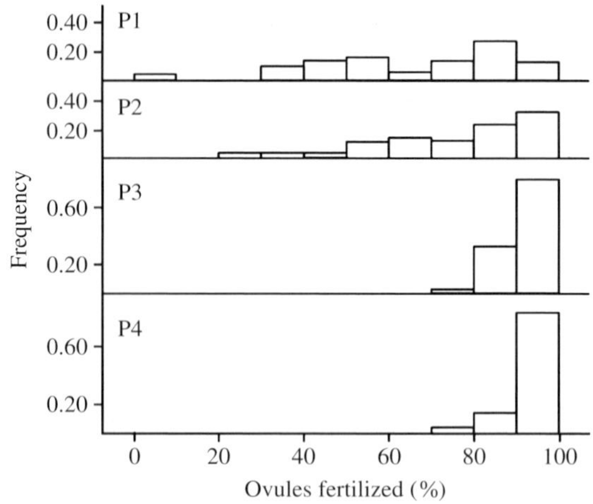

Fig. 13.3 Frequency of occurrence of fruit of *Cassia fasciculata* containing various percentages of fertilized ovules at four consecutive stages of fruit development numbered P1 to P4. Fruits with low percentages of fertilized ovules are preferentially aborted. Redrawn with permission from Lee and Bazzaz (1982), *Ecology*.

result in selective abortion. Third, developing embryos may produce hormones that regulate abortion. Several studies indicated that developing embryos excrete a substance, related to the plant hormone indoleacetic acid, that promotes the abortion of other younger fruits or seeds (Raju *et al.* 1996; Krishnamurthy *et al.* 1997). If some viable embryos excrete more hormones than others this may lead to selective abortion. In this scenario embryos with an interference strategy are selected rather than those with a high genetic quality.

The mechanisms above imply that the developing embryos compete and the worst competitor is aborted, while the mother plant may set the level of abortion. Alternatively, selective abortion may be under maternal control and can be seen as part of female choice over potential partners. The endosperm may be decisive for the outcome of the conflict of interest between the maternal plant and the developing embryo (Westoby and Rice 1982; Shaanker *et al.* 1996; Chapter 15). The problem of maternal choice versus competition among embryos is as yet unresolved.

13.3 Are otherwise viable embryos aborted?

Central in the SEA hypothesis is the idea that otherwise viable embryos are aborted and that selective abortion leads to higher offspring quality. This implies that genes that are expressed in the embryonic stage are also beneficial later in life. Likely candidates are genes controlling, for instance, metabolic processes. Goldberg *et al.* (1989) estimated that the vast majority of the mRNAs present in mid maturation stage embryos are also present in cotyledons and mature plants. Experimental evidence for the abortion of otherwise viable embryos comes largely from experiments in which abortion levels were manipulated. A number of experiments showed that destruction of ovules led to higher seed set in the remaining ovules, indicating that without this experimental treatment otherwise viable embryos would have been aborted (Casper 1988; Rocha and Stephenson 1991; Melser and Klinkhamer 2001). In a number of species with linearly arranged ovules, developing ovules at the basal end of the fruit are more likely to abort. Nakamura (1988) could, however, successfully culture such embryos in *Phaseolus vulgaris* (common bean), indicating that they were potentially viable. In addition, Ganeshaiah and Shaanker (1988) removed two distal seeds in pods of *Dalbergia sissoo* (Indian rosewood). The basal seeds, which were aborted in the control, now developed normally. Note that in these species the basal seeds normally have the highest abortion rate in contrast to *Raphanus sativus* discussed earlier. In many species of the Boraginaceae most of the four ovules in each flower do not produce mature seeds. In *Cryptantha flava* (yellow hiddenflower) (Casper 1988) and in *Cynoglossum officinale* (Melser and Klinkhamer 2001) random destruction of three ovules resulted in doubling of the chance of maturation in the remaining ovule compared to a control that was not manipulated. This again indicates that under normal conditions otherwise viable embryos are aborted.

13.4 Selection among pollen donors

By now there is quite some evidence for selection among pollen donors in plants. Examples include *Aquilegia caerulea* (blue columbine) (Montalvo 1992) and *Dalbergia miscolobium* (Gibbs and Sassaki 1998). Most apparent is selection against selfing. For instance, Rigney (1995) removed embryos that were in the process of being aborted and analysed their paternity in *Erythronium grandiflorum* (glacier lily). Selfed embryos were more likely to abort than outcrossed ones while nearby donors were less successful than distant ones. In *Echium vulgare* Melser *et al.* (1997) found that in some individuals self pollen sired most seeds, while in other individuals outcross pollen was most successful. Because pollen donors did not differ in pollen viability, germination and pollen tube growth, these differences most probably occurred in the embryonic stage. Early selection among embryos can potentially be an efficient way of coping with effects of inbreeding depression, while maintaining the advantages of recombination and passing on two copies of the plant's own genome to the offspring (the Fisherian advantage from Chapter 12). Selection among outcross donors has also been documented. For instance, in *Campsis radicans* (trumpet creeper) (Bertin 1982, 1985), *Raphanus raphanistrum* (Mazer *et al.* 1986), *Raphanus sativus* (Ellstrand and Marshall 1986; Marshall and Ellstrand 1988) and *Chamaecrista fasciculata* (partridge pea) (Fenster 1991). In *Raphanus sativus* (Marshall 1988, Marshall and Whittaker 1989) and *Asclepias speciosa* (Bookman 1984) pollen donors that sired most seeds also sired the largest ones.

13.5 Do higher abortion rates increase the quality of the remaining offspring?

Most evidence for the fact that developing embryos with a potential low fitness later in life are preferentially aborted comes from pruning experiments. When part of the ovules is destroyed, the remaining ovules have a higher chance of developing a seed. The idea behind these experiments is that the effects of random abortion (through destruction of ovules by the experimentator) are compared with the effects of selective abortion by the plant. In *Crypthanta flava* offspring from the control group (with natural abortion) showed higher seedling emergence and survival during two years of growth compared to offspring from the ovule removal treatment (Casper 1988). Superiority of the control group was also found for *Lotus corniculatus* (bird's-foot trefoil) (Stephenson and Winsor 1986), *Phaseolis coccineus* (Rocha and Stephenson 1991) and *Cynoglossum officinale* (Melser and Klinkhamer 2001). Casper (1988) cautioned that negative effects of mechanical damage itself might cause differences among treatments. Therefore Melser (2001) sought ways to manipulate seed set and abortion rate without causing damage. She compared, for *Cynoglossum officinale*, the quality of offspring produced early in the flowering season (1.23 seeds per flower with four ovules, i.e. low abortion levels), with that of offspring produced late in the flowering season (0.36 seeds per flower, i.e. high abortion levels). In line with the SEA hypothesis, performance was better for offspring produced late in the season when abortion levels were high. Taking the product of survival rate and plant size after one year as a measure of offspring performance, high abortion levels almost doubled performance (Melser 2001).

13.6 Using molecular markers to study SEA

A way to overcome the problems involved in manipulation experiments is to study abortion at the level of molecular markers. When SEA takes place, this should show up as a non-Mendelian segregation of markers, closely associated with the genes upon which this selection acts. Korbecka *et al.* (2002) reviewed published genetic maps to study the frequency of non-Mendelian segregation. In such maps it is common practice to test by means of a χ^2 test at a 5% significance level if a certain marker deviates from the expected segregation ratio. Thus without selection 5% of all markers should show a significant test result only by chance. Korbecka *et al.* (2002) showed for 59 maps that the percentage of markers showing non-Mendelian segregation was significantly higher than 5%. Average values were 14.6% for cultivars and 11.5% for wild species. The authors discussed that these averages are probably an underestimate because an unknown percentage of markers is located in regions of the genome that are not or only loosely linked to functional regions and for such markers no selection is expected. This literature study indicated that non-Mendelian segregation is more common than textbooks suggest. Many processes other than SEA, however, can result in non-Mendelian segregation and without additional analysis it is not possible to tell whether it results from pre- or post-fertilization selection (Fig. 13.4). Korbecka *et al.* (2002) applied the scheme of Fig. 13.4 to 53 loci with non-Mendelian segregation and found that post-fertilization selection took place in 71% of the cases. There are several advantages to the use of molecular markers. Markers avoid limitations of pollination experiments because multiple donors can be compared that compete within the same flower and even selection among offspring of a single pollen donor (including selfing) can be detected. A second advantage is that the presence or absence of markers that are under- or over-represented in the offspring can be related to offspring fitness later in life. This makes a genetic link between SEA and offspring quality. So far no such studies have been published, but undoubtedly application of new molecular techniques will revitalize the debate on SEA.

13.7 Aborting fruits with too few seeds: Burd's model

So far we have mainly discussed the effects of embryo abortion on the quality of the remaining seeds. Plants may also retain fruits with the highest numbers of seeds, regardless of quality (Burd 1998). Martin Burd's idea was that a fruit is costly to produce when it is poorly pollinated and contains only few seeds. In such cases it may be more advantageous to abort the fruit and produce a new fruit, possibly with more seeds. In his model the number of fruits that fully develop is fixed and fruits initiated have different numbers of seeds drawn from a specified probability distribution. He then asked the question: 'How does total seed number increase with flower number through the wider choice the plant has to select fruits with the highest number of seeds?' A floral surplus, and subsequent selective abortion, can generate a large increase in total seed

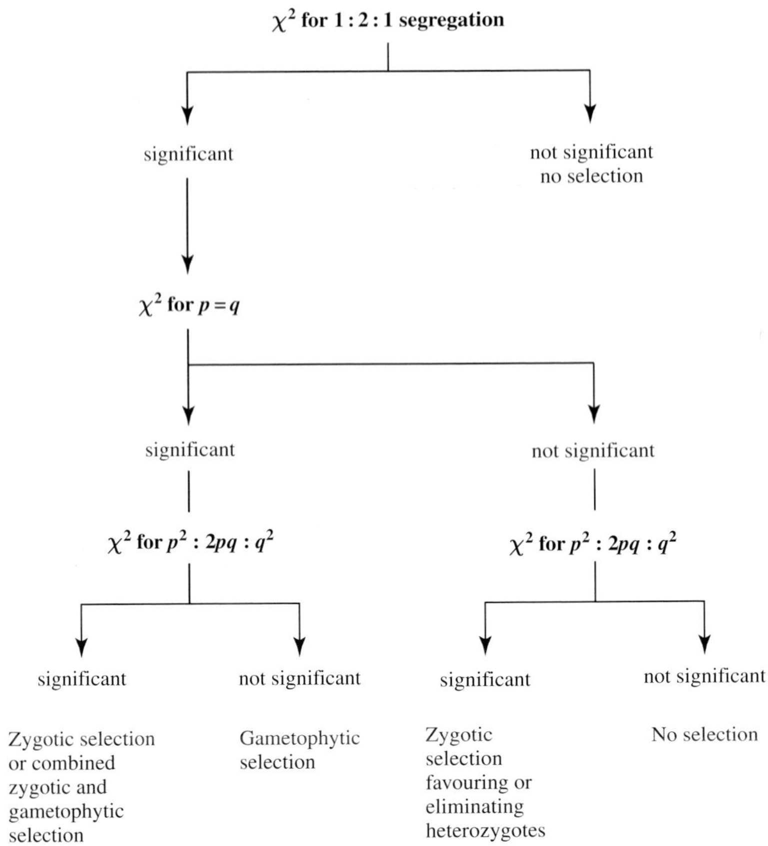

Fig. 13.4 Key for determining whether selection takes place before or after fertilization in a segregating F_2 population analysed with co-dominant markers. (After Korbecka *et al.* 2002, based on Pham *et al.* 1990.) Markers are called codominant if one can count the exact number of copies present in an individual. Genetic maps are usually made by producing an F_1 generation by crossing two inbred lines. Individuals from the F_1 generation can then either be crossed among each other or backcrossed with one of the parents to produce an F_2 population. The expected genotype frequencies for a particular polymorphic marker, if no selection occurs are 1 : 2 : 1 and 1 : 1, respectively. The gene frequencies p and q represent the frequency of the presence and absence of the marker. The expected frequencies are 1 : 1 and 3 : 1, respectively for crossing between F_1 individuals and backcrossing with a parent. The example depicted is for a cross betweeen F_1 individuals. In the first step of this key it is determined whether genotypes show Mendelian segregation or not. If they do, no selection has taken place at all. In the second step it is determined whether both alleles occur at equal frequencies. If they do, no selection before fertilization (gametophytic selection) has taken place. In the third step it is determined whether the genotypes segregate according to the frequencies of the alleles. If they do, no selection after fertilization (zygotic selection) has taken place.

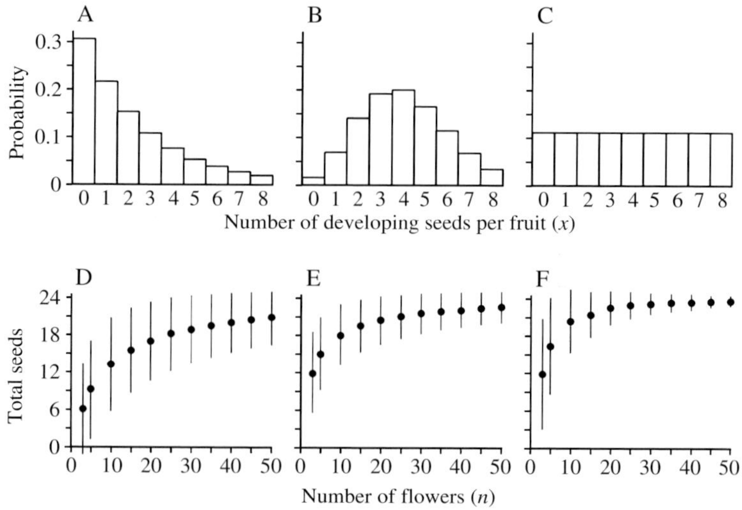

Fig. 13.5 Total seed production (D–F) of a hypothetical plant producing three fruits and different numbers of flowers. In this model flower production is not at the cost of fruit production and the more flowers are produced the wider the choice for the plant to select the three fruits with the highest number of seeds. Three possible probability distributions of seeds per fruit are represented by A–C. Total seed production is most strongly increased by flower number if the probability of fruits with no seeds or low seed numbers is high and that of fruits with many seeds is low. Redrawn with permission from Burd (1998), *Ecology*.

production (female fitness). This potential increase depends on the variance in number of seeds among initiated fruits and on how well plants can select the fruits with most seeds. In all cases total seed production was a decelerating function of flower number (Fig. 13.5). Note that this method does not allow calculation of the optimal fruit-to-flower ratio because fruit numbers are assumed to be fixed, with no trade-off between fruit and flower production.

13.8 Optimizing the seed-to-flower ratio in plants with SEA

We will first analyse the optimal seed-to-flower ratio for a female individual in a dioecious population or a female (male-sterile) individual in a gynodioecious population because this enables us to look at the effects of SEA without introducing sex allocation as well. Note that optimizing seed-to-flower ratios in females is a straightforward optimization problem and not an ESS problem because the optimal strategy of the female is independent of the strategy played by the other plants in the population. Finally we will discuss an ESS model for seed-to-flower ratios in hermaphrodites. This model will include both effects of SEA and sex allocation.

13.8.1 Optimal seed-to-flower ratio in female plants

Melser and Klinkhamer (2001) concluded that, within a rather broad range, large differences in the seed-to-flower ratios could be compensated for by a relatively small increase in quality per seed. This can be demonstrated by a numerical example. We assume the number of ovules per flower to be fixed at 4, as in the family of the Boraginaceae. Set the cost of producing a flower at 1, and the cost to produce a seed at 3 (which is close to the trade-off found for *Cynoglossum officinale*). With a total budget of 450 units, a plant would produce 34.5 flowers and 138.5 seeds if it produces the maximum of 4 seeds per flower. With the same budget a plant could produce 112.5 flowers and 112.5 seeds if the seed-to-flower ratio is decreased to 1. This means that, when only counting the female component of fitness, an increase in quality of 23% [(138.5 – 112.5)/112.5] would compensate for the lower seed production. In other words an increase in seed quality of a female plant of 23% would select for an abortion level of over 75%. Now we will turn to a mathematical analysis. We assume that the costs of producing a flower are fixed and that flowers have a fixed number of ovules. We assume that all ovules are fertilized and produce potentially viable embryos in single-seeded fruits. Without costs the flower can abort one or more of the developing embryos. The total budget of a plant is T resources, the cost of producing a flower is 1 unit of resource, and the cost of producing a seed is c units of resources. A plant producing n flowers can then produce $(T-n)/c$ seeds. Seed quality Q is a function of the number of seeds per flower (x). For simplicity, we assume a linear relationship between $Q(x)$ and x. Melser (2001) also explored other shapes of this relationship. With increasing seed numbers per flower and consequently lower levels of selective seed abortion the quality of the seeds decreases. $Q(x) = a + bx$, with $x = (T-n)/cn$. Because x closely corresponds to flower number n we can rewrite the seed quality of a plant as a function of the number of flowers (n): $Q(n) = b(T-n)/cn + a$ (note that $b < 0$). With linear gain curves we write the fitness of a female plant with n flowers as:

$$W(n) = Q(n)(T-n)/c \tag{13.1}$$

We find the optimum by differentiating fitness with respect to flower number (n) and setting the derivative to zero. This gives:

$$\frac{\mathrm{d}W(n)}{\mathrm{d}n} = -Q(n) + \frac{\mathrm{d}Q(n)}{\mathrm{d}n}(T-n) = 0 \tag{13.2}$$

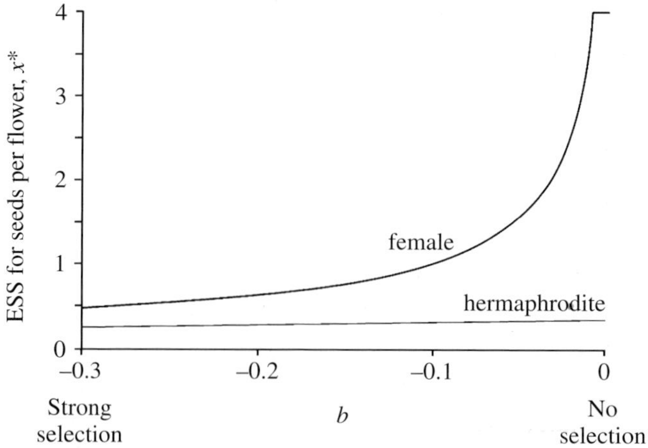

Fig. 13.6 Optimal seed-to-flower ratio for females and ESS seed-to-flower ratio for hermaphrodite plants, as a function of the slope of the regression between offspring quality (Q) and number of seeds per flower (x). When $Q(x) = a + bx$, $b = 0$ there is no effect of abortion on offspring quality; more negative values of b correspond to a stronger positive effect of abortion on offspring quality. For the calculations for the hermaphrodite it was assumed that there were no male-sterile plants in the population (Eq. 13.10).

with $\mathrm{d}Q(n)/\mathrm{d}n = -bT/cn^2$. Substituting this in Eq. (13.2) yields $n^2 = bT^2/(b - ac)$, or:

$$n^* = T\sqrt{b/(b - ac)} \tag{13.3}$$

This optimal n^* corresponds to optimal number of seeds per flower according to $x^* = (T - n^*)/cn^*$, so that the optimal number of seeds per flower is:

$$x^* = \frac{1}{c\sqrt{b/(b - ac)}} - 1/c \tag{13.4}$$

We would like to apply this to female plants of *E. vulgare*. However, we have no estimates available for the parameters a, b and c for female plants. Instead we will use the estimates for hermaphrodite individuals of *E. vulgare*: $a = 0.59$, $b = -0.085$, $c = 1.87$ (Melser 2001). Melser (2001) estimated the parameters a and b from a regression of seed quality on seeds per flower in experiments in which seeds per flower varied considerably. Without selective seed abortion ($b = 0$) the optimal number of seeds per flower for male-sterile individuals becomes very large (Fig. 13.6). In *E. vulgare* this number is constrained, however, to a maximum of 4 seeds per flower. Without selective abortion female plants should produce this maximum of 4 seeds per flower. With selective abortion and assuming the parameter values given above, the optimal number of seeds per flower for male-sterile individuals predicted by Eq. (13.4) was 1.46. This is somewhat higher than found for females in natural populations of *E. vulgare* (1.1 seed per flower). The calculation nevertheless illustrates that SEA can, in itself, select for a considerable reduction in seed-to-flower ratios, which corresponds to an abortion level as high as 63% [100(4 – 1.46)/4].

13.8.2 ESS for seed-to-flower ratios in hermaphrodites

Next we will examine the ESS for the seed-to-flower ratio for an outcrossing hermaphrodite plant with linear gain curves with no female plants in the population. Chapter 8 discussed how the ESS for sex allocation of hermaphrodites is affected by the presence of females: in that case the ESS for number of seeds per flower decreases but never to zero (hermaphrodites are selected to become more male but never completely male). We assume hermaphrodite flowers that produce a standard amount of pollen and that have a fixed number of ovules. We assume that all ovules are fertilized and produce potentially viable embryos in single-seeded fruits. Without costs the flower can abort one or more of the developing embryos. We extend the analysis from Chapter 4 to include selective abortion. The quality of the seeds of the mutant with m flowers is Q_m, and we write the seed quality of the common type with n flowers as Q. With linear gain curves the absolute fitness of a rare mutant producing m flowers in an outcrossing population of plants producing n flowers, is:

$$W_m = Q_m(T - m)/c + \frac{m}{n}Q(T - n)/c \tag{13.5}$$

This is the Shaw–Mohler equation in which the first term represents the female component of fitness (number of seeds produced) and the second term the male component of fitness (number of seeds sired on other plants). In this case m/n is Lloyd's (1984) competitive share because each flower contributes the same amount of pollen to the outcross pollen pool. Note that by differentiating Eq. (13.5) with respect to flower number of the mutant and setting the derivative equal to zero we find that, without SEA (Q is a constant, not a function of flower number), the ESS for number of flowers is $0.5T$ and the corresponding number of seeds is $0.5T/c$. For seeds per flower at the ESS this gives $x^* = 1/c$. The ratio of allocation to male (flowers) and female (seeds) reproduction is then $c : 1$. This corresponds to Fisher's (1930) result that the cheapest sex should be overproduced (Chapter 4). To find x^* for a situation with SEA, we follow again Parker's method of differentiation of the fitness function of the mutant. We start from the idea that the plant can vary the number of flowers and due to the trade-off the number of seeds varies as well. We could also have taken the number of seeds per flower (x) as the evolutionary variable in all equations; this would give the same result. To compute the ESS flower number, we differentiate the absolute fitness of the mutant (Eq. 13.5) with respect to m and set the derivative equal to zero.

$$\frac{\mathrm{d}W_m}{\mathrm{d}m} = -Q_m/c + \frac{\mathrm{d}Q_m}{\mathrm{d}m}(T-m)/c + Q(T-n)/cn = 0 \qquad (13.6)$$

With $\mathrm{d}Q_m/\mathrm{d}m = -bT/m^2$, setting $n = m$, $Q = Q_m$ and some algebra this gives the candidate ESS for flower number n^*:

$$\frac{\mathrm{d}W_m}{\mathrm{d}m} = (bT - bn + acn)(T - 2n) - bT(T - n) = 0 \qquad (13.7)$$

or

$$T(ac - 2b) = n(2ac - 2b) \qquad (13.8)$$

which we can rewrite as:

$$n = n^* = \frac{T(ac - 2b)}{2(ac - b)} \qquad (13.9)$$

From Eq. (13.9) it can be seen that without selective abortion ($b = 0$), n^* equals again $T/2$, a familiar result. Furthermore it can be seen that n^* increases, and thus the number of seeds per flower decreases, when the effect of abortion on seed quality is stronger (b is more negative). Substituting Eq. (13.9) in the formula for the number of seeds per flower, $x = (T - n)/(cn)$, gives:

$$x^* = \frac{2(ac - b)}{c(ac - 2b)} - 1/c \qquad (13.10)$$

Without SEA ($b = 0$) this would lead for *E. vulgare* to $1/c = 1/1.87 = 0.53$ seeds per flower. In this case the low value for the ESS number of seeds per flower is entirely due to sex allocation. With SEA included as well ($a = 0.59$, $b = -0.085$), the ESS for number of seeds per flower was further reduced to 0.46. This shows that sex allocation alone has a strong effect on the ESS for number of seeds per flower in a hermaphrodite (for *E. vulgare* from the maximum of 4 to 0.53) and that including both sex allocation and SEA leads to a relatively small further decrease (from 0.53 to 0.46) (Fig. 13.6). The estimated values are somewhat lower than the value of 0.96 seeds per flower that was observed for hermaphrodite plants in natural populations of *E. vulgare*. At this point it is unclear what causes this discrepancy. Including female individuals in the population would lower the predictions even further. Including some selfing and costs of abortion would increase the predicted values (Chapter 8).

Note that although the ESS number for seeds per flower is not strongly affected in hermaphrodites it is still of selective advantage if you do not mature all your seeds to abort the ones with the lowest quality.

To sum up: some plant species have the ability to abort low-quality embryos selectively, which raises average quality of the surviving offspring. The use of molecular markers now enables us to study SEA without manipulating the mother plant and makes it possible to relate abortion and offspring quality directly to the presence or absence of a marker. One of the evolutionary consequences of SEA is that it selects for lower seed-to-flower ratios. The effect is strong in females and male-steriles, but may be weak in hermaphrodites, because in the latter the ESS seed-to-flower ratio is primarily determined by sex allocation. SEA is most pronounced for selfed seeds. In this way plants avoid the consequences of inbreeding at low costs while maintaining the Fisherian advantage.

Questions

(1) Does abortion of multiple-seeded fruits always increase offspring quality?
(2) Which three evolutionary hypotheses were given for low seed-to-ovule ratios?
(3) What does sex allocation theory predict for seed set in female plants?
(4) What does the SEA hypothesis state and why are female plants useful in the study of this hypothesis?

14

Attractiveness to pollinators

A widespread misunderstanding in popular scientific literature is the idea that plants and pollinators have evolved to cooperate. Any biology student should know better. Plants have evolved to receive and export pollen as efficiently as possible at the lowest cost. Pollinators have evolved to collect the highest rewards with the lowest possible effort and within the shortest possible time. These conflicting interests gave rise to several interesting features of plant–pollinator systems. For example, nectar-robbing insects in *Symphytum officinale* (comfrey) sometimes bite a hole in the corolla to 'steal' nectar instead of collecting the nectar in a legitimate way, depositing and exporting pollen. As another example, pollinating insects of the South African water lily *Nymphaea capensis* (Meeuse and Morris 1984) are trapped and drown in the liquid that is contained within the flower. The pollen grains on their bodies sink to the bottom of the flower where the stigmas are situated. In the male phase of the flower, stamens cover the trap so that the visiting insects are not at risk and can depart to another (female) flower. Less dramatic are flowers that mimic the scent of insect food sources, like dung or mushrooms. Although the pollinators are attracted to these flowers they are apparently not rewarded. Some species (for instance, *Chiloglottis* and *Ophrys* orchids) even mimic the scent or shape of female insects in order to attract male insects that pollinate the flowers through pseudo-copulation. It is beyond the scope of this book to even begin to try to give a complete overview of the ways that plants and pollinators can interact but the examples given above should be sufficient to make our point clear: interests of plants and their pollinators do not necessarily run in parallel.

14.1 Coevolution

Coevolution between plants and their pollinators is driven by their conflict of interest. Darwin already suggested that the evolution of deep corolla tubes was a kind of 'race' between plants and pollinating insects. Pollinating insects with long tongues forage more efficiently for nectar than insects with short tongues; they can hover in front of the flower and do not need to enter. At the same time, however, this makes them less efficient pollinators from the plant's point of view because of their reduced contact with the anthers or stamens. Deeper corolla tubes in plants are advantageous because more pollen is deposited on a standard position on the pollinator's body when the pollinator is forced to thrust its proboscis deep into the flowers thereby making more contact with stigma and anthers. Deeper corolla depth in turns drives selection for longer tongues and so on. Experimental support for the plant part of this hypothesis comes from Nilsson (1988) who studied removal and receipt of pollen packages (pollinia) in *Platanthera bifolia* and *P. chlorantha*, two moth-pollinated orchid species with long, nectar-containing floral spurs. Shortening the spurs reduced pollen removal and receipt. The reduction was smallest for male function, indicating that individuals with short tubes were relatively better at being male than female. Field studies in the *Disa draconis* complex (Orchidaceae) showed that artificial shortening of spurs resulted in a significant decline in female function (pollen receipt and fruit set), while pollen removal was not affected (Johnson and Steiner 1997). Because fruit set in the studied populations was pollen limited, selection on spur length occurred mainly through female function in this species.

Wasserthal (1997) suggested that the long tongues have evolved for predator avoidance, rather than for foraging efficiency. Pollinators with long tongues keep a distance from predators that ambush visited flowers. Wasserthal (1997) argued that the long-tubed flowers are not the result of a coevolutionary arms race, but were selected for after a pollinator with already a long tongue shifted to a new plant species. Unfortunately, too little experimental evidence is available to settle this debate.

14.2 Effects of nectar reward on pollinator behaviour

Floral traits such as style length, anther length, flower size, petal colour and nectar volume influence plant fitness through their effects on pollinator behaviour (Waser and Price 1981; Galen and Plowright 1985; Stanton *et al.* 1986; Campbell 1991; Mitchell 1993). In particular, the effects of nectar on pollinator behaviour have been studied extensively. A high nectar reward may increase the number of pollinator approaches to inflorescences (Pleasants 1981; Real and Rathcke 1991; Klinkhamer *et al.* 2001), the number of flowers visited in sequence on the same inflorescence (Pyke 1981; Galen and Plowright 1985; Mitchell 1993; Pappers *et al.* 1999) or the time spent in individual flowers

(Zimmerman 1983; Galen and Plowright 1985; Mitchell and Waser 1992; Pappers *et al.* 1999). In addition, low nectar volumes have been shown to cause bumble-bees to increase inter-plant flight distances (Heinrich 1983; Zimmerman and Cook 1985). The response of the pollinator is likely to depend on the experimental design of the experiment or the small-scale population structure. A pollinator is likely to stay longer and visit more flowers in a sequence if nectar standing crop is higher. However, bumble-bees may learn the exact location of plants with a high nectar production rate (NPR) and approach these plants more often. In such a case differences in nectar standing crop may be small and NPR has no effect on the number of flowers visited in sequence on the same plant (bout length). If, on the other hand, pollinators do not learn the exact location of individual plants and do not match approaches to NPR, nectar standing crop will be higher on plants with high NPR, inducing longer bout lengths on these plants.

14.3 Attractiveness: a plant's dilemma

It is generally expected that a flower on an attractive plant is more frequently visited and receives and exports more pollen. However, although perhaps counterintuitive, pollen export may even be reduced if plants are too attractive. This may occur when bout lengths are longer on attractive plants. Geitonogamy then sets a limit to pollen export. When a bee visits n flowers on a plant, n units of pollen are removed from the anthers of that plant (Chapter 3).When n is very large, the bee exports approximately $E=1/k$ units of pollen (k was the fraction of pollen on the bee that was lost between flowers) on its body when it flies off (Eq. 3.8). Consequently with large n the efficiency of pollen export, $E/n=1/(nk)$, is a decreasing function of n. To give an example: for $k=0.1$ (Chapter 3) and a bee visiting 30 flowers in sequence, only approximately one-third of the pollen removed from the anthers is actually exported.

So on the one hand attractiveness leads to less efficient pollen transfer in one approach. On the other hand attractive plants are more often approached so that a greater fraction of the pollen is removed from the anthers. When pollinators are scarce the latter effect of attracting more pollinators may dominate. When pollinators are abundant most pollen is removed anyway, and in such a case only the efficiency of pollen export matters. In this situation plants are ideally pollinated by an infinite number of pollinators that each visit only a single flower (Klinkhamer *et al.* 1994b). In the intermediate situation there is a dilemma for the plant: being more attractive increases pollen removal but reduces the efficiency at which this pollen is exported (Fig. 14.1).

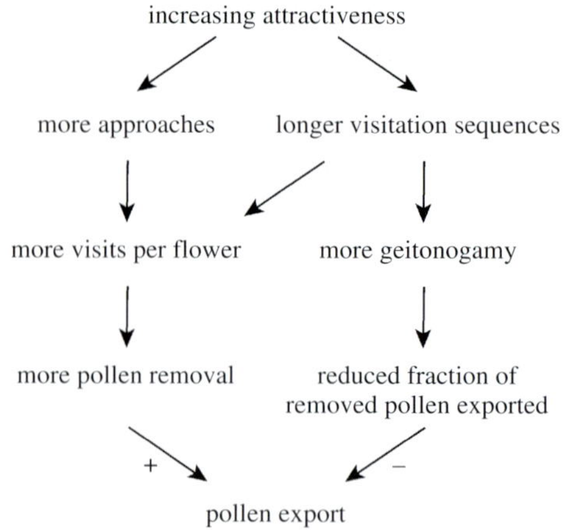

Fig. 14.1 The plant's dilemma. Redrawn with permission from Klinkhamer and de Jong (1993), *Oikos*.

The dilemma only exists with a positive correlation between the number of approaches and bout length. Such a correlation always existed when we compared plants with different flower numbers; a larger flower number increased both the number of approaches to plants and the bout length (Chapter 3). For NPR the dilemma may not exist if pollinators learn the exact location of high NPR plants and match approaches to reward. In such a case differences in nectar standing crop would be small and bout length could be more or less equal in plants with high and low NPR (Klinkhamer *et al.* 2001).

14.4 How to escape the dilemma?

Without intra-plant variation in nectar standing crop, a pollinator should depart to another plant, as soon as the reward drops below the expected average reward of the patch. The pollinator should continue visiting flowers if the reward is higher than expected. With intra-plant variation in nectar standing crop the best foraging strategy for the pollinator is not obvious.

The effects of nectar reward on the decision of the pollinator to stay on the plant and visit another flower or to leave have been studied extensively (Pyke 1982; Hodges 1985; Cresswell 1990; Kadmon and Shmida 1992; Pappers *et al.* 1999). Some of these studies suggested that pollinators use a threshold departure rule. If the pollinator encounters a reward that is below this threshold it leaves the plant. Cresswell (1990) discussed the proposition that a strict departure rule can be efficient with a clear gradient of nectar reward, as in the vertical inflorescences of *Digitalis purpurea* (foxglove) that bear the oldest flowers at

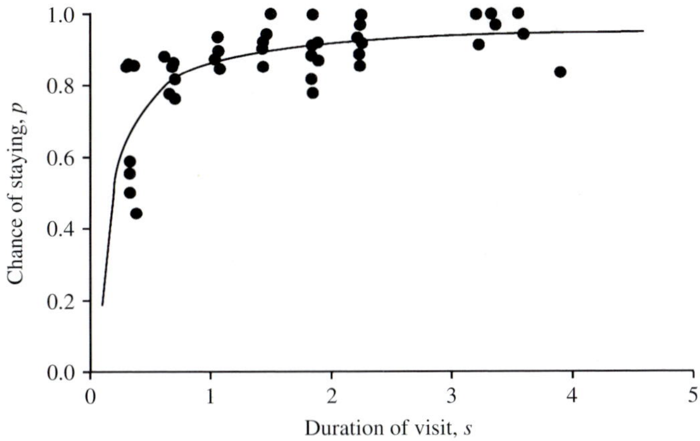

Fig. 14.2 The chance that a bumble-bee stays on a flowering plant of *Echium vulgare* increases with the duration of the current flower visit (which was highly correlated with nectar reward). The line is based on 2064 flower visits, during 159 visitation sequences, pooled in 49 groups. Redrawn with permission from Pappers *et al.* (1999), *Oikos*.

the bottom and the youngest flowers at the top (Best and Bierzychudek 1982). For more random patterns of nectar distribution within the plant, a probabilistic departure rule is likely to be better than a threshold departure rule (Cresswell 1990). With a probabilistic departure rule the chance to visit another flower on a plant increases with nectar reward in the previous flower. Even after encountering a small reward, there is a slight chance that the pollinator stays on the plant whereas with a strict rule it would always leave (Cresswell 1990; Kadmon and Shmida 1992). Pappers *et al.* (1999) found that the time bumble-bees spent in a flower of *Echium vulgare* was a good measure for the nectar standing crop of that flower. They then studied the chance of the bee departing from a plant as a function of the duration of the current and previous flower visits. Bumble-bees used the probabilistic departure rule (Fig. 14.2); their decision to leave the plant was based solely on the reward in the current flower. Other studies showed that bees could process the information of at least three flowers (Dukas and Real 1993). Pappers *et al.* (1999) used their decision rule to simulate flower visitation sequences for plants with the same mean nectar reward per flower, but with different variance. In simulations, more variation in nectar content led to shorter bout lengths. Perhaps variation in nectar reward can be an adaptive character in itself. Clearly more experimental work is needed on the link between variation in nectar content within plants and pollen export and receipt.

14.5 Effects of population structure on plant–pollinator relations

14.5.1 Population density

The density of a plant population can influence both the number of pollinators attracted and the quality of their visits. Often populations that vary in size also differ in density. In an experiment with *Brassica kaber* (wild mustard), Kunin (1997) tried to uncouple effects of population size from those of density. He found that the number of individuals had no effect on pollinator visitation, while increasing density had strong effects on both visitation and reproductive success. In another experiment with the same species he examined the effect of a background population on pollination quality (Kunin 1993). Pollinators visiting *B. kaber* behaved as generalists in backgrounds of *Brassica hirta* (mustard cress) and behaved as specialists where no alternative flowers were provided. When pollinators behaved as generalists, the quality of their visits was dramatically reduced, as indicated by reduced seed set of *B. hirta* plants. Population structure and density may therefore have a pronounced effect on natural selection. Selection on reproductive traits in rare species seems to have taken two different directions (Kunin and Shmida 1997). Rare self-incompatible species that depend on cross-pollination typically have evolved specialized flowers that are unusually large and deep. Rare self-compatible species, in which selfing provides reproductive assurance, have evolved generalized, small, shallow flowers (Kunin and Shmida 1997).

14.5.2 Small-scale population structure

Experiments with bumble-bees foraging on *Aralia hispida* (bristly sarsaparilla) (Thomson 1988) showed that bees were capable of learning the exact location of individual inflorescences at scales varying from tens of centimetres to hundreds of metres. The ability to learn to discriminate between rewarding and non-rewarding flowers with different colours was lower if the reward was variable within flower morph (Dukas and Real 1993).

It is well documented that the variation in nectar production among flowers of single plants is usually high. Feinsinger (1983) referred to this as a bonanza–blank pattern. Because of this variation it may take a lot of sampling effort for the pollinator to identify (and remember) the exact location of the most rewarding inflorescences. Manning (1956) found that bumble-bees only remembered the location of individual plants if inflorescences were at least 3.5 metres apart. In addition, the presence of a rewarding plant that is often approached by pollinators reduces the travelling time to

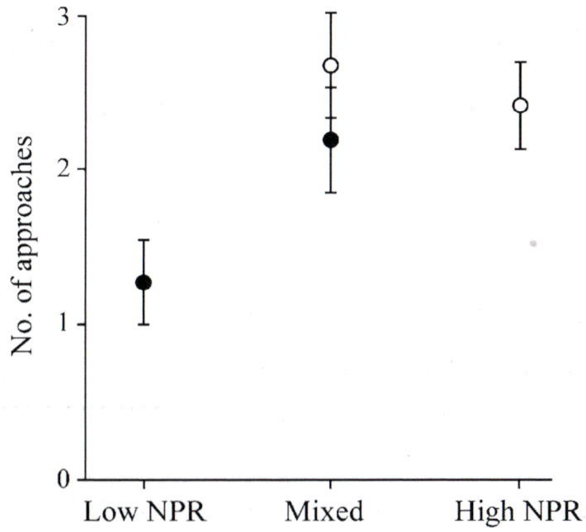

Fig. 14.3 Number of approaches per 5 minutes to individuals in a group of plants with low nectar production rates (NPR), a mixed group and a group of plants with high NPR. In the mixed group results are given for plants with high NPR (open circles) and the plants with low NPR (solid circles). Plants with low NPR profit from nectar-rich neighbours. Redrawn from Fig. 3 in Klinkhamer *et al.* (2001), *Trends in Ecology and Evolution*, with permission from Elsevier.

its less rewarding neighbours, making it worthwhile to approach the neighbours as well. Foraging decisions of pollinators in natural populations may therefore be taken at larger spatial scales than the inflorescence or the individual plant (Johnson *et al.* 2003) and this has consequences for selection on NPR.

In experiments with bumble-bees foraging on small groups of *E. vulgare* plants differing in NPR, Klinkhamer *et al.* (2001) found that bumble-bees approached groups of plants with high NPR more often than groups with low NPR, if the groups were separated by at least 6 metres (Fig. 14.3). Pollinators did not discriminate within groups when plants with high and low NPR were placed next to each other. Therefore, plants with low NPR profited from the attraction of pollinators by their close neighbours with high NPR. Klinkhamer *et al.* (2001) suggested that this could lead to frequency-dependent selection and to the maintenance of genetic variation in NPR (see also Chapter 17).

14.6 Rewardless plants

Many orchid species do not offer a reward to their pollinators. Most of these orchids do not resemble other rewarding species but some do. Dafni (1983, 1984) suggested that pollinators find it hard to distinguish between mimics and rewarding species. For example, the non-rewarding flowers of the South African orchid *Disa pulchra* are very similar to the flowers of a sympatric nectar-producing iris (*Watsonia lepida*). Both species are pollinated by a long-tongued fly (*Philoliche aethiopica*). In choice experiments inflorescences of both species were approached at equal rates (Johnson 2000). Similarly, the red flowers of the rewardless orchid *Disa ferruginea*, which depends on a single butterfly species for its pollination, mimic the nectar-producing flowers of *Tritoniopsis triticea* (Iridaceae). In a mixed stand pollinators did not discriminate between the mimic and the model species and fruit production of the mimic was lower in populations where it grew without its model (Johnson 1994). Gigord *et al.* (2002) observed naive bumble-bees foraging on arrays of plants with a rewarding model species and a rewardless mimic (elder-flowered orchid, *Dactylorhiza sambucina*) with two colour morphs. Inexperienced bees visited both rewardless morphs randomly, while bees that had gained experience with the model species preferred the rewardless morph that most resembled the model species. Likewise Gumbert and Kunze (2001) found in a natural situation with experienced pollinators (*Apis mellifera* and *Bombus* spp.) that colour similarity to rewarding model plants increased pollination in the rewardless orchid *Orchis boryi*. On the basis of their findings, Gigord *et al.* (2002) argued that mimicry could explain the maintenance of rewardlessness but only under restricted ecological conditions. Early-flowering orchids have only a slight advantage of resembling a mimic.

14.6.1 Costs and benefits of being rewardless

Nectar-producing orchids are more successful in setting fruit than nectarless species; on average the presence of nectar doubled the probability of fruit set in both temperate and tropical areas (Neiland and Wilcock 1998). In a comparative analysis of pollen limitation of flowering plants Larson and Barrett (2000) also found that pollen limitation was less intense in nectar-producing species. Being rewardless thus impairs a reproductive cost on a per-flower basis. In experimental studies, however, artificially enhanced nectar levels did not always lead to higher seed set per flower. In the orchid *Comparettia falcata*, which presents only small amounts of nectar, no overall differences in reproductive success between plants with enhanced rewards and unmanipulated controls were found (Salguero-Faria and Ackerman 1999). Smithson (2002) supplemented inflorescences with artificial nectar in the rewardless orchid *Anacamptis morio* and found no effect on fruit set, while pollinia removal even tended to be reduced. We must be careful, however, with the interpretation of these nectar enhancement experiments. The instantaneous response of pollinators to instant nectar enhancement may differ from their response to nectar production on an evolutionary timescale.

What are the possible advantages of rewardlessness that can compensate for the apparent costs? Several hypotheses have been put forward. One hypothesis is based on the observation that pollinators leave quickly after visiting non-rewarding flowers so that geitonogamy is reduced (Johnson and Nilsson 1999; Johnson 2000). Ferdy and Smithson (2002) modelled pollen transfer on the basis of actual foraging sequences on rewarding and non-rewarding artificial inflorescences and indeed predicted higher selfing rates on rewarding inflorescences. Direct evidence for reduced deposition of self pollen or lower selfing rates in rewardless individuals is, however, weak. This led Johnson and Nillson (1999), Salguero-Faria and Ackerman (1999) and Smithson (2002) to conclude that in their study systems reduced geitonogamy is not an adequate explanation for selection on rewardlessness. As a second hypothesis Smithson and Gigord (2001) suggested that pollinators try harder to find the nectar when visiting rewardless flowers, thereby increasing the probability of removing pollinia. Supplementing nectar to flowers of the bumble-bee-pollinated orchid *Barlia robertiana* decreased the probability that a pollinator removed pollinia by a factor of ten. In populations where inflorescences were supplemented with nectar for 20 days, pollinia removal was reduced by more than 50%. Although this presents the strongest experimental case of an evolutionary advantage of rewardlessness in the Orchidaceae, it remains to be tested how general this explanation is. The third hypothesis for the evolution and maintenance of rewardlessness states that not presenting a reward saves resources. Recent studies show that many of the genes that are expressed in the nectary involve defence against fungal and microbial infection (Thornburg *et al.* 2003; Carter and Thornburg 2004) which seems to suggests that the total production cost of nectar exceeds that of sugars alone. Whether or not nectar production is costly is still an open question. We have found only one study on this topic. By manipulating nectar production in flowers of *Blandfordia nobilis* (Christmas bells), Pyke (1991) found that plants with increased NPR produced fewer seeds than controls.

14.6.2 Some consequences of rewardlessness

Many rewardless species depend on pollinators that have not yet learned to avoid the unrewarding flowers. After each unrewarding flower the pollinator visits, the chance increases that it will visit a different morph. Consequently non-rewarding plants may only receive visits as long as they are rare. This may explain why non-rewarding orchids are so much rarer than rewarding ones in the British Isles (Neiland and Wilcock 1998). An encounter with a

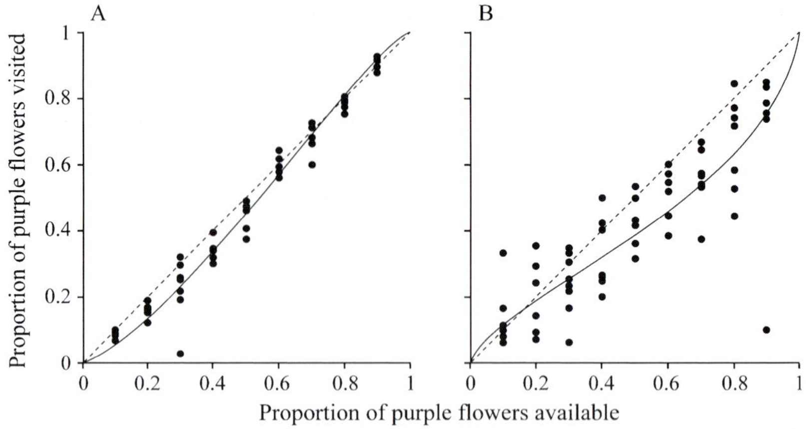

Fig. 14.4 Visitation of bumble-bees to purple flowers in relation to the proportion of purple flowers presented. Always three colours were present: yellow, blue and purple. In all the experiments one-third of the available flowers were yellow and were always rewarding. The proportion of blue and purple flowers varied and they were either rewarding or non-rewarding. The proportion of purple flowers present was taken as the proportion from the total number of blue and purple flowers present. Similarly the proportion of purple flowers visited was taken as the proportion from the total number of blue and purple flowers visited. (A) When blue and purple flowers were also rewarding, purple flowers were visited less than expected when rare. (B) When purple and blue flowers were both unrewarding, purple flowers were visited more than expected when rare and less than expected when common. Redrawn with permission from Smithson and Macnair (1997), *Evolution*.

non-rewarding colour morph increases the probability that a pollinator will switch to a different colour on the next visit. Hence, a rare morph advantage arises because a pollinator samples rare morphs more than randomly. This advantage may also explain why many species of non-rewarding orchids that do not mimic a rewarding species are polymorphic for corolla colour. Smithson and Macnair (1997) experimentally tested the hypothesis that pollinator behaviour results in a preference for rare colour morphs of deceptive species. Bumble-bees were trained to forage in the lab on arrays of artificial flowers with different corolla colour morphs. Morphs were varied in frequency, and bumble-bee preferences were recorded on arrays where morphs did and did not contain sucrose solution rewards. Bumble-bees preferred the most common colour morph when flowers contained sucrose solution rewards, but over-visited rare morphs when sampling flowers with no reward (Fig. 14.4). A negative frequency-dependent selection through pollinator preference for the rare morph was demonstrated in field experiments with the rewardless orchid *Dactylorhiza sambucina* (Gigord *et al.* 2001). Such frequency-dependent pollinator behaviour can allow stable polymorphisms in corolla colour to be maintained in non-rewarding species. In the experiments of Smithson and Macnair (1997), bumble-bees repeatedly switched colour morphs when these were unrewarding. If such a shift occurred between different orchid species this might lead to hybridization. Indeed, the analysis of Neiland and Wilcock (1998) suggested that hybridization occurs more frequently among nectarless orchid species in Britain and Europe than among nectar-producing ones.

14.7 Bateman's principle and Bell's principle

Until the 1970s selection on flower traits was assumed to act through female function. The importance of male function first gained widespread attention through the papers of Horovitz and Harding (1972), Janzen (1977) and Willson (1979). Willson (1979) and Charnov (1979) applied Bateman's principle to plants. Bateman's (1948) principle states that male fitness is usually limited by the number of matings, while female fitness is limited by the amount of resources available for reproduction. For hermaphrodite plants this would imply that female fitness decelerates faster with increased levels of attraction than male fitness. We call this Bell's principle (Fig. 14.5) (Bell 1985). Under Bell's principle variation in floral characters should primarily affect male fitness. Experimental support for this hypothesis came from Queller (1983), Bell (1985) and from a series of papers by Stanton and collaborators on *Raphanus* (Stanton *et al.* 1986, 1989, 1991). Stanton showed that plants with white instead of yellow petals attracted fewer pollinators. This resulted in a reduced male reproductive success (siring seeds on other plants) but not in a reduction of seed set on the white-flowered plant itself (Fig. 14.6). Bell's principle can easily be tested, by quantifying how flower visitation relates to male function (pollen removal or export) and female function (seed set). Surprisingly few data have been collected in this context. Figure 14.5B shows that for *Echium vulgare* seed set decelerated strongest with the number of bumble-bee visits, but the difference with pollen removal was only small. The idea that flowers primarily serve a male function spread rapidly in the literature during the 1980s (e.g. Sutherland and Delph 1984). In fact the idea became so generally accepted that two independent papers advised cautiousness (Burd 1994; Wilson *et al.* 1994). Bell's principle would lead one to believe that pollen limitation of fruit and seed set will be uncommon. Burd's (1994) review on pollen limitation showed, however, that limitation of female success due to inadequate pollen receipt is a common phenomenon in plants. In his review he found cases of pollen limitation in 159 out of 258 species. Burd writes:

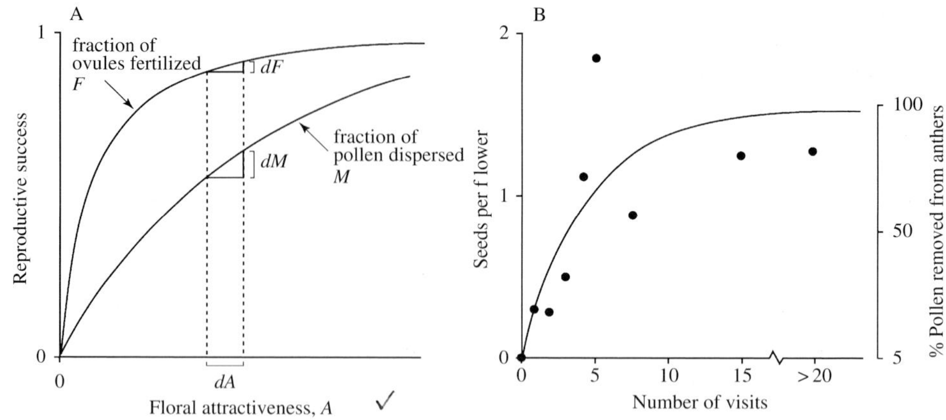

Fig. 14.5 (A) The effect of floral attractiveness on female and male reproductive success. At the dotted lines an increase in attractiveness (*dA*) leads to a steeper increase in male (*dM*) than in female success (*dF*) (Bell 1985). Redrawn with permission from Fig. 1 in Bell (1985), *Proceedings Royal Society of London B*. (B) The number of seeds per flower in *Echium vulgare* (as a measure of female reproductive success; solid circles) and the percentage of pollen removed from the anthers (as a measure of male reproductive success; line) in relation to visits per flower. After five to ten visits seed set was saturated, while additional visits still dispersed some more of the remaining pollen from the anthers (T. J. de Jong and P. G. L. Klinkhamer, unpublished data).

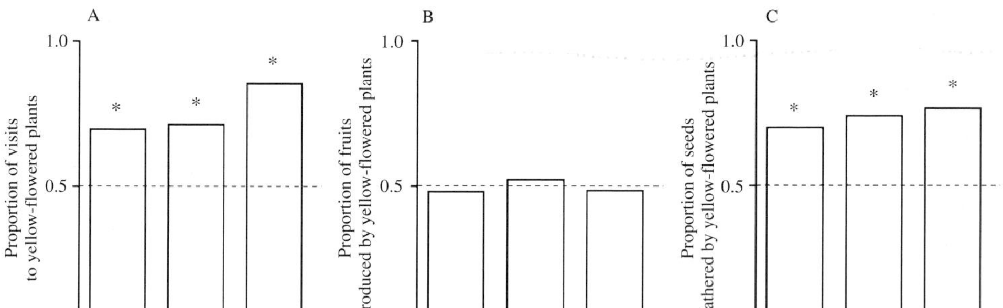

Fig. 14.6 (A) Pollinator discrimination, (B) maternal function and (C) paternal function for yellow-flowered plants in three populations with equal numbers of white- and yellow-flowered plants of *Raphanus raphanistrum*. Pollinators (mainly *Pieris rapae* butterflies) preferred yellow flowers over white ones. This preference did not affect female reproductive success, while male reproductive success benefited from more visits. Reproduced with permission from Stanton *et al.* (1986), *Science*, American Association for the Advancement of Science.

Because flowering plants often allocate considerable resources to pollinator attraction, both export and receipt of pollen could be limited primarily by resource investment in floral advertisement and rewards. But whatever investment is made in attraction, pollinator behavioural stochasticity usually produces wide variation among flowers in reproductive success through both male and female functions.

Independently Wilson *et al.* (1994) raised several concerns against what they called 'floral Batemania'. First, the common practice of additional hand-pollinating can be detrimental to the supplemented flowers. Second, many of these studies were done in large, dense populations that present situations that are especially favourable for pollination. One might also argue that if pollen limitation is shown by hand-pollinating a small fraction of the flowers on an inflorescence it does not immediately follow that pollen is limiting seed set at the plant level. The experimental evidence shows that in many situations Bell's principle does not apply to plants while there are also well-documented cases where it does (as in Stanton's experiments on *Raphanus*). Perhaps one of the few generalizations possible is the suggestion by Burd (1994) that Bell's principle is more likely to apply to self-compatible species.

14.8 How does attractiveness influence sex allocation?

Does investment in attractive structures change the earlier results of our sex allocation models in which we accounted only for investment in male and female function? Initially several authors argued that because investment in attraction is a cost shared by male and female function this stabilizes hermaphroditism. In an article entitled 'Why be an hermaphrodite?' Charnov *et al.* (1976) suggested that these shared costs possibly explain why so many plant species are hermaphrodite and so few are dioecious. Under Bell's principle, however, investment in attractiveness should not be seen as shared costs but primarily as an investment enhancing male reproductive success.

14.8.1 Attractiveness only influences male function

A useful starting-point for making these ideas more exact is a paper by Charnov and Bull (1986), which used the Shaw–Mohler equation (Chapter 4) without any selfing. Resources available to reproduction are scaled to 1 and divided over three functions: pollen production (r), seed production (f) and attractiveness (a) with $r+f+a=1$. The first assumption is that seed set is not affected by investment in attractiveness, i.e. that it is not pollen limited, while pollen export to other plants does increase with the investment in attractiveness.

Charnov and Bull (1986) furthermore assumed that the female component of fitness (F) is a linear function of the investment in seeds (f) but we will relax this restriction. If we set the total budget for reproduction at unity we obtain: $F = cf^{\beta}$, in which c is a scaling factor and β is the exponent of the female gain curve. The chance h of a pollen grain being picked up by a pollinator, transported and deposited on a stigma of a conspecific is a function of the amount of resources invested in attractiveness: $h = a^{\gamma}$. This is a convenient mathematical description, which is close to mechanistic functions of pollen export from Chapter 3 (Fig. 3.4) (see also Iwasa *et al.* 1995). Note that the shape of this function is assumed to be independent of the allocation strategy of the resident population. This assumption is not necessarily correct and at this point the consequences of different assumptions for the outcome of the model are unclear. Charnov and Bull (1986) further assumed that pollen exported is a decelerating function of r, such that the total effective number of pollen grains exported (E) is given by:

$$E = br^{\alpha}h = br^{\alpha}a^{\gamma} \tag{14.1}$$

in which b is a scaling factor. For a rare mutant with allocations r_m, f_m and a_m in a population of a common type with allocations r, f and a, absolute fitness is given by: $W_m = F_m + M_m$.

$$W_m = cf_m^{\ \beta} + \frac{br_m^{\ \alpha}a_m^{\ \gamma}}{br^{\alpha}a^{\gamma}}cf^{\beta} \tag{14.2}$$

in which $r_m^{\ \alpha}a_m^{\ \gamma}/r^{\alpha}a^{\gamma}$ represents Lloyd's competitive share, i.e. the ratio of seeds sired by mutant and resident. With $f + r + a = 1$, we can rewrite this as

$$W_m = c(1 - a_m - r_m)^{\beta} + \frac{r_m^{\ \alpha}a_m^{\ \gamma}}{r^{\alpha}a^{\gamma}}c(1 - a - r)^{\beta} \tag{14.3}$$

If investment in attractiveness is negligible and equal for mutant and resident, this equation reduces to Eq. (4.7). We follow Parker's method to find the candidate ESS for allocation. We differentiate W_m, set the derivative equal to zero and set equal values for the evolutionary variable for the mutant and the resident. The difference with previous calculations is that we now have two parameters (a_m and r_m) determining the strategy, instead of one. The ESS is at the point where a shift in each of both parameters leads to a reduction of W_m. This means that to find the ESS we have to set both partial derivatives equal to zero ($\mathrm{d}W_m / \mathrm{d}r_m = \mathrm{d}W_m / \mathrm{d}a_m = 0$).

$$\frac{\mathrm{d}W_m}{\mathrm{d}r_m} = -c\beta(1 - a_m - r_m)^{\beta-1} + \alpha\frac{r_m^{\ \alpha-1}a_m^{\ \gamma}}{r^{\alpha}a^{\gamma}}(1 - r - a)^{\beta}c = 0 \tag{14.4}$$

Setting $r_{\mathrm{m}} = r$ and $a_{\mathrm{m}} = a$ gives:

$$-\beta/(1 - r - a) + \alpha/r = 0 \tag{14.5}$$

$$\frac{\mathrm{d}W_{\mathrm{m}}}{\mathrm{d}a_{\mathrm{m}}} = -c\beta(1 - a_{\mathrm{m}} - r_{\mathrm{m}})^{\beta-1} + \gamma\frac{{r_{\mathrm{m}}}^{\alpha}{a_{\mathrm{m}}}^{\gamma-1}}{r^{\alpha}a^{\gamma}}(1 - r - a)^{\beta}c = 0 \tag{14.6}$$

Setting $r_{\mathrm{m}} = r$ and $a_{\mathrm{m}} = a$ gives:

$$-\beta/(1 - r - a) + \gamma/a = 0 \tag{14.7}$$

Combining Eqs. 14.5 and 14.7 and substitution gives the allocations at the candidate ESS:

$$a^* = \gamma/(\alpha + \beta + \gamma) \tag{14.8a}$$

$$r^* = \alpha/(\alpha + \beta + \gamma) \tag{14.8b}$$

$$f^* = \beta/(\alpha + \beta + \gamma) \tag{14.8c}$$

The relative allocations at the ESS are determined by the shape coefficients and relate to each other as: $r^*{:}f^*{:}a^* = \alpha{:}\beta{:}\gamma$. As in the models from Chapter 4, where attractiveness was not included, the result $r^*/f^* = \alpha/\beta$ holds (Eq. 4.8).The stability of the candidate ESS can be tested by taking the second derivatives with respect to r_{m} and a_{m} (Chapter 4). In this case the maths becomes rather complex so we follow another line of reasoning. A necessary condition for hermaphroditism to be stable is that pure males and females cannot invade the population. By writing down the fitness equations for the hermaphrodite and the pure male and female types, under the assumption that they all optimized their allocation strategy, we can calculate these conditions. The total fitness of the common hermaphrodite type equals two times its female fitness, which amounts to: $2cf^{\beta}$ (Eq. 14.2). Equation (14.2) shows that the absolute fitness of a rare female mutant ($r_{\mathrm{m}} = 0$) equals $c{f_{\mathrm{m}}}^{\beta}$, and the fitness of a pure male mutant ($f_{\mathrm{m}} = 0$) equals $({r_{\mathrm{m}}}^{\alpha}{a_{\mathrm{m}}}^{\gamma}/r^{\alpha}a^{\gamma})cf^{\beta}$ with a, r and f for the hermaphrodite at the ESS given by Eq. (14.8) and with optimal investment in male function and attractiveness for the male mutant $r_{\mathrm{m}} = \alpha/(\alpha + \gamma)$ and $a_m = \gamma/(\alpha + \gamma)$. When the hermaphrodite is at the ESS, a female mutant can invade if:

$$\left(\frac{\beta}{\alpha + \beta + \gamma}\right)^{\beta} < \frac{1}{2} \tag{14.9}$$

Without any effects of attractiveness ($\gamma = 0$) this equation is a reformulation of Eq. (4.11). A male mutant can invade if:

$$\left(\frac{\alpha+\beta+\gamma}{\alpha+\gamma}\right)^{\alpha+\gamma} > 2 \tag{14.10}$$

Charnov and Bull (1986) examined the simplified case where female fitness was a linear function of the investment in female reproduction ($\beta = 1$). At the ESS the conditions for stability of hermaphroditism (Eqs. 14.9 and 14.10) reduce to $\alpha + \gamma < 1$. With a linear female gain curve ($\beta = 1$), for hermaphroditism to be stable male fitness must be a strongly decelerating function of both investment in male reproduction and attractiveness. If there is an effect of attractiveness on male fitness ($\gamma > 0$), hermaphroditism may not be stable even if the shape coefficient for the investment in male function (α) is smaller than 1. Combining this condition for stability of the hermaphroditism with Eq. (14.8) shows that the ESS for investment in female reproduction (f^*) must be smaller than the total investment in male reproduction and attractiveness together. Note, however, that this result is brought about by the a priori assumption that the shape parameter of female reproductive success equals 1. The more general case is given by Eqs. (14.9) and (14.10). The left-hand side of Eq. (14.9) decreases and that of Eq. (14.10) increases when α, β and γ become larger. This means that both pure males and females are more likely to invade when exponents of the curves for male and female function and attractiveness are higher. A high value of γ seems likely when pollinators are scarce and the number of approaches to plants limits pollen export. A high value of γ will also lead to a higher investment in attractiveness when the hermaphrodite plays an ESS. A high investment in attractiveness may destabilize hermaphroditism, which is in contrast to what was initially expected. Note that in this model there are no shared costs for attractiveness and the result depends on the assumption that female fitness is not related to attractiveness, as would be the case if Bell's principle applies in its most extreme form. Although this may be true sometimes, there are also examples where seed production is pollen limited and where we must assume that being more attractive results in more pollinator visits and consequently leads to higher female fitness.

14.8.2 Attractiveness influences both male and female functions

Including the effect of attractiveness on female fitness makes the model more complicated. Moreover, the way it should be modelled is not straightforward. Attractiveness not only reduces the amount of resource available for male and female functions, but also affects the fraction of ovules fertilized and thus the number of seeds that can be matured. Two models have included an effect of attractiveness on female fitness. Morgan (1992) included this in his model by writing female fitness as a power function of both the initial investment in female function and the investment in attractiveness ($F = bf^{\beta}a^{\delta}$). It is questionable whether such a model formulation is realistic because it assumes that investment in attractiveness affects seed production without any repercussions for the total investment in seed production. In other words the costs of producing more seeds if a plant is more attractive is not at the expense of the investment in male reproduction or attractiveness. Morgan (1992) concluded that a higher investment in attractive structures at the ESS does not necessarily destabilize hermaphroditism. He writes:

> Accelerating gains associated with pollinator attraction may contribute to the evolution of separate sexes only if male function benefits from increased investment in attractive structures at a disproportionate rate compared to female function.

This is exactly the situation we analysed above in Charnov and Bull's model. Charlesworth and Charlesworth (1987) divided total reproduction into two periods. In the first period flowers with pollen, ovules and attractive structures are formed and resources are traded off among these flower parts. In their model a certain fraction of the ovules is self-fertilized and the fraction of non-selfed ovules that is fertilized is determined by the investment in attractiveness. In the second period resources set a limit to the number of seeds that can actually mature. Seed number is a decelerating function of the total number (selfed and outcross) of fertilized ovules. Furthermore they assumed that pollen export is proportional to the investment in male reproduction. The chance that a pollen grain becomes part of the outcross pollen pool is a power function of the investment in attractiveness. Charlesworth and Charlesworth (1987) analysed the ESS allocation by means of numerical simulations. They concluded that the spread of females was unlikely unless there was high inbreeding depression and a rather high selfing rate. With a linear relation between numbers of fertilized ovules and seeds matured, invasion by females was more likely when investment in attraction was low than when it was high. The spread of male mutants was never likely in their models.

To sum up: even if attractiveness costs nothing, being more attractive is not always best; it may result in more pollen removal but also in less efficient pollen export if pollinators stay longer. Avoidance of geitonogamy has been put forward as an explanation for the evolution of nectarless flowers. Alternatively, pollinators try harder to find a reward in flowers without nectar, which results in closer contact and more removal and deposition of pollen. Bell's principle states that attractiveness affects male reproductive success more than female reproductive success. When Bell's principle is included in sex allocation models, the relative allocations to male and female functions and attractiveness are controlled by the coefficients of the gain curves. Accelerating male gains associated with pollinator attraction destabilize hermaphroditism and may contribute to the evolution of separate sexes. Bell's principle should, however, be treated with caution. We included allocation to attractiveness in the sex allocation models. How to include such effects is not straightforward and more theoretical work should be done on this.

Questions

(1) What is Bell's principle?
(2) In Charnov and Bull's (1986) model a high investment in attractiveness may destabilize hermaphroditism. On which assumption does this result depend?
(3) Give an explanation for Neiland and Wilcock's (1998) finding that hybridization occurs more frequently among nectarless orchid species than among nectar-producing ones.
(4) What is the plant's dilemma regarding attractiveness?

15

Parent–offspring conflict

> Clearly, if the gene in the offspring exacts too great a cost from the parent, that gene will be selected against even though it confers some benefit to the offspring. To specify precisely how much cost an offspring should be willing to inflict on its parents in order to gain a benefit, one must specify how the offspring is expected to weigh the survival of siblings against its own survival. The problem of specifying how an individual is expected to weigh siblings against each itself . . . has been solved in outline by Hamilton.
>
> *R. L. Trivers (1974)*

15.1 Introduction

Trivers' (1974) paper clearly pointed out that genes that affect how much resource is given to each offspring and genes that affect how much each offspring can take from its mother are under different selection pressures, i.e. they are in conflict. This concept had a huge impact on evolutionary biology and optimality thinking (Trivers 2002). Although several authors, especially Lloyd, Westoby and Haig, have pointed out that parent–offspring conflict applies to plants, these ideas have not diffused well into mainstream plant ecology.

As with any new scientific theory, it did not take long before objections were raised against Trivers' concept of parent–offspring conflict (reviewed in Godfray 1995).

First, it was argued that if offspring take too much resource from their mother, this advantage would be nullified by the fact that their offspring inherit the same selfish genes so that in the next generation their children also take too much from their mother. Such a lineage would then perform poorly and would be selected against: an intuitive and seemingly entirely reasonable statement, which can, however, be shown to be false in explicit genetic models. We illustrate such a genetic model in Box 15.1 for the example of seed size.

Second, it was stated that the mother is always in control so the offspring cannot affect the decisions she makes. This is an important point that may well be true in many cases. However, it is now realized that in many instances the offspring has control too and what we see in nature might be a compromise between the interests of the parents and the offspring. It all depends on the specific situation if the offspring can also jump into the control seat. In some situations it seems reasonable that it can. For instance, in Cohen's (1966) seed bank model (Chapter 7) it was implicit that the mother controls the germination fraction of her offspring. Risk-spreading was optimal for the mother; not all seeds should germinate at the same time in the first year, but a fraction should delay germination for one or more years. For an individual seed, however, it would be best to germinate as soon as possible, avoiding mortality connected with residing in the soil seed bank. This is the basis for a conflict over germination fraction (Shmida and Ellner 1981; Ellner 1986). The mother may place germination-inhibiting hormones in the seed coat so that germination is delayed. The offspring may be selected to break this control. On the one hand are genes for hormone production in the mother that result in high hormone concentrations in the seed coat (maternal tissue). On the other hand are genes that are expressed in the embryo and that reduce the sensibility to germination-inhibiting hormones or that simply promote leakage of such hormones from the seed coat. The genes expressed in the embryo and in the flowering plant are selected in opposite directions.

Apart from differences in germination time, conflict is now considered to play a role in many characters for which parent and offspring interact. For plants these include, among others, seed size (Hamilton 1964b), the functioning of the endosperm (Queller 1984; Härdling and Nilsson 1999), reproductive effort (Charnov 1982b), seeds per fruit (Shaanker *et al.* 1988), self-compatibility (Lloyd 1992) and the seed sex ratio in dioecious plants (Beukeboom *et al.* 2001).

15.2 Optimal seed size under offspring control

Although parent–offspring conflict was typically a product of the 1970s, following the papers of Trivers, Hamilton (1964b) already made an insightful comment on this: 'For one seed to expand selfishly at the expense of its neighbours may or may not be advantageous to the inclusive fitness of its bearer, but it is almost certainly not in the interest of that of the parent.'

To prove this point and to quantify it, it is most straightforward to construct an explicit population genetics model in which the character we are interested in is under offspring control. We revisit the Smith–Fretwell model for optimal

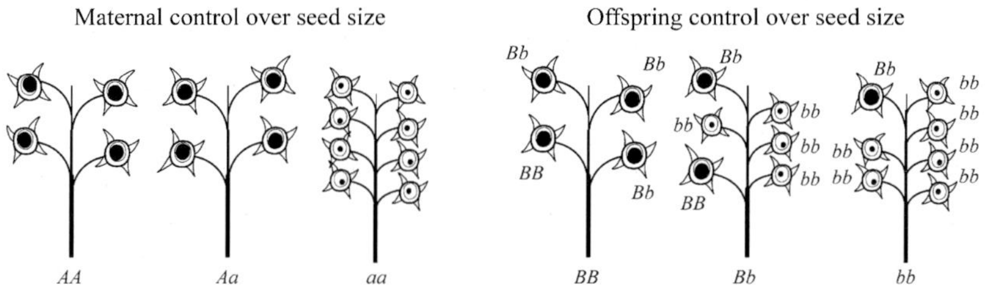

Fig.15.1 Seed size under maternal and offspring control. Under control of the mother a gene with alleles *A* and *a*, expressed in the parent during seed formation, determines the size of each seed. In that case all seeds on the same plant have the same size. Under control of the offspring a gene with alleles *B* and *b*, expressed in the developing embryo or the associated endosperm, has an effect on the amount of resource taken from the maternal parent and therefore on its own seed mass.

seed size that we dealt with in Chapter 7. In that model the question was what is the best seed size for the maternal parent. The terminology 'best for the parent' and 'best for the offspring' is easy to use, but might also be confusing. In the first case, we refer to a gene with different alleles, *A* or *a*, that is expressed in the maternal parent during seed formation. Each allele of that gene corresponds to a certain seed size. In this model the genotype of the mother determines seed size, she controls the size of each seed and all seeds on the plant are of the same size. Alternatively, the genotype of each individual seed with alleles *B* and *b* has control and determines its own size (Fig. 15.1). In this second case, seeds differ in size due to their genotype. Under maternal control we examine changes in frequency of *A*, under offspring control we examine the change in frequency of *B*.

Now assume there are two alleles for a gene that is expressed in the seed and that affects seed size. Dominant allele *B* takes more resources than allele *b*. If all seeds on a mother plant were of *BB* genotype and on another plant the seeds were of *bb* genotype, then most copies would be transferred of the allele that corresponded most closely to the optimal seed size in the Smith–Fretwell model. However, when *BB* seeds and *bb* seeds are in competition on one maternal plant, the *BB* seeds take more resources so that fewer other seeds can be formed. This competition between the two alleles on one plant selects for larger seed sizes than predicted by the Smith–Fretwell model from Chapter 7. If plants were 100% selfing than in no time all plants would be homozygous and also their selfed seeds would be genetically identical. The competition between alleles then does not exist and we are back at the Smith–Fretwell model. When plants are fully outcrossing the competition between alleles is strongest. Several authors have pointed to the conflict between mother and seed, but the conflict has not been quantified (Hamilton 1964b; Queller 1984; Haig and Westoby 1988; Shaanker *et al.* 1988; Lloyd 1992). Since alleles are competing against each other, genetic models are most straightforward. However, they are also a brute-force method that is rather long-winded; if you are already convinced that parent–offspring conflict exists you might want to skip Box 15.1 and go immediately to Hamilton's method, which gives a better idea of the essentials. The results of the two approaches are identical.

Box 15.1 A population genetics model for seed size under offspring control

Consider a gene for seed size that is expressed in the offspring. The gene has many alleles so that many seed sizes are possible, but in one simulation we always compare two alleles: B is dominant and b is recessive. Seed phenotypes are M (genotype BB and Bb) and m (genotype bb). As in the Smith–Fretwell model $f(m)$ denotes the survival of the seed as a function of its mass. Now there are three types of maternal parents in the population: bb, Bb and BB. The frequencies of these genotypes are denoted as f_1, f_2 and f_3, respectively. To calculate the number of seeds each mother makes it suffices to know the average seed size Z of all seeds on one mother plant: from T resource T/Z seeds can be made. The average sizes of seeds from maternal genotypes bb, Bb and BB are denoted as Z_1, Z_2 and Z_3, respectively. Now first consider plants that are fully selfing. Next we consider full outcrossers and in the end we combine the previous results for partial selfers. The average seed sizes for mother plants with 100% selfing are $Z_1 = m$, $Z_2 = 0.75M + 0.25m$ and $Z_3 = M$. Homozygotes only generate themselves and are produced by the heterozygote mother after selfing. With full selfing the new frequencies (denoted by the prime) depend on the frequencies the year before as:

$$bb\colon\ f_1' = f_1 f_m R/Z_1 + 0.25 f_2 f_m R/Z_2 = U_1 \tag{15.1a}$$

$$Bb\colon\ f_2' = 0.5 f_2 f_M R/Z_2 = U_2 \tag{15.1b}$$

$$BB\colon\ f_3' = f_3 f_M R/Z_3 + 0.25 f_2 f_M R/Z_2 = U_3 \tag{15.1c}$$

Since the three frequencies always need to add up to 1, we still have to divide by $U_1 + U_2 + U_3$ so that their sum is 1. One can compare different seed sizes using these equations. If you simulate this on a computer, it turns out that the allele that corresponds most closely to the Smith–Fretwell seed size is always the winner.

Now consider fully outcrossing plants. If pollen of all plants in an infinitely large population is well mixed in the outcross pollen pool, the frequency of the b allele in the pollen is $p_b = f_1 + 0.5f_2$ and the frequency of the B allele is: $p_B = 1 - p_b$.

An outcrossing *bb* plant now makes a fraction p_b seeds that are of genotype *bb* and size m and a fraction p_B seeds of genotype *Bb* and size M. The average seed size is then $Z_1 = p_b m + p_B M$. Similarly for the heterozygous *Bb* mother plant $Z_2 = 0.5M + 0.5p_b M + 0.5p_b m$. For the dominant homozygous mother, each seed has at least one *B* allele and the seed mass is simply $Z_3 = M$. The new frequencies for the full outcrossers are:

$$bb\colon\ f_1' = f_1 f_m(R/Z_1)p_b + 0.5f_2 f_m(R/Z_2)p_b = V_1 \tag{15.2a}$$

$$Bb\colon\ f_2' = 0.5f_2 f_M(R/Z_2) + f_1 f_M(R/Z_1)p_B + f_3 f_M(R/Z_3)p_b = V_2 \tag{15.2b}$$

$$BB\colon\ f_3' = f_3 f_M(R/Z_3)p_B + 0.5f_2 f_M(R/Z_2)p_B = V_3 \tag{15.2c}$$

We still need to divide by $V_1 + V_2 + V_3$ to obtain new frequencies that add up to 1.

Now consider the situation with partial selfing. The fraction of selfed seeds is denoted S. In this case the average seed sizes on the three maternal genotypes are:

$$bb\colon\ Z_1 = (1 - S)(p_b m + p_B M) + Sm \tag{15.3a}$$

$$Bb\colon\ Z_2 = (1 - S)(p_B M + 0.5p_b M + 0.5p_b m) + S(0.75M + 0.25m) \tag{15.3b}$$

$$BB\colon\ Z_3 = M \tag{15.3c}$$

For partial selfers we can combine Eqs. (15.1) and (15.2) to obtain:

$$bb\colon\ f_1' = SU_1 + (1 - S)V_1 \tag{15.4a}$$

$$Bb\colon\ f_2' = SU_2 + (1 - S)V_2 \tag{15.4b}$$

$$BB\colon\ f_3' = SU_3 + (1 - S)V_3 \tag{15.4c}$$

We divide all new frequencies by $S(U_1 + U_2 + U_3) + (1 - S)(V_1 + V_2 + V_3)$ so that the new genotype frequencies always add up to 1. With these equations we can then compare alleles for different seed sizes, and examine if there exists an ESS seed size that wins against others that are slightly different. For a chosen shape of the survival function, $f(m) = (m - 1)/(m + 5)$ (Fig. 15.2) we can do these calculations with a computer. For full selfing the optimal seed size under offspring control turns out to be 3.45 (point 1 in Fig. 15.2), which is equal to the optimal size for the parent, i.e. with full selfing there exists no parent–offspring conflict. For full outcrossing the optimal size under offspring control becomes 20.24 (point 3 in Fig. 15.2), much larger than the optimal size with maternal control. For intermediate selfing rates we can calculate values between 3.45 and 20.24 with the computer; we will now show in the text that it is possible to derive an exact equation for optimal seed weight under offspring control using Hamilton's method of inclusive fitness.

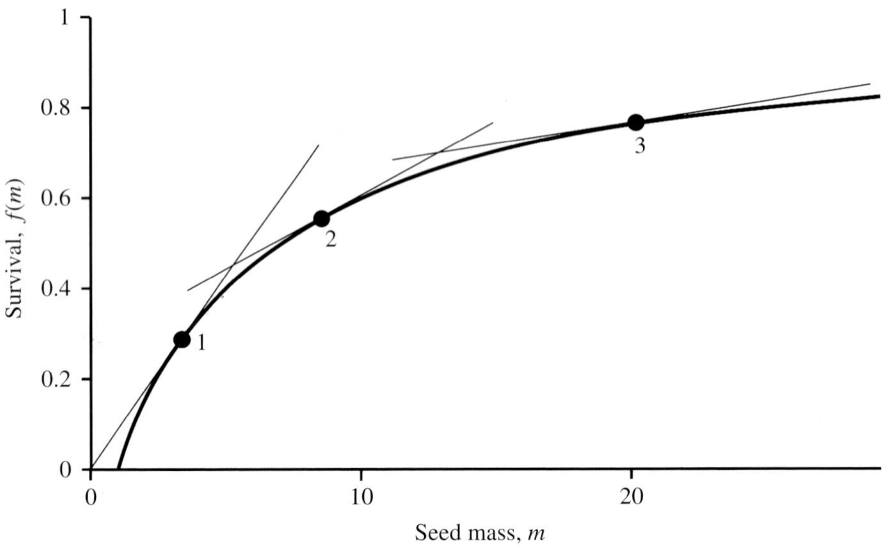

Fig.15.2 The Smith–Fretwell model of optimal seed size. Under maternal control the optimum seed size is the point where the line through the origin touches the survival curve as a function of seed mass. In this point $df(m)/dm = f(m)/m$ (point 1). If genes in the offspring control seed size, optimal seed size for the offspring is at the point where $df(m)/dm = rf(m)/m$ (point 2 for $r = 0.5$ and point 3 for $r = 0.25$). Here r measures the average genetic relatedness between seeds. If $r < 1$ the optimal seed size for the offspring is larger than for the mother. The optimal seed size under offspring control can be calculated by a genetic model (Box 15.1) or directly from Eq. 15.7

15.3 Hamilton's method of inclusive fitness

The population genetics model in Box 15.1 demonstrated existence of conflict between autosomal genes, depending on where and when they are expressed: in the maternal parent during seed production or in the developing embryo. The genetics in Box 15.1 can all be handled with a computer, but is quite complicated and it is now appropriate to see if there are simpler ways to tackle the same problem. Hamilton's inclusive fitness method (Hamilton 1964a; Reeve and Keller 1999) can be applied in many cases, including this one. In trying to explain altruistic behaviour Hamilton realized that one should not focus on the organism but rather on the genes it contains, taking social actions into account. Different alleles of these genes are transmitted to the next generation through the actor but also through the related individuals that are affected by the behaviour and that share some of the alleles. Take a random allele in individual J. Hamilton's coefficient of relatedness r between individuals J and K now estimates the probability that individual K received the same allele as present in J by identical descent and passes it on to a gamete, *divided by the probability that J itself passes on the allele to a gamete*. The importance of the latter part of the sentence will soon become clear (see Box 15.2, below). Identical by descent (IBD) means that the identical allele comes from a common ancestor. For outbreeding diploids, relatedness simply means what fraction of the genes is shared by identical descent. For two identical twins all genes are IBD $r = 1$, the coefficient of relatedness cannot be higher than this value. Fathers and mothers pass on 50% of their genes to their daughters, so that the relatedness to their children is 0.5. Full brothers and sisters share 50% of their genes, so that relatedness between full sibs is 0.5. For two half sibs $r = 0.25$. In Chapter 17 we discuss that IBD is not the same as having a concrete allele in common (IIS or identical-in state). Unrelated individuals may share concrete alleles by chance, especially if these alleles are common in the population. The calculation of relatedness is straightforward for outcrossers in a very large population. With self-fertilization or mating between relatives (inbreeding) the computation of the coefficient of relatedness is more complicated (see Box 15.2, below).

The rationale behind the definition of the coefficient of relatedness is clear: we count the allele copies passed on to the next generation rather than only the reproductive success of a focal individual. Even if an allele is rare in the population then not only the focal individual but also its relatives have that allele and we have to take both into account to calculate if the allele can invade. We encountered the same logic already in the example of Haldane's drowning brothers from Chapter 1.

To predict the outcome of selection Hamilton defined inclusive fitness of an allele. In the seed size example, inclusive fitness consists of the fitness of the focal seed and fitness of the other seeds on the plant, weighted with respect to their relatedness to the focal seed. The fitness of the focal seed is a function $f(m)$ of its seed mass. If the focal seed claims m resource from the total of T, then $T - m$ remains for the others, so that only $(T - m)/m$ other seeds of size m can be formed. The more resource the focal seed claims, the less is left for the other seeds on the plant. In equation the inclusive fitness W of the common strategy of claiming m resource per seed is:

$$W = f(m) + rf(m)\frac{(T - m)}{m} \tag{15.5}$$

For a selfish type with heavier seeds that takes $m + \Delta m$ resource per seed from the mother plant we can write a similar equation. The survival of the focal seed, $f(m + \Delta m)$, is now higher, but the other seeds together have fewer resources $(R - m - \Delta m)$. How many seeds can a plant still make after provisioning the focal seed? This depends on the average seed size, and this depends in its turn on the fraction of seeds with the selfish allele. How many seeds on the same plant are selfish if the selfish allele is still rare in the population? Conveniently this fraction is also equal to r (see bottom of Box 15.2). Thus a fraction r of these seeds also has the allele for bigger seeds and a fraction $1 - r$ has not. The average seed size is then $r(m + \Delta m) + (1 - r)m$. The inclusive fitness of the rare mutant strategy of claiming $m + \Delta m$ resource per seed is:

$$W_{\text{mut}} = f(m + \Delta m) + rf(m + \Delta m)\frac{(T - m - \Delta m)}{r(m + \Delta m) + (1 - r)m} \tag{15.6}$$

According to Hamilton (1964a) the type (allele) with the highest inclusive fitness wins. With some rewriting of $W_{\text{mut}} > W$ the selfish type wins if: $[f(m + \Delta m) - f(m)]/\Delta m > rf(m)/m$. For small Δm the left-hand side of this equation is the derivative of $f(m)$ with respect to seed size m. Optimal seed size under offspring control then follows from:

$$\frac{\mathrm{d}f(m)}{\mathrm{d}m} = r\frac{f(m)}{m} \tag{15.7}$$

With full selfing, corresponding to $r = 1$ between seeds, we return to the Smith–Fretwell equation (Eq. 7.2) that we analysed before. In that case the optimum is where the line through the origin touches the survival curve in Fig. 15.2. If the relatedness between the seeds is 1, no parent–offspring conflict exists. With full outcrossing ($S = 0$) with many possible fathers, seeds are half sibs with relatedness $r = 0.25$. The optimal seed size can then be glanced from the survival curve as the point where the slope of the tangent is $1/4$ of $f(m)/m$.

In Fig. 15.2 we used as a function for fitness per seed: $f(m) = (m - 1)/(m + 5)$. The derivative of this function is $6/(m + 5)^2$. Substituting this in Eq. (15.7) allows us to solve it for m. This yields: $m^* = -2 + 3/r + 0.5\sqrt{(4 - 6/r)^2 + 20}$.

So we have an explicit equation for the optimal seed size m^* for the offspring that replaces the genetic model from Box 15.1. For an outcrossing plant with many fathers ($r = 0.25$), we find $m^* = 20.24$, the same result as produced by the genetic simulation model. Hamilton rulez! Remains the matter of how to calculate the coefficient of relatedness between random seeds as a function of the selfing rate and the number of competing fathers (Box 15.2).

Box 15.2 Computation of relatedness between seeds on partially selfing plants

For more background the reader is referred to Bulmer (1994). Here we just illustrate how to calculate the coefficient of relatedness r between seeds for a partially selfing plant. The calculation is straightforward but it is well worth going through it step by step as this procedure is rarely applied to cosexual plants in the literature and a mistake is easily made. To calculate relatedness between individual J and K, we need as a tool the coefficient of consanguinity F_{JK} between individual J and K. The coefficient of consanguinity is the probability that two alleles that we draw at the same locus *with replacement* are IBD. One could also calculate this coefficient for an individual with itself (F_{JJ}). If J is a diploid outbred individual AB in which alleles A and B are not IBD, then F_{JJ} is 0.5. We call this individual allozygous. If individual K has genotype AA with two alleles IBD, we call this individual autozygous and $F_{kk} = 1$. The coefficient of consanguinity for J and K, $F_{JK} = 0.5$: with 50% probability we draw A from the first individual and we always draw A from the second. This coefficient of consanguinity is symmetrical: $F_{JK} = F_{KJ}$. Following the notation of Bulmer (1994) the coefficient of relatedness r between J and K is then defined as:

$$r_{JK} = F_{JK}/F_{JJ} \tag{15.8}$$

To see the rationale behind this definition, note that our text definition was that relatedness is the probability that individual K received the same allele as present in J by identical descent and passes it on to a gamete, *divided by the probability that J itself passes on the allele to a gamete*. This is reflected in Eq. (15.8). If we focus on individual J the chance that it passes on a randomly chosen allele A is F_{JJ}. The chance that if we pick an allele in J, we pick an identical allele from K is F_{JK}. The chance that the identical allele is present among the two alleles in K is F_{JK}/F_{KK}; this chance is twice as high if K has two non-identical alleles ($F_{KK} = 0.5$) as compared to two identical alleles ($F_{KK} = 1$). The chance that individual K passes on allele A to a gamete equals the chance that it has the allele multiplied by the chance K passes it on, which is $(F_{JK}/F_{KK})\, F_{KK} = F_{JK}$. So if we focus on allele A, then F_{JJ} copies are passed on to a gamete by individual J and F_{JK} copies by individual K. The factor $F_{JJ} + F_{JK}$ is proportional to $1 + F_{JK}/F_{JJ} = 1 + r$, showing the rationale behind Hamilton's inclusive fitness (Eq. 15.5) and the definition of relatedness (Eq. 15.8).

All this detail does not matter for outcrossers in a large population for which $F_{JJ} = F_{KK} = 0.5$ and there are simpler ways to estimate relatedness (see text).

Note the coefficient of relatedness is not symmetrical if the F_{JJ} and F_{KK} are different, which they are in our example (invidual J is AB and individual K is AA). For allozygous individual J, $F_{JJ} = 0.5$ and for autozygous individual K, $F_{KK} = 1$. The coefficients of relatedness are asymmetrical: $r_{JK} = 0.5/0.5 = 1$ and $r_{KJ} = 0.5/1 = 0.5$.

To calculate relatedness we have to distinguish between selfed and outcross seeds (embryos) and between allozygous and autozygous parent. All possibilities are written down in Table 15.1 with the corresponding coefficient of consanguinity

Table 15.1 *Coefficients of consanguinity* (F_{JK}) *when comparing focal* (J) *and other* (K) *seeds (embryos) on a partially selfing, cosexual plant*

Mother allozygous (*AB*)							
Focal seed J selfed		Other seeds K selfed				Other seeds K outcrossed	
	F_{JJ}	AA	AB	AB	BB	AY	BY
Possible types AA	1	1	0.5	0.5	0	0.5	0
AB	0.5	0.5	0.5	0.5	0.5	0.25	0.25
AB	0.5	0.5	0.5	0.5	0.5	0.25	0.25
BB	1	0	0.5	0.5	1	0	0.5
Focal seed J outcross		Other seeds K selfed				Other seeds K outcrossed	
	F_{JJ}	AA	AB	AB	BB	AY	BY
Possible types AX	0.5	0.5	0.25	0.25	0	$0.25 + \frac{1+F}{8n}$	$\frac{1+F}{8n}$
BX	0.5	0	0.25	0.25	0.5	$\frac{1+F}{8n}$	$0.25 + \frac{1+F}{8n}$
Mother autozygous (*AA*)							
Focal seed J selfed		Other seeds K selfed				Other seeds K outcrossed	
	F_{JJ}	AA				AY	
Possible type AA	1	1				0.5	
Focal seed J outcross		Other seeds K selfed				Other seeds K outcrossed	
	F_{JJ}	AA				AY	
Possible type AX	0.5	0.5				$0.25 + \frac{1+F}{8n}$	

Notes: F_{JJ} denotes the probability that in the embryo two randomly chosen alleles (with replacement) at the same locus are identical by descent (IBD). A and B are two alleles at the same locus, which are not IBD. X and Y denote alleles from an unknown sire, which have probability $(1+F)/n$ to be IBD. The parameter F is the inbreeding coefficient of the population and n is the number of competing fathers. See text for further explanation.

between two seeds of known genotype. The corresponding table of the coefficients of relatedness of J to other seeds K (r_{JK}) then follows by dividing each F_{JK} value by the F_{JJ} of the focal seed (Eq. 15.8). First we compare selfed and outcrossed seeds in an allozygous plant. There are four possibilities.

(1) For a selfed seed (embryo) on an allozygous mother average relatedness of the embryo to that in another selfed seed then follows from summing all 16 possible relations in the table. We thus need to compute all 16 coefficients of relatedness from Table 15.1. This yields four values for the top row $1/1 = 1$, $0.5/1 = 0.5$, $0.5/1 = 0.5$ and $0/1 = 0$. In the second and third row all eight values are $0.5/0.5 = 1$. For the fourth row we have $0/1 = 0$, $0.5/1 = 0.5$, $0.5/1 = 0.5$ and $1/1 = 1$. The coefficient of relatedness between two selfed seeds on an allozygous mother plant is found by averaging over these 16 values, which yields 0.75.

(2) Next calculate the relatedness of a focal selfed seed to an outcrossed seed on the same plant. The paternally derived allele of the outcrossed seed is denoted as Y in the table and since the mother and father are assumed to be unrelated, Y is not identical by descent to A or B. Average relatedness of the focal selfed seed to the outcrossed seed is now for the top and bottom row $0.5/1 = 0.5$ and $0/1 = 0$. For the second and third row it is $0.25/0.5 = 0.5$. Combining all possibilities gives $r = 0.375$.

(3) Now calculate the relatedness of a focal outcross seed to a selfed seed. The value of F_{JJ} of a focal outcross seed is always 0.5. The average relatedness of a focal outcross seed to a selfed seed is found by dividing all values in Table 15.1 by 0.5, which yields an average r of 0.50. Note that there is an asymmetry here as compared to point 2.

(4) When we compare two outcross seeds, the chance that we draw the identical maternal allele in both seeds is 0.25 if they have a maternal allele in common (compare AX and AY) and zero when the maternal allele is not shared (AX and BY). For randomly chosen outcrossed seeds on the same plant the probability of drawing the same maternal allele is then 0.125. Now if there are n fathers that sire seeds on one plant, the chance is $1/n$ that two seeds on one mother plant share the same father. The chance that the same allele is drawn from this one father is 0.5 when he is allozygous. The father could also be autozygous and the probability that this is the case is F. When we combine this, the probability of drawing the same allele from the other father is $(1 + F)/8n$. The coefficient of consanguinity is then $0.125 + (1 + F)/8n$ and the corresponding average relatedness between two outcross, allozygous seeds ($F_{JJ} = 0.5$) becomes $0.25 + (1 + F)/4n$.

For the second case, that the maternal parent is autozygous, the calculation is shorter. If the focal seed is selfed it has relatedness 1 to all other selfed seeds and 0.5 to all other outcross seeds. If the focal seed is outcrossed it is related to other selfed seeds with coefficient 1 and to other outcross seeds with coefficient $0.5 + (1 + F)/4n$.

All values for coefficients of relatedness are summarized in Table 15.2. Now we need to combine these values to calculate the average relatedness between seeds.

Table 15.2 *Coefficients of relatedness between seeds of a partially selfing plant. The average relatedness* r *can be split up in a matrilineal (*m*) and patrilineal component (*p*)*

	Other seed selfed			Other seed outcross		
	r	m	p	r	m	p
Maternal parent allozygous						
Focal seed selfed	0.75	0.75	0.75	0.375	0.375	0.375
Focal seed outcross	0.50	1	0	$0.25+(1+F)/4n$	0.5	$(1+F)/2n$
Maternal parent autozygous						
Focal seed selfed	1	1	1	0.5	0.5	0.5
Focal seed outcross	1	2	0	$0.5+(1+F)/4n$	1.0	$(1+F)/2n$

We need to give a weight to the probabilities at which all combinations in Table 15.2 occur. With probability S^2 the focal seed and the other seed are both selfed. With probability $S(1-S)$ the focal seed is selfed and the other seed is outcross. With probability $(1-S)S$ the focal seed is outcross and the other seed is selfed, and finally with probability $(1-S)^2$ both seeds are outcrossed. The average relatedness between seeds if the maternal parent is allozygous is then:

$$\begin{aligned} r_{\text{allo}} &= 0.75S^2 + 0.375S(1-S) + 0.5(1-S)S + 0.25(1-S)^2 + \frac{(1+F)(1-S)^2}{4n} \\ &= 0.25 + 0.375S + 0.125S^2 + \frac{(1+F)(1-S)^2}{4n} \end{aligned} \tag{15.9}$$

If the maternal parent is autozygous the average relatedness is:

$$\begin{aligned} r_{\text{auto}} &= S^2 + 0.5S(1-S) + (1-S)S + 0.5(1-S)^2 + \frac{(1+F)(1-S)^2}{4n} \\ &= 0.5 + 0.5S + \frac{(1+F)(1-S)^2}{4n} \end{aligned} \tag{15.10}$$

The inbreeding coefficient F denotes which fraction of the plants is autozygous. In equilibrium, in an infinitely large population with selfing rate S and random mating: $F = S/(2-S)$ (Hartl and Clark 1989). The mean relatedness between two randomly chosen seeds is then found by combining the probabilities F and $1-F$ with the corresponding relatedness coefficients from Eqs. (15.9) and (15.10):

$$r_{\text{average}} = Fr_{\text{auto}} + (1-F)r_{\text{allo}} \tag{15.11}$$

With some algebra we finally arrive at:

$$r_{\text{average}} = (0.5 + 0.5S)^2 + \frac{(1+F)(1-S)^2}{4n} = (0.5 + 0.5S)^2 + \frac{(1-S)^2}{2n(2-S)} \quad (15.12)$$

Relatedness between seeds declines with n, the number of fathers that sire seeds on one plant. If this number of fathers is large, as assumed in the population genetic model of Eqs. (15.1) to (15.4), the second term becomes zero and cancels. Average relatedness increases with increasing S from $r = 0.25 + 1/4n$ for $S = 0$ to $r = 1$ for $S = 1$.

We noted above that the chance that K has an allele present in J equals F_{JK}/F_{KK}. We can go through the whole table again to calculate this chance, which turns out to be r for two random seeds. This convenient result is due to the symmetry in the table.

As a final note it should be pointed out that, only under certain assumptions (Rousset 2002), r can be calculated empirically from molecular genetic data for two individuals without knowing the exact family relations (Box 17.1). Such methods are based on the fact that related individuals have more concrete alleles in common than two randomly chosen plants.

15.4 Haig's kinship theory of genomic imprinting

In Box 15.2 we calculated the coefficient of consanguinity between two seeds by picking a random allele in the focal seed and then estimating the probability that such an allele is also present in other seeds on the plant. You could apply the same argument to the paternally or maternally inherited allele. The relatedness of these alleles in the focal seed to alleles in other seeds is not the same. Compare, for instance, two outcross seeds on an allozygous mother plant. The chance that the other seed received the identical allele from the mother is 0.5. If you pick the paternally derived allele in one seed, the chance you pick the identical allele in another seed is $(1 + F)/4n$, which is close to 0 when seeds on one plant are sired by a great number of fathers so that n is large. Haig's (1997, 2000) original idea was to break up the coefficient of relatedness in a patrilineal and a matrilineal component. This is shown in Table 15.2 for the seed example.

For all trios it holds that the average relatedness r is the average of the matrilineal (m) and patrilineal component of relatedness (p) (Haig 1997). With full selfing the mother is autozygous and $r = m = p = 1$ but for any other selfing rate the average relatedness of the matrilineal allele to the other seeds is always larger than for the patrilineal allele: where the values differ in Table 15.2, m is always larger than p. While the coefficient of relatedness lies between 0 and 1 this is not necessarily true for its component m. Consider the autozygous mother plant AA. If we take the maternal allele (A) of a focal outcross seed

(AX), then the focal seed passes this on to the gametes with probability 0.5. Another seed on the same parent that is selfed has genotype AA and passes allele A on with probability 1. Once you have selected maternal allele A in the focal outcross seed, the probability that you draw the identical allele in the selfed seed is 1 and the corresponding estimate of relatedness is $1/0.5 = 2$. In other words, if allele A in an outcross seed on an autozygous plant sacrifices itself for the survival of a selfed seed, two copies of A, instead of one copy, are passed on, explaining the odd 2 in Table 15.2.

Because the maternal allele is more likely to be present in other seeds on the plant, the paternally derived allele in the offspring is selected to take more resource from the maternal parent than the maternally derived allele. However, an allele that is maternally derived in one generation may be paternally derived in the next, so how does this work? The answer to this question is a conditional strategy: take as much as you can when you are the paternally derived allele, but hold back when you are the maternally inherited allele. This phenomenon is called imprinting (Mochizuki *et al.* 1996). There is a difference between an allele (an imprint) if it passes through the pollen or through the female gametophyte. Paternally derived alleles that take more resource from the mother are switched on while the same allele is silenced when transmitted through the mother. Parent-specific gene expression (PSGE) is another term used in this context. The silencing of gene expression occurs through DNA methylation or is mediated through the chromatin (Alleman and Doctor 2000). Both processes are entirely reversible so that a gene that is silent in one generation may be active again in the next.

15.5 The endosperm

To see how the endosperm may modify our conclusions we must first have a clear understanding of how meiosis works in plants. Angiosperms have double fertilization. The pollen contains three haploid nuclei that all arise from mitotic division of the haploid microspore mother cell and therefore are genetically identical. One nucleus functions only during pollen tube growth. The second one fertilizes the egg nucleus, the third one fertilizes the central cell. The central cell is diploid, when it fuses with a pollen nucleus it forms a $3n$ endosperm, the so-called *Polygonum*-type endosperm. The haploid megaspore mother cell in the mother goes through three mitotic divisions, and from the eight nuclei that are formed two identical nuclei fuse to form the central cell. Consequently, if an embryo receives allele A from the mother and X from the father then the endosperm receives AA from the central cell in the mother and X from the father. These alleles A and X are identical to the alleles in the embryo. In calculating relatedness measures in the previous section, we drew random alleles at one locus from the genome of two individuals. Because the two maternal alleles in the endosperm are identical we run into trouble when applying Hamilton's r, which is based on drawing random alleles, to endosperm genotypes (Lloyd 1992).

Does it then matter if seed size is under control of the endosperm or of the embryo? It does if effects of alleles are additive. You may want to check for yourself by modifying the model in Box 15.1 for endosperm control. With two alleles b_1 and b_2 corresponding with seed sizes m_1 and m_2 there are four possible endosperm genotypes and four corresponding seed sizes in the model: endosperm $b_1b_1b_1$ with seed size m_1, endosperm $b_2b_2b_2$ with seed size m_2, endosperm $b_1b_1b_2$ with seed size $(2m_1 + m_2)/3$ and endosperm $b_2b_2b_1$ with seed size $(m_1 + 2m_2)/3$. The effect of these complications is that the genotype of the mother is more important for seed size and this selects for smaller seeds, closer to the Smith–Fretwell size. Härdling and Nilsson (1999) argued that stronger maternal control is one of the advantages of having endosperm in the first place.

15.6 Model predictions

Dr Pangloss would have said that each plant has the best seed size for the environment in which it lives. In calculating the different optima for genes that are expressed in different life phases and tissues and not a single optimal seed size, we have taken quite a different route. But what does the theory actually predict and can we compare it with observations?

First, if the offspring has some control the predicted seed sizes in the Smith–Fretwell model should be typically much too low.

Second, if the offspring has some control over seed size and the realized seed size is a compromise between parent and offspring, then seeds from plants that are normally outcrossing should be larger than seeds from selfing plants. In many species this seems to be the case. To give just one example, *Arabidopsis thaliana* (length of seed 0.3–0.5 mm) and *A. suecica* (Swedish thale cress, a hybrid species between *A. thaliana* and *A. arenosa*, length of seed 0.5–0.8 mm) are selfers. The sister species *A. arenosa* (seed 0.8–1.2 mm), *A. halleri* (0.8–1.2 mm) and *A. lyrata* (lyrate rockcress; 0.8–1.2 mm) are all outcrossing (data from Al-Shehbaz and O'Kane 2002 and Mitchell-Olds 2001). The lower seed sizes in the selfers are consistent with our prediction, but of course many more species pairs need to be compared before we can reach any definite conclusion.

Third, because the paternal allele is less related to other seeds on the plants, it should take more resource from the mother than the maternal allele. This prediction was tested by making crosses between plants with different ploidy levels, varying the number of paternal and maternal genomes in the seed (Haig

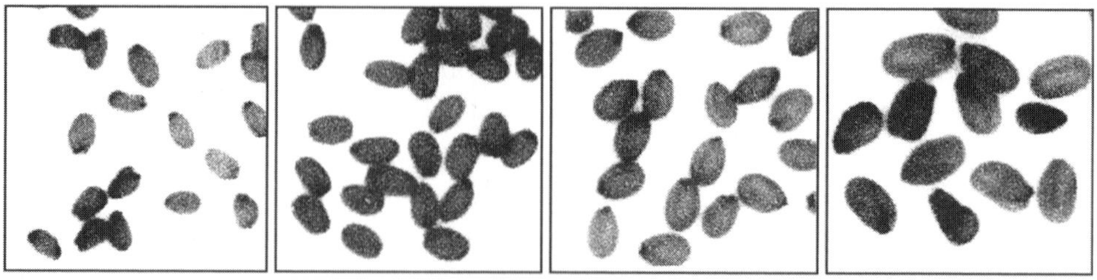

Fig.15.3 By reciprocally crossing *Arabidopsis thaliana* plants with different ploidy levels ($2n$ or $4n$) one can vary the relative contribution of the maternal and paternal parent to the endosperm or embryo. If the father contributes $2n$ chromosomes to the embryo and the mother n, seeds become large and if the mother contributes $2n$ and the father n, seeds remain small. Redrawn with permission from Scott *et al.* (1998), *Development*.

and Westoby 1991; Alleman and Doctor 2000). For instance in *A. thaliana* $2n$ lines were crossed with $4n$ lines (Fig. 15.3). A higher ploidy level leads to bigger seeds (compare the two frames in the middle). A much larger effect is however due to the parent of origin. From seeds with a $3n$ embryo, those which received a double dose of chromosomes from the father (right) were much larger than those which received a double dose from the mother (left) (Scott *et al.* 1998). This is consistent with Haig's kinship theory and the mechanism is considered to be genomic imprinting (Alleman and Doctor 2000). Such experiments suggest that endosperm or embryo genotype have a large influence on seed size so that the models in this section make sense; we cannot simply say that the mother is in full control. Nevertheless the degree of control by mother and offspring remains an open question for both ecologists and plant physiologists.

15.7 Sporophytic–gametophytic conflict

Lloyd (1992) stated: 'Self-incompatibility is tantamount to a suicide–infanticide pact.' He went on to show that selection on rejection genes, expressed in style or stigma, is different from selection on genes expressed in the pollen, that signal pollen identity and thereby seal their own fate. When a gene in the style can allow or block the germination of *all* self pollen, selfing or outcrossing is selected depending on the level of inbreeding depression. Inbreeding depression of 0.5 is the critical level (Chapter 11).

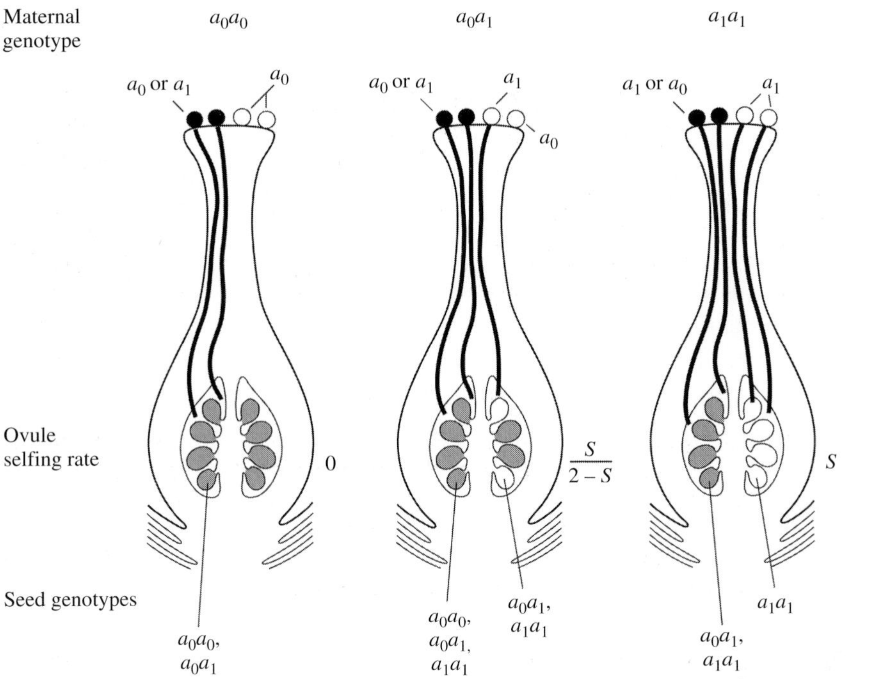

Fig.15.4 Conflict between genes expressed in the sporophyte (the mother plant) and the gametophyte (pollen). Imagine there exists gametophytic self-incompatibility that normally allows all outcross pollen to germinate but prevents germination of self pollen on the stigma. On top of that a different gene is expressed in the pollen that may prevent recognition of self pollen. Allele a_0 is common and does nothing; the plant recognizes self pollen, an SI reaction occurs on the stigma and the plant is SI (left). Pollen with the allele a_1 is not recognized and is able to fertilize ovules on the plant on which it was formed. The fraction of self pollen on the stigma is denoted as S. Selfed pollen and selfed seeds are blank (right-hand side).

With gametophytic self-incompatibility (gSI) the situation is, however, different. To show this it is easiest to start with a plant that already has a well-established gametophytic self-incompatibility system with many so-called S alleles (Richards 1997). Now suppose an entirely different gene at locus a (not an S allele) is expressed in the pollen (see also Porcher and Lande 2005). The common allele a_0 does nothing. Essentially a_0 is a suicide allele; the S allele in the pollen is recognized and the pollen is rejected. A new allele a_1 makes the recognition or the blocking of pollen germination by a stigmatic reaction more difficult. The proportion of self pollen on the stigma is S (Fig. 15.4). For a mother plant of common a_0a_0 genotype, no self pollen germinates and its selfing rate is then 0; this plant is SI. For a plant with genotype a_0a_1, half the self pollen grains (those with the a_1 allele) germinate on the stigma. All outcross pollen germinates provided they have a different S allele than the mother plant. This plant then has $0.5S$ self pollen on a total of $1 - 0.5S$ germinating pollen grains. Its new selfing rate is thus $0.5S/(1 - 0.5S) = S/(2 - S)$. From the selfed seeds 50% is of genotype a_1a_1 and 50% is genotype a_0a_1. If the mother is homozygous for a_1, all self pollen of genotype a_1 germinates and the selfing rate is S; all selfed seeds have genotype a_1a_1.

Can the rare a_1 allele, which merely sabotages the maternal recognition system, invade into the population? The calculation is similar to the population genetic model of seed size in Box 15.1 in the beginning of this chapter. It is actually simpler. Let f_1 be the frequency of a_0a_0 plants, f_2 the frequency of heterozygote a_0a_1 plants and f_3 the frequency of a_1a_1 plants. The frequency of the a_0 allele in the pollen is p_{a_0} and the frequency of the a_1 allele in pollen is p_{a_1}: $p_{a_0} = f_1 + 0.5f_2$ and $p_{a_1} = 1 - p_{a_0}$. The frequencies in the next generation are denoted by the primes. The seed genotypes are outlined in Fig. 15.4. We then have:

$$a_0a_0\colon\ f_1' = f_1p_{a_0} + f_2 0.5p_{a_0}\left(1 - \frac{S}{2 - S}\right) = U_1 \tag{15.13a}$$

$$a_0a_1\colon\ f_2' = f_2 0.5\left(\frac{S}{2 - S}\right)(1 - \delta) + f_2 0.5\left(1 - \frac{S}{2 - S}\right) + f_3(1 - S)p_{a_0} + f_1p_{a_1} = U_2 \tag{15.13b}$$

$$a_1a_1\colon\ f_3' = f_3S(1 - \delta) + f_2 0.5\left(\frac{S}{2 - S}\right)(1 - \delta) + f_3(1 - S)p_{a_1} + f_2 0.5\left(1 - \frac{S}{2 - S}\right)p_{a_1} = U_3 \tag{15.13c}$$

For calculating the frequencies in the next year we still have to divide the new frequencies by $U_1 + U_2 + U_3$ so that they sum to 1. Using matrix algebra Charlesworth and Charlesworth (1979a) solved mathematically when a rare allele can invade into the population. We can also write a simple computer program that calculates the frequencies of the genotype over a number of generations. This program comes up with the result that if inbreeding depression δ exceeds approximately 0.64 (the exact value increases with S from 0.618 when $S = 0$ to 0.667 when $S = 1$), the suicide allele a_0 wins from the allele a_1. At lower levels of inbreeding depression the sabotage allele a_1 can invade. When $0.5 < \delta < 0.64$ conflict exists between the sporophytic mother and the gametophytic pollen. Mothers are selected to recognize all their own offspring and exclude them from siring ovules if inbreeding depression is higher than 0.5. When inbreeding depression is not too high, self pollen is selected to sabotage recognition and to self-fertilize ovules.

15.8 Possible examples of conflict

The evidence for gametophyte–sporophyte conflict is at best circumstantial. If the interests of gametophyte and sporophyte coincide you would expect honest signalling by the offspring and recognition systems that work in a simple and effective way. If the interests do not coincide you would expect an evolving system that never works perfectly for one of the parties. The sporophyte may have the upper hand at one time, but the gametophyte at another. Such labile systems seem to be more the rule than the exception.

For instance, the genes for gametophytic SI systems are ancient and are considered to be much older than the origin of the species in which they reside (Richman and Kohn 1996). These genes are well studied and characterized by molecular biologists. However, Stone (2002) noted that modifier loci are also abundant but that virtually nothing is known about identity and function.

Richards (1997) detailed the sporophytic SI systems of 52 *Primula* species. There was no simple SI system; failure of pollen germination, failure to penetrate the stigma and failure of the pollen tube to grow down the style all contribute to SI, depending on the species. Richards (1997) stated that a small amount of selfing is usually possible in most *Primula* species.

Finally, in *Brassica* the molecular details of sporophytic SI have been studied (Hiscock and Tabah 2003). An SI reaction on the stigma blocks self pollen. It has been shown that the proteins that are involved in the stigmatic rejection are present only in the pollen coat. It is significant then that the pollen coat is of maternal origin. This allows the mother full control and leaves no opportunity for the gametophyte to sabotage the recognition system imposed by the sporophyte.

To sum up: nuclear genes may be in conflict, depending on the stage in which they are expressed. Such conflict only becomes apparent if we examine explicit genetic models. In a conflict situation there is not a single optimum and genes with different effects may be involved in a tug-of-war. To predict the outcome of such a tug-of-war it is important to know which genes are in control and in a few cases this is known. We detailed conflict over germination fraction, seed size and self-compatibility, but this list is far from complete.

Questions

(1) Over what sort of characters do you expect parent–offspring conflict?

(2) Would you expect parent–offspring conflict over generation time in monocarps?

(3) Should a seed, if it had the possibility, always take as much resource as possible from the parent?
(4) Why are paternally derived alleles in the offspring selected to take more resource from the mother plant than maternally derived alleles?
(5) What does the coefficient of relatedness indicate?
(6) Under what conditions will signalling of identity by the pollen to the sporophyte (Lloyd's infanticide–suicide pact) evolve?
(7) Shouldn't pollen always try to germinate, even at very high level of inbreeding depression?
(8) What is imprinting and when is it favoured by selection?

16

Intragenomic conflict

Cell organelles like mitochondria and chloroplasts contain DNA. This is a remnant from the time, some 2 billion years ago, that bacteria fused to form the first unicellular algae, ancestors of all higher plants. It is now a generally accepted theory that the nucleus in each cell is derived from a spirochaete, the mitochondrion from a purple bacterium and the chloroplast from a cyanobacterium (Margulis and Sagan 2002; Willis and MacElwain 2002). Although there has been transfer of many organelle genes to the nucleus, these cell organelles still contain their own DNA. There is a great degree of cooperation between organelles. Groups of genes, so-called coreplicons (Cosmides and Tooby 1981), have similar interests. However, between coreplicons interests may diverge, depending on the way the DNA is inherited. Cytoplasmic genes (mitochondrion, chloroplast) are usually transmitted through the seeds, while nuclear genes can be passed on through both pollen and seeds (Mogensen 1996). Most cytoplasmic genes are transmitted to the next generation when seeds are formed, while the transmission of nuclear genes depends on both seeds and pollen. In this chapter we first illustrate, at length, the conflict between cytoplasmic and nuclear genes in hermaphrodite plants. This conflict is considered to be of major importance for understanding the evolution of gynodioecy. Second, we consider the same conflict in dioecious plants. Third, we consider meiotic drive, the situation where one of the chromosomes is over-represented in the gametes at the expense of the homologous chromosome.

16.1 Gynodioecy

Gynodioecy, the occurrence of cosexuals and females (= male-steriles, MS) within the same population, is a prime example of intragenomic conflict. In MS plants male function is disrupted (Fig. 16.1) and, because of reallocation of some of the resources (compensation), usually more seeds are produced. The disruption of

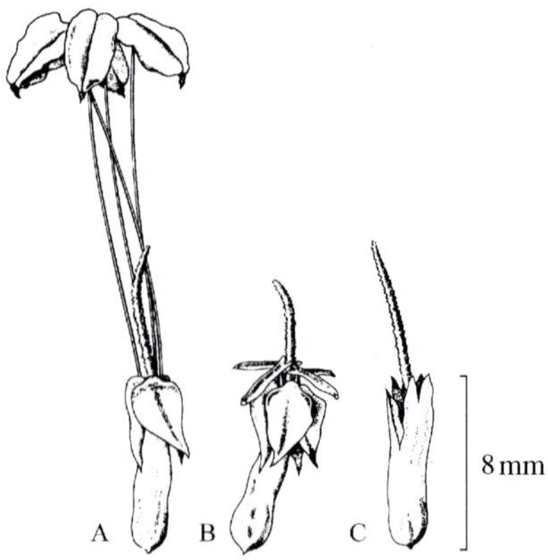

Fig. 16.1 Normal hermaphrodite flowers (A) and two types of male-steriles (B, MS1 and C, MS2) that can be distinguished visually in the field in *Plantago lanceolata* (van Damme 1983).

male function can occur in an early stage or in a late stage. *Echium vulgare* is an example of the latter; in MS plants the anther and filament are formed but the yellow pollen grains are empty and not viable (Klinkhamer *et al.* 1991).

Male sterility is often caused by a cytoplasmic factor. Cytoplasmic male sterility (CMS) has been demonstrated in a number of well-studied plant species in nature, like *Plantago lanceolata*, *P. coronopus*, *Thymus vulgaris* and *Beta vulgaris* and in crop species like *Zea mays*. The paper by Koelewijn and van Damme (1995) is a good example of the genetic crossing scheme necessary to demonstrate cytoplasmic inheritance and Kaul (1988), Samitou-Laprade *et al.* (1994) and Frank and Barr (2001) have reviewed CMS. The mechanism behind CMS has been studied in detail by plant physiologists; in all known cases sterility factors were located in the mitochondrion. Occasionally, new mitochondrial genes arise through recombination. These new genes are open reading frames that code for novel polypeptides and proteins, which may disrupt pollen production (Samitou-Laprade *et al.* 1994; Budar *et al.* 2003).

A CMS factor that would lead to more seeds, at the cost of pollen production, would transmit more copies of itself to the next generation than a neutral cytoplasm. Remember *Plantago lanceolata*, which was self-incompatible. Females produced more and heavier seeds, and 60% was the highest estimate of the relative advantage of females over hermaphrodites (Kuiper and Bos 1992). The nuclear model of gynodioecy (Chapter 8) predicted that females would rapidly be eliminated from the population. With CMS the maintenance of females is easily explained; the MS cytoplasm in *P. lanceolata* passes on 60% more copies of itself than a neutral cytoplasm. On the contrary, it is the maintenance of hermaphrodites that requires explanation.

16.2 Epidemics of females

When a female with CMS produces more seeds than a hermaphrodite plant with a neutral cytoplasm, the MS cytotype will increase in frequency. What factors can stop this ever-increasing number of females in plant populations?

First, Lewis (1942) suggested that if females increased in frequency, then at some point there would be so few pollen-producing hermaphrodites left that the seed production of the females becomes pollen limited. Self-compatible hermaphrodites have reproductive assurance through self-pollination and will have less of a problem with pollination. Once hermaphrodites and females produce equal numbers of seeds of the same quality there exists equilibrium; the same number of copies of the neutral and of the MS cytotype is transmitted. While there are some reports that seed production in isolated females suffers when they are too far away from pollen-producing hermaphrodite neighbours (for instance, Widen and Widen 1990), there are surprisingly few data that support Lewis's (1942) idea of an equilibrium under pollen limitation. Females systematically produced more seeds than hermaphrodites in 32 species examined by Shykoff *et al.* (2003) (see also Fig. 8.10). Also Lewis (1942) does not apply to self-incompatible species like *P. lanceolata*. There must exist another mechanism than pollen limitation that prevents the further increase of females.

Second, nuclear genes can restore the male function. In a female-biased population, the pollen of plants with nuclear restorer genes will be successful on many female plants where they produce male-fertile offspring. As a result the restorer allele will rapidly increase in frequency. These nuclear genes, dominant or recessive, are known to work specifically against certain CMS types (Samitou-Laprade *et al.* 1994). In *P. lanceolata* at least two cytotypes, MS1 and MS2, usually coexist within the same population (see also de Haan *et al.* 1997). Separate nuclear genes restore MS1 and MS2, say at locus *A* and *B*. Genotypes MS1/*AA* and MS1/*Aa* are then hermaphrodite and MS1/*aa* is female, regardless of the genotype at the *B* locus. Genotype MS2/*BB* or MS2/*Bb* is hermaphrodite and MS2/*bb* is female, regardless of the genotype at the *A* locus. Why do several CMS types coexist in *Plantago lanceolata*? It was suggested that there were costs to having a useless restorer gene against a cytoplasm that is not present in the plant. For instance, having *BB* (or *Bb*) is useless in a plant with the MS1 cytotype and there may be costs when such a genotype performs worse than the genotype MS1/*bb*. With high costs of restoration, at least 30%, Gouyon *et al.* (1991) calculated that the two cytotypes cycle and occur in dynamic equilibrium (Gouyon *et al.* 1991). However, de Haan *et al.* (1997) found that costs of restoration are much lower (0–13.4%). With low costs the models predict that both restorer genes go to fixation, eliminating male sterility from the population. The coexistence of several CMS types in *P. lanceolata* populations therefore remains unexplained with these models of a single well-mixed population. Several authors (for instance, Couvet *et al.* 1998) have suggested that maintenance of such a nucleocytoplasmic polymorphism can be more easily explained in a metapopulation context. In the above example, with small costs of restoration allele *B* will slowly disappear from a local population of 100% CMS1, which can then be invaded again by CMS2 from a neighbouring local population, after which *B* increases in frequency and so on. Such models would have to account for the fact that restorer genes are transported first by pollen and then by seeds, and are therefore better mixed in the metapopulation than cytoplasmic genes which are transported by seeds only (Murayama *et al.* 2004). While metapopulation models may indeed explain the maintenance of genetic variation in these systems, it remains a puzzle why species like *P. lanceolata* show such variation and other species do not.

Frank (1989) studied a computer model in which plants have several CMS types and matching restorer alleles. This model showed that the population dynamics of MS is highly unpredictable: after a period of little change, unrestored cytotypes may suddenly sweep through the population. Such sudden epidemics of females may well occur in the field. The *Plantago* studies showed that MS plants grow typically in clumps that expand in size (Kuiper and Bos 1992). However, the scale at which these female epidemics occur still needs to be substantiated by long-term demographic studies.

Box 16.1 Restorer alleles in *Plantago coronopus*

To illustrate the concept of restorer alleles it is instructive to examine the crosses that Jos van Damme (1991) made with *P. coronopus*. Male-sterile plants in this species have small brown anthers (Fig. 8.8) that resemble the MS1 type in *P. lanceolata*. The purpose of this study was to examine whether a single dominant restorer allele was present. Van Damme started his experiments with a single male-sterile plant that was open-pollinated. Since this plant is MS, the female parent itself had no dominant restorer allele; the hypothetical nuclear genotype of the plant is *aa*. The pollen coming from outside may or may not contain the restorer allele *A*. If the pollen contains *A*, seeds grow up to become a hermaphrodite with genotype *Aa*. If the pollen contains *a*, the seeds produce MS plants (*aa*). Van Damme found both types in the first generation. Then in experiment 1 the MS plants (*aa*) of the first generation were crossed with the first-generation hermaphrodites (*Aa*). The expectation was that this would produce 50% MS (*aa*) and 50% hermaphrodite (*Aa*) offspring. As Fig. 16.2

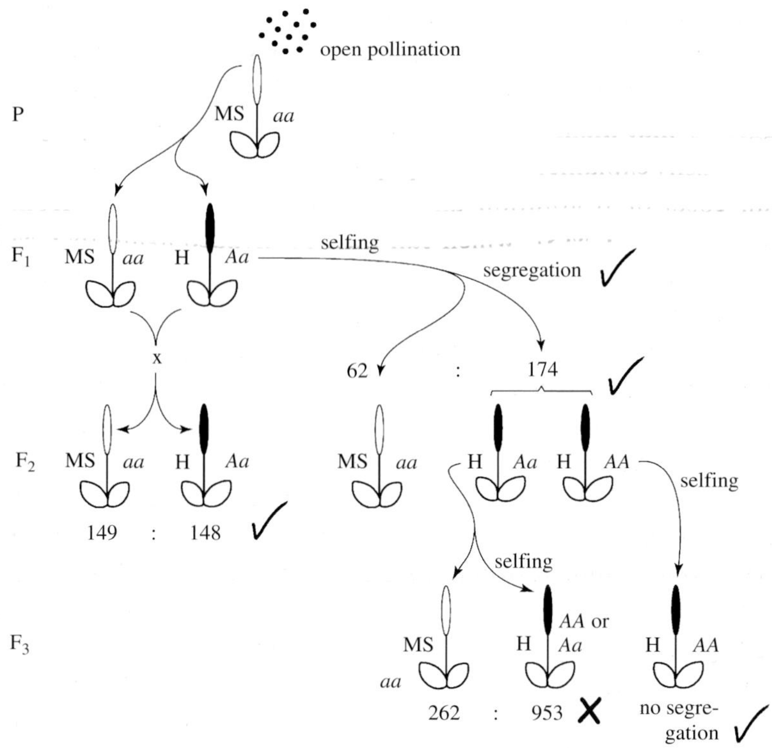

Fig. 16.2 Starting with one male-sterile plant of *Plantago coronopus* that was open-pollinated, van Damme (1991) tested the hypothesis that male fertility was restored by a single dominant nuclear allele *A*. H, hermaphrodite; MS, male-sterile. The tick marks (✓) indicate results consistent with the hypothesis of one restorer gene with two alleles, the cross (×) indicates a result inconsistent with the hypothesis.

shows the results were close to this expectation. In a second experiment the hermaphrodite plants from the first generation (genotype *Aa*) were selfed. With Mendelian inheritance this should yield one MS plant (*aa*) to three hermaphrodites (genotype either *Aa* or *AA*). Again the results were close to the theoretical ratio. In a third experiment van Damme then self-pollinated these second-generation hermaphrodites. In one-third of the hermaphrodite mother plants segregation among the seeds is not expected: we expect that these are the mother plants with the *AA* genotype that produce 100% hermaphrodite offspring. In two-thirds of the mother plants (genotype *Aa*) segregation is expected. In fact 24 families segregated and 14 did not, close to the prediction. The offspring in the segregating families are again expected to have a ratio of one MS to three hermaphrodites. When pooling over different series in the experiment, the fraction of MS plants was actually 21.5%, significantly lower than the expected 25%.

The system is therefore more complicated than a single dominant restorer allele for MS. Van Damme (1991) proposed that an unplanned drought may have caused different mortality for the two types. Despite this complication these results show that his approach goes a long way to explaining the genetics of nuclear restoration of MS.

16.3 Calculation of the frequency of a restorer allele in a population with CMS

The simple model developed by Frank and Barr (2001) is very useful for understanding the dynamics of CMS. This model is still far removed from any real plant species with CMS; it does not include selfing and inbreeding depression, both common features of gynodioecious plant species (Chapter 8). Nevertheless it is a good starting-point.

Assume the population contains only one MS cytotype. Females have a relative fecundity of $1+k$ compared to cosexuals. The restorer gene has two alleles: the dominant allele A restores male function whereas a does not. We thus have three nuclear genotypes AA, Aa and aa with corresponding frequencies f_1, f_2 and f_3 in the population, respectively. Genotypes AA and Aa are hermaphrodite, genotype aa is female.

For simplicity, Frank and Barr (2001) assumed full outcrossing and assumed that seed set of females was assured (no pollen limitation). The frequency p of the A allele in the pollen (all pollen produced by plants of the AA genotype and half the pollen of the Aa genotype) is $p = (f_1 + 0.5f_2)/(f_1 + f_2)$. The frequency of the alternative a allele in pollen is then $1 - p$. The frequency q of the A allele in the ovules is $q = (f_1 + 0.5f_2)/[f_1 + f_2 + (1 + k)f_3]$ and because $f_1 + f_2 + f_3 = 1$ this reduces to $q = (f_1 + 0.5f_2)/(1 + kf_3)$. The frequency of the a allele in the ovules is $1 - q$. Now in the next generation (denoted by the prime) the frequencies are:

$$
\begin{aligned}
AA&: f'_1 = pq \\
Aa&: f'_2 = p(1-q) + (1-p)q \\
aa&: f'_3 = (1-p)(1-q)
\end{aligned}
\qquad (16.1)
$$

When frequencies no longer change $f'_1/f_1 = f'_2/f_2 = f'_3/f_3 = 1$. We can then solve the three equations (16.1) with only three unknown variables (f_1, f_2 and f_3). With some calculus we find that the equilibrium genotype frequencies of the two hermaphrodites are $f_1^* = [(k+1)/2k]\left(k - \sqrt{k^2-1}\right)$ and $f_2^* = [(k+1)/2k]\left(1 - k + \sqrt{k^2-1}\right)$, respectively. Most important is the equilibrium frequency f_3^* of the genotype with the unrestored MS cytotype which equals the frequency of females in the population:

$$aa\text{: } f_3^* = \text{frequency females} = (k-1)/2k \qquad (16.2)$$

To remain in the population, females have to produce at least twice as many seeds as hermaphrodites ($k > 1$). If not, the restorer allele A goes to fixation, rendering all plants hermaphroditic again. For self-incompatible *P. lanceolata*, for which the highest estimate was $k = 0.6$, one expects that the restorer gene spreads readily through the population, returning the species to hermaphroditism. Equation (16.2) matches the prediction from Chapter 8 exactly (Eq. 8.15). We have gained something, however, since we now have an explicit statement about the frequency of the restorer allele and the different genotypes.

16.4 The effect of CMS females on sex allocation in the hermaphrodite

In the phenotypic model of Chapter 8 the presence of females increased the number of seeds that a hermaphrodite plant sired and this drove selection towards relatively more male allocation. Does this result change with CMS? We can extend the Frank and Barr (2001) model to include sex allocation. Assume another gene with two alleles (r_1 dominant and r_2 recessive) for sex allocation; the gene is not linked to the nuclear restorer gene. The effect of this sex allocation gene could be to increase the number of anthers per flower, the size of the anthers or pollen grains per anther. Genotypes r_1r_1 and r_1r_2 have male allocation r_1 and consequently female allocation $1 - r_1$. Genotype r_2r_2 has male allocation r_2 and female allocation $1 - r_2$. With two genes, one controlling MS and the other controlling sex allocation, each with two alleles, we have nine possible genotypes (Table 16.1).

All genotypes (1 to 3) that lack the dominant restorer allele are female. They produce no pollen but their seed production is a factor $1 + k$ higher than the seed production of the hermaphrodite, which is proportional to $1 - r$. This is the standard assumption in all the models from Chapter 8, which we will later see is not satisfactory here. The genotypes (4 to 9) with a dominant restorer allele are hermaphrodite: they trade off pollen production against seed production. As in the Frank and Barr (2001) model from the previous paragraph, we may proceed by calculating the allele frequencies of A versus a and r_1 versus r_2 in pollen and in ovules. From the allele frequencies, we calculate the genotype frequencies in the next year. For instance, the new frequency of genotype 1 (aa, r_1r_1) is the chance that it is aa multiplied by the chance that it is r_1r_1. This equals the frequency of a in ovules, times the frequency of a in pollen, times the frequency of r_1 in ovules,

Table 16.1 *Nuclear genotypes in a population that is fixed for a single cytotype that causes male sterility*

Genotype	Restorer genotype	Sex allocation genotype	Frequency	Phenotype	Seed production	Pollen production
1	*aa*	r_1r_1	f_1	Female	$(1+k)(1-r_1)$	0
2	*aa*	r_1r_2	f_2	Female	$(1+k)(1-r_1)$	0
3	*aa*	r_2r_2	f_3	Female	$(1+k)(1-r_2)$	0
4	*Aa*	r_1r_1	f_4	Hermaphrodite	$1-r_1$	r_1
5	*Aa*	r_1r_2	f_5	Hermaphrodite	$1-r_1$	r_1
6	*Aa*	r_2r_2	f_6	Hermaphrodite	$1-r_2$	r_2
7	*AA*	r_1r_1	f_7	Hermaphrodite	$1-r_1$	r_1
8	*AA*	r_1r_2	f_8	Hermaphrodite	$1-r_1$	r_1
9	*AA*	r_2r_2	f_9	Hermaphrodite	$1-r_2$	r_2

Notes: The dominant allele *A* restores male function, whereas allele *a* does not. Alleles r_1 (dominant) and r_2 (recessive) denote relative male allocation (pollen production). Pollen production trades off against seed production: if a fraction r of the resources is allocated to pollen $1-r$ remains for seed production.

times the frequency of r_1 in pollen. The same routine holds for genotypes 2 to 9. The model is not inherently difficult; it is only a stiff amount of work to write all the recurrence equations and type them into the computer.

What is the result of this exercise? It turns out that under these assumptions equal allocation to pollen and seed, $r=0.5$, is an ESS! There is no effect of females on the sex allocation of hermaphrodites, contrary to the prediction of the phenotypic models. At first glance this result is surprising but on second thoughts it is clear how it arises. In the phenotypic model we counted the number of successful gametes and stopped there. However in this model, as soon as the gene for more male allocation occurs in combination with *aa* in an MS plant, it has a disadvantage. The r_1r_1 genotype may have more anthers per flower or bigger anthers with more pollen. If male function is then aborted at a late stage, only part of the investment costs can be recuperated. In that case it is a disadvantage if the plant had initiated the formation of many anthers or many pollen grains. Apparently in the ESS the advantage of producing more pollen in the hermaphrodite cancels out the disadvantage when the allele occurs in a female plant. The ESS stays where it was (Chapter 4), 50% allocation to male and 50% to female.

We can ask what happens with sex allocation of hermaphrodites if not one but more cytotypes are present within the same population (Maurice *et al.* 1993). Consider the Frank and Barr model above with one CMS type, CMS1, and a restorer gene that restores CMS1. Assume $k > 1$, so that females are being maintained (Eq. 16.2). Now a new CMS type, CMS2, arrives that is not restored. New females arrive in the population. Do these extra females lead to higher reproductive success through pollen and therefore to an increase in the frequency of hermaphrodites? No, they lead to a decrease! CMS1 is competing against CMS2 and the winner is the type that produces most seeds. CMS1 produces females and hermaphrodites and CMS2 produces only females. CMS2 is not 'wasting', from the viewpoint of the cytoplasm, any resource on male function. CMS2 pushes CMS1 out of the population and as CMS1 goes down the hermaphrodites go down also. Without the above model, one would be surprised to see this result.

We do not claim that this is what happens exactly in the real world. Our purpose is merely to illustrate that even the simple models with one or two cytotypes are quite tricky and the direction of selection on sex allocation genes is difficult to predict and sometimes counterintuitive. There is a small amount of data in support of our conclusion that females do not necessarily cause selection for maleness in hermaphrodites: in *Thymus vulgaris* hermaphrodites in populations that usually have a high female frequency (sometimes up to 90%) are not more male-biased than hermaphrodites in populations with low female frequencies (Thompson and Tarayre 2000). This observation is at odds with the phenotypic models from Chapter 8 but could fit in with the models in this chapter.

Another example in *Triticum aestivum* (wheat) varieties showed that, even with CMS, male allocation of hermaphrodites can rapidly change in response to selection for outcrossing (David and Pham 1993). Starting with partially selfing lines David and Pham (1993) set up experimental populations. In these populations they used CMS plants, which were recessive homozygotes for the restorer gene (*aa*). Upon natural pollination these plants produced seeds that became female (*aa*) or hermaphrodite (*Aa*), as in the *Plantago coronopus* example from Box 16.1. These seeds were sown and after growth and seed set, only the seeds from the MS plants were harvested and resown. Seeds of hermaphrodites were discarded so that hermaphrodites only passed on genes to the next generation by pollinating the MS plants. The design is partly similar to the scheme in Table 16.1, with the difference that hermaphrodites pass on genes through pollen and seeds in the scenario of Table 16.1 but not in David and Pham's (1993) set-up, in which wheat behaves as if it has suddenly become dioecious and is fully outcrossing. David and Pham (1993) documented that pollen production of the hermaphrodites greatly increased in a six-year period (Fig. 16.3), as expected. Data on relative sex allocation, involving both pollen and seed numbers and mass, are not available for

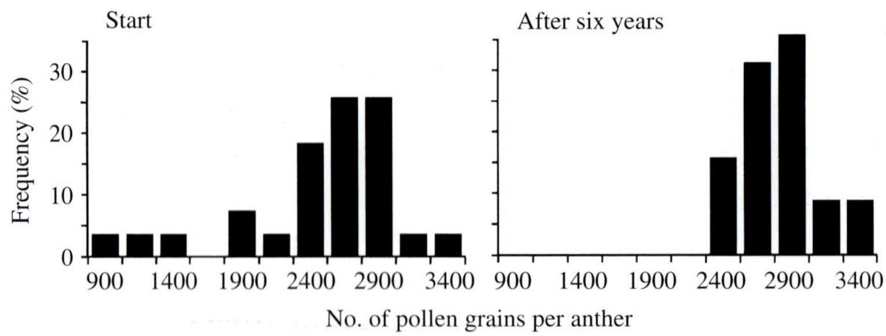

Fig. 16.3 Due to selection, pollen grains per anther increased when partially selfing lines of *Triticum aestivum* (wheat) (left) were exposed to six generations of outcrossing (right). Redrawn with permission from David and Pham (1993), *Journal of Evolutionary Biology*, Blackwell.

this study so it is not clear whether relative male allocation converged to a certain value.

16.5 A better way to model compensation

The way compensation is incorporated into the model is not satisfactory. We used the common assumption in the literature that hermaphrodites make $1-r$ seeds and females produce $(1+k)$ times the seed production of the hermaphrodites. It would seem more logical to assume that a plant could reallocate a fraction of its male investment to make extra seeds. In that case a female produces $1-r$ seeds plus she produces extra seeds due to compensation for aborting anthers and pollen, which is a fraction ν times r. When $\nu=0$ there is no compensation, $\nu=1$ means full resource compensation; $0<\nu<1$. We can equate the alternative ways of writing seed production of females, $(1-r)(1+k)=1-r+\nu r$, so that,

$$\nu=\frac{k(1-r)}{r} \quad or \quad k=\frac{\nu r}{1-r} \tag{16.3}$$

The parameter k is now under constraints and depends on the sex allocation of the plant; it is no longer a constant. If $r=0.5$ then with full compensation $k=1$, i.e. plants can at best double their seed production. One could modify the model to include selfing and inbreeding depression but we will not pursue this point any further. It would seem important to measure for a given species exactly how the compensation occurs and relate k to allocation, rather than consider it as a constant. Clearly much can still be done on this point, both with respect to the physiological background of compensation and the modelling aspects. For instance, simulations using the scheme of

Table 16.1 showed that when we change the assumption about compensation, genes for more male allocation can be selected for, even if they are not coupled to the restorer alleles (results not shown). The paper by Jacobs and Wade (2003) is an excellent example of how evolution in such systems can be further analysed.

16.6 Sex determination can evolve from CMS

We have assumed that the restorer genes are inherited independently of the sex allocation genes. This would be the case if the two genes were on different chromosomes. However, if a sex allocation gene is linked to a dominant restorer allele, it will only be expressed in hermaphrodites, not in MS plants. A gene for more male allocation then has no disadvantage, only a selective advantage and can spread (see also Charlesworth and Gutmann 1999; Jacobs and Wade 2003). One would then expect a rapid increase in frequency of male allocation genes when tightly linked to the restorer allele (Fig. 16.4). If a gene r_1, linked to A, turns the hermaphrodite (AA or Aa) into a male, which can only cross with females (aa), all offspring will be either Aa males or aa females. This is essentially a sex determination system with males as the heterogametic sex. This is an interesting finding since heterogametic males are most common in dioecious plant species and it is now well established that the Y chromosome contains a number of active genes that are not present on the X chromosome (Roldan and Gomendio 1999).

16.7 Intragenomic conflict in dioecious plants

We have sketched for gynodioecious plants the conflict between nuclear genes and cytoplasmic genes with uniparental inheritance. In dioecious plants this

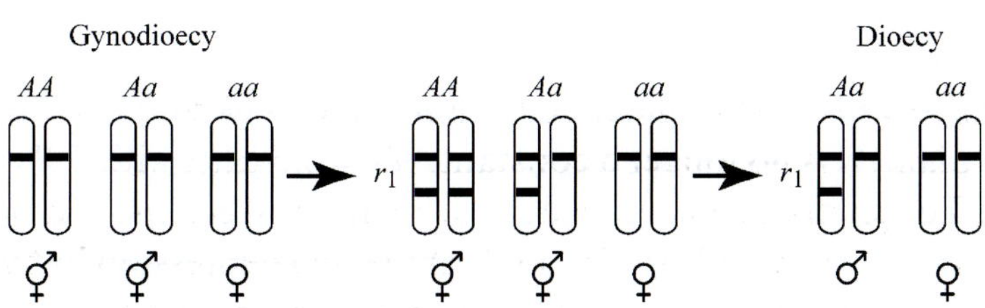

Fig. 16.4 Hypothetical evolution of a sex determination system from CMS. Once a gene for high male allocation becomes linked to the restorer allele A, it is only expressed in hermaphrodites and is selected. If gene r_1 fully masculinizes the plant, selfing is no longer possible and after a while the population will consist of heterogametic males and homogametic females.

conflict is more complex. Cosmides and Tooby (1981) made some predictions that are soundly based on theory but have hardly been tested in plants. They noted that the transition from cosexuality to dioecy transfers the locus of intragenomic conflict from the allocation of pollen versus seeds to the sex ratio of the seeds that females produce. If males do not pass on cytoplasmic genes, the production of males is a fitness loss from the point of view of cytoplasmic genes. Cytoplasmic genes in the female gametophyte may then be selected to kill androgenic pollen (pollen carrying a Y chromosome or, more generally, pollen with a male factor) or resist fertilization. Fertilization by androgenic sperm is genetic death for cytoplasmic genes in the ovule. Once the ovule is fertilized, the cytoplasmic genes in the developing embryo may be selected to cause it to degenerate if this frees resources that can be used by other developing seeds, some of which are females. Finally, a cytoplasmic gene in a male plant will be selected to kill the plant if this has even the slightest positive effect on its female kin. Cosmides and Tooby (1981) pointed out that in a dioecious plant species there is no selection at all on cytoplasmic genes in males to function properly. Whether the male is successful or dies before flowering, in both cases zero copies of the cytoplasm are transmitted. Cytoplasmic genes in males are then expected to behave in a suicidal way, killing themselves and their bearer, whenever this promotes the transmission of their kin through females. Nuclear genes are selected to stop these subversive actions.

As noted these ideas have hardly been tested but they could well apply. For instance, in *Rumex acetosa* (sorrel) males invest less in reproduction but are still less vigorous than females. In vegetation succession the females persist but the feeble males rapidly disappear from the vegetation (Korpelainen 1991). In general, male plants of dioecious species are much more susceptible to disease and herbivory (Ashman 2002). The hypothesis that males do poorly because of enhanced intragenomic conflict, leading to reduced resistance, is speculative but well worth considering.

It seems that CMS is unique to plants, but still this problem has a clear parallel in animals (see Hurst *et al.* 1996 and Frank 2000 for reviews). *Wolbachia* bacteria living inside insects are only transmitted through the female line; when these bacteria are in a male they are at a dead end. *Wolbachia* has been shown to manipulate the sex ratio of its host to its own advantage so that only daughters are born (Hurst *et al.* 1996). An analogy also exists with respect to endophytic fungi living in plants (Clay and Schardl 2002). Because sometimes the fungus is transmitted through seeds, but never through pollen, one could imagine the fungus promoting seed production at the cost of pollen production. We do not know if such examples exist.

Silene?

16.8 Mice, maize and meiotic drive

According to Mendel's laws each allele has a 50% chance of being represented in the offspring. Imagine a mouse species, that is monogamous; male and female stay together for life. An allele that increases its representation in the gametes by killing half of the other sperm cells with the other allele, would transmit twice as many copies as the alternative allele in the male mouse. The driving allele will rapidly replace the other allele in the population. This phenomenon is called meiotic drive. Drive also occurs in plants; for instance, in maize, knob-containing chromosomes are more likely to be incorporated in the egg cell than normal chromosomes (Hiatt and Dawe 2003). When pollen is well mixed in the population, the advantage for the driving allele is smaller: when you never compete with the other pollen grain of the same plant, there is no advantage of killing them. Only with local mate competition, when related pollen grains compete for access to ovules, may the selective advantage for the driving allele be substantial, although never as high as in the example of the monogamous mice. In the plant kingdom Mendel's laws usually apply, suggesting that meiotic drive is a rare phenomenon. On the other hand, the replacement of one allele by another driving allele will go so fast that it will be difficult to observe in nature and once the driving allele has gone to fixation, Mendelian inheritance prevails again. So perhaps drive is quite common but difficult to detect (Taylor and Ingvarsson 2003). Drive becomes apparent when it affects the X or Y chromosomes because it then leads to aberrant sex ratios. In principle, a driving Y chromosome could lead to a very high percentage of male offspring, followed by population extinction (Hamilton 1967). In nature the driving chromosome frequently has some other disadvantage that stops it from completely eliminating the other non-driving chromosome. *Silene latifolia* has been intensively studied in this respect. This is a dioecious species with sex chromosomes; the male is XY and the female XX. As in the best-studied animal systems, fruit flies and mice, drive seems to affect the male gametes. In crosses the genotype of especially the father was important for the sex ratio of the offspring and not so much that of the mother (Taylor 1994). Males with strongly female-biased offspring had a high percentage of aborted pollen (Fig. 16.5), suggesting that pollen carrying the Y chromosome was aborted. Taylor *et al.* (1999) found that the surviving pollen of males with female-biased offspring performed poorly in competition with pollen from plants with a 50% sex ratio. They produced fewer sons, as expected from Fig. 16.5, but they also produced fewer daughters. The pollen of driving males is thus generally of poor quality and this may help explain why the driving X chromosome fails to take over completely in *S. latifolia*. Such strong

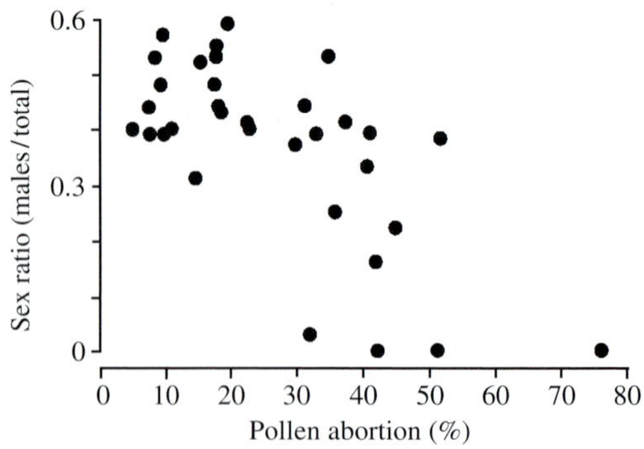

Fig. 16.5 In *Silene latifolia*, males that produced a female-biased sex ratio in their progeny have a higher fraction of unviable pollen. Presumably the ratio of X- to Y-bearing pollen is initially 1 : 1 but some of the Y-bearing pollen grains are killed during meiosis. Redrawn from Taylor and Ingvarsson (2003), *Genetica*, with kind permission of Springer Science and Business Media.

negative side-effects of segregation distorters have also been reported for fruit flies and mice (Taylor and Ingvarsson 2003).

To sum up: intragenomic conflict in plants involves several sets of genes with overlapping interests. We examined nuclear–cytoplasmic conflict over sex allocation in hermaphrodites, nuclear–cytoplasmic conflict over the seed sex ratio in dioecious plants and meiotic drive. Cosmides and Tooby (1981) remarked:

> the particular sequence of intragenomic events may explain major trait sets in a way that the simple appeal to ecological circumstances cannot. This gives an unstable, interactive, and historical character to the evolutionary process.

This is reflected in some of the models presented here. Nevertheless it seems possible to build predictive models that amend traditional sex allocation theory.

Questions

(1) What is the difference between parent–offspring conflict and intragenomic conflict?
(2) Which two factors are suggested to stop a cytoplasmic male-sterility factor from increasing in frequency?
(3) What is meiotic drive?
(4) A mutation in a chloroplast gene results in higher photosynthesis. Would you expect selection on this character in male plants?
(5) How were costs of restoration defined?

17
Group and kin selection

17.1 What is group selection?

Imagine a competition experiment with three genotypes of the same annual plant species, differing at a single locus. The genotypes are *AA*, *Aa* and *aa*, with allele *A* dominant over allele *a*. After each round of competition we harvest, mix and resow all the seeds. If plants of genotype *AA* and *Aa* produce more seeds than genotype *aa*, then allele *A* will increase in frequency. There are various means by which allele *A* can outperform *a*. Plants with this allele could simply convert their resources more efficiently into seeds. However, genotypes *AA* and *Aa* could also be less efficient than genotype *aa*, but still do relatively better due to a 'selfish' growth strategy based on interference with neighbours. Such interference strategies include producing a tall stem (Falster and Westoby 2003), side branches, a dense root system or even a toxin that poisons the neighbours. Making long stolons can also be considered as an interference strategy in clonal plants (Lenssen *et al.* 2005). In a mixture aggressive individuals could acquire resources at the expense of their neighbours. The downside of this behaviour is that the aggressive genotypes will do poorly when planted together. This is well known in agriculture. Goodnight and Stevens (1997) noted that plant breeders typically grow plants in monocultures of one genotype and select the group with the highest yield, rather than the high-yielding individuals from a mixed-genotype group. Seed yield of wheat can greatly be increased by dwarfing genes, which lower the centre of gravity of the plant and reduces the risk of lodging (Ellis *et al.* 2004). Of course, such genes have not been selected in this species in nature, since they would have reduced competitiveness against neighbours of the same or different species. Opposite to the 'selfish' strategy in which individuals increase their own success at the expense of others is the 'altruistic' strategy. By altruistic we mean that plants grow less aggressively, allowing more space for their neighbours, even at a cost for their own success.

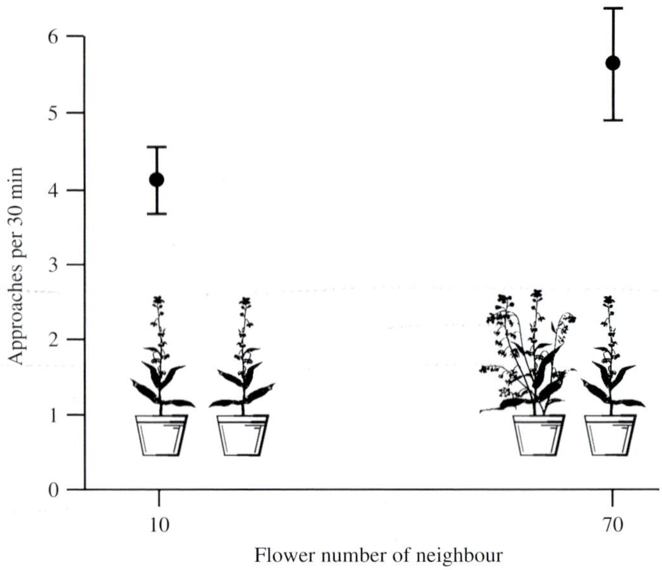

Fig. 17.1 During the 2004 field course in ecology at Leiden University students counted bumble-bee approaches to potted *Echium vulgare* plants with 10 flowers, immediately after placing them 50 cm from a similar plant with 10 flowers or 50 cm from a much larger plant with *c*.70 flowers. Plants next to large neighbours received more approaches ($n = 18$, $P = 0.05$ in a sign test).

Many characters place a genotype somewhere on a scale from selfish to altruistic, including some characters that you might not immediately think of. For instance, a genotype releasing volatiles after being damaged by herbivores sends out information that the neighbouring plants may use to raise their defence level. Such a strategy could be regarded as altruistic. When a plant strategy of producing many flowers attracts not only more pollinators to the plant itself but also to the neighbours, it could also be regarded as partly altruistic (Fig. 17.1).

Are selfish genotypes then always the winners? Not necessarily, when there is some grouping within the population. A well-known animal example given by Wilson (2002) illustrates this point. Some birds call out when they spot a predator and by doing so they attract attention to themselves. This increases the chance that they are themselves attacked, while lowering the chance that other group members fall prey. In a group of 20 birds, 10 callers and 10 non-callers, selection always works against calling. Now imagine that the group breaks up in two subgroups of 10 birds, with 1 caller and 9 non-callers in the first group and 9 callers and 1 non-caller in the second. In the first group the survival rate is 25% for callers and 50% for non-callers. Within this group there is selection against calling. Now examine the second group, with 9 callers and 1 non-caller. Because many birds are watching out for each other their survival is high, say 75% for the callers and 100% for the non-caller. Within this group also there is selection against calling. However, the second group with many callers fares better than the first group and sends out more offspring. The callers have 0.25 offspring in the first group and $9 \times 0.75 = 6.75$ in the second. This adds up to 7 offspring for the callers for the two groups together. The non-callers have $0.5 \times 9 = 4.5$ survivors in the first group and 1 survivor in the second. This adds up to 5.5 survivors for the two groups. Callers leave more offspring despite within-group selection against calling. We say that there exists group or interdemic selection for calling (deme = group). This selection counteracts within-group selection (intrademic selection) for non-calling. The effect is due to the assumed group structure. Because of this group structure, benefits of certain behaviour do not fall randomly on all genotypes in the population, as would be the case in a large unstructured, well-mixed population (the example we started off with). There may be two simple reasons why the groups are often not random draws from the population. First, groups may differ in their survival probabilities or the number of offspring they send out depending on their genotypic composition. This is the case in the bird example. It is also the case in, for instance, Charnov's barnacle model that we analysed in Chapter 4. Second, if the behaviour is heritable, young will resemble their parents. If they stay in each other's vicinity, this also results in a group structure. The benefits of altruistic behaviour then fall mostly on the relatives, the kin, and not randomly on all genotypes in the population. This form of group selection is called kin selection.

We now first sketch how group structure can affect selection for altruistic characters even if individuals are unrelated. Next we examine kin selection in a spatially structured population and review the evidence for kin selection. Finally we consider selection in metapopulations.

17.2 Altruism between unrelated plants

Imagine a desert in which water is a severe limiting factor for plant growth. One genotype uses water in the soil sparingly so that it can use the full growing season and produce many seeds. Another genotype from the same species takes up as much water as it can so that the supply in the soil is already depleted early in the growing season and few seeds can be produced. The first genotype can be considered as altruistic when the water spared is also available to neighbouring plants. The second genotype wastes water and is therefore selfish. In competition between two genotypes, the selfish plant wins. Now assume that in the field plants interact in pairs. If two selfish plants (players) grow together the basic yield Y for each plant is P seeds; P stands for

Table 17.1 *Pay-offs for plants playing an altruistic or selfish strategy against different opponents*

A			B		
	Opponent			**Opponent**	
Player	Altruist	Selfish	**Player**	Altruist	Selfish
Altruist	R	S	Altruist	$Y-c+b+s$	$Y-c$
Selfish	T	P	Selfish	$Y+b$	Y
C: $b=10,\ c=5,\ s=0,\ Y=0$			D: $b=10,\ c=5,\ s=10,\ Y=0$		
	Opponent			**Opponent**	
Player	Altruist	Selfish	**Player**	Altruist	Selfish
Altruist	5	−5	Altruist	15	−5
Selfish	10	0	Selfish	10	0

Notes: Synergistic effects that may occur when both players are altruistic are denoted as s. Costs of altruism are denoted as c and benefits from having an altruistic opponent as b. Y denotes the yield in a population of selfish plants.

punishment (Table 17.1A). If one player is selfish while the other is altruistic the former receives the high pay-off T, the so-called temptation to defect, while the latter receives the sucker's pay-off S. Assume that the cost of altruism is c and the benefit of having an altruistic neighbour is b. Then $T=Y+b$ and $S=Y-c$ seeds. Finally two altruistic plants have the reward R as pay-off; they each make $R=Y-c+b$ seeds. One could even imagine that altruism sometimes has an extra premium, in that case yield is $Y-c+b+s$, in which s is the so-called synergistic effect (Maynard Smith 1998). The pay-offs for the different strategies are given in Table 17.1. In board games like chess the two players can only divide a single point; such games are called zero-sum games. Soccer used to be a zero-sum game but after the rules were changed, so that a won game counted for 3 points and a draw only for 1 point, this was no longer true. Probably biological games are often not zero-sum, so that the entries in the matrix can take on different values. Nevertheless there is certain ranking in the pay-offs. With $s=0$ (Table 17.1C) the ranking is always $T>R>P>S$. This ranking of the pay-offs as in Table 17C is known in game theory as the prisoner's dilemma. Actually the prisoner's dilemma is not a dilemma at all; $T>R$ and $P>S$ so regardless of what your opponent does, being selfish is always best (Sigmund 1993). If $s>0$, i.e. with strong synergistic effects of altruistic behaviour, the reward R, that two players receive when they behave altruistically, increases and the ranking may change into $R>T>P>S$. This is known as the stag hunt game (Hauert 2002). The name refers to a hunt for a large prey with a high pay-off, such as a deer, which can only be caught if two hunters cooperate. If they do not cooperate they can only catch a small prey such as a rabbit with a low pay-off. There are many more possible rankings of the matrix elements (Hauert 2002), but those in Table 17.1, and in Table 17.2 that we deal with later in this chapter, seem particularly relevant for plants.

Can a rare altruistic mutant invade a large selfish population when plants are teamed up in groups of two? No, it cannot with the pay-offs in Table 17.1. When it is common the fitness of the selfish type is Y. When the mutant is rare it will always be teamed up with a selfish plant; it receives the sucker's pay-off, which is $Y - c$ and this is always smaller than Y.

Is there no possible equilibrium between altruistic and selfish plants in the population? Rather surprisingly an unstable equilibrium exists. If the altruistic type is more common, altruistic plants will sometimes grow together in pairs and reap the benefits of their behaviour.

Denote the frequency of altruists as p and that of selfish plants as $1 - p$. In a large population each plant has probability p of being teamed up with an altruist and $1 - p$ of being with a selfish plant. For simplicity set $Y = 0$. The fitness W of a player is the sum of the pay-off in its rows in Table 17.1B, averaged for the probability at which the situations occur. Fitness is then:

$$W_{\text{altruist}} = p(b - c + s) + (1 - p)(-c) = pb + ps - c \qquad (17.1a)$$

$$W_{\text{selfish}} = (1 - p)(0) + pb = pb \qquad (17.1b)$$

It follows that $W_{\text{altruist}} - W_{\text{selfish}} = ps - c$; without synergistic benefits ($s = 0$) the selfish type always wins. With synergistic benefits an equilibrium exists at $p = c/s$; this equilibrium is however unstable. When we start at a frequency below c/s, the selfish type wins and p goes to zero. When we start at a higher frequency the altruist wins and p goes to 1. So there exists a paradoxical situation. Mutants that are altruistic cannot invade a population when rare, but they take over when they are common enough. In this context population size could well be important. A rare mutant has no chance in a large well-mixed population, but if it is locally more common than frequency p and the population is not well mixed, it may locally increase. Also when many plants interact, one selfish plant may spoil it for everybody and pure altruistic groups will be rare. Altruism is then most likely to be selected in small groups.

17.3 The free rider

When small groups of flowering plants of *Echium vulgare* were more than 6 metres apart, bumble-bees discriminated between groups with low and high

Table 17.2 *(A) Bee approaches to plants per 5 minutes in* Echium vulgare, *for groups containing 6 high-nectar plants, 6 low-nectar plants and a mixed group with 3 low and 3 high nectar plants; (B) corresponding pay-offs if each bee approach contributes one unit to fitness and the extra nectar production costs* c

A			B		
	Opponent			**Opponent**	
Player	High nectar	Low nectar	**Player**	High nectar	Low nectar
High nectar	2.4	2.4	High nectar	$2.4 - c$	$2.4 - c$
Low nectar	2.4	1.2	Low nectar	2.4	1.2

Source: Klinkhamer *et al.* (2001).

nectar production rate (NPR) (Table 17.2; Fig. 14.3) (Klinkhamer *et al.* 2001). The bees made fewer approaches to the group with low nectar. However, bees approached the mixed group with some plants with low and others with high NPR, at about the same rate as the high nectar group and did not discriminate within the group (Table 17.2). Probably the bees learned where the good patches were, but once arrived at a patch they just went from flower to flower without discriminating between individual plants.

The pay-offs for the high-nectar plants and low-nectar plants are shown in Table 17.2B. In the mixed group the low-nectar plants were winning because they received visits without any costs in terms of nectar production. Their fitness was 2.4 whereas that of the type with high nectar production is $2.4 - c$. When rare, these free riders can always invade a population of plants with high nectar production. However, when the low-nectar plants are common the reverse is also true. If the population consists only of low-nectar plants, there are few bee visits and each plant has a gain of only 1.2 (Table 17.2B); the rare altruist then has a fitness of $2.4 - c$, which, for small values of costs c ($c < 1.2$), is larger than the fitness of the common low-nectar type. Note that the ranking of the pay-offs in the matrix has changed as compared to Table 17.1; now $T > R = S > P$. This frequency-dependent situation, which is called a hawk–dove game in animal ecology, results in a stable equilibrium in which low and high nectar-producing plants coexist. Free riders remain in the population but cannot take over. In section 17.5 we examine whether intermediate strategies can evolve.

17.4 Altruism with relatedness

Wilson (1987) constructed a chessboard model of the evolution of altruistic characters in plants. In his simulations pollen or seeds stayed close to home, so that there existed some spatial genetic structure in his model populations. Spatial genetic structure means that neighbours have a higher statistical probability of sharing the same alleles than two randomly selected plants (see Box 17.1, below). The benefits of an altruistic allele then go more to relatives that share this allele than to others in the population. We can analyse such problems by examining the inclusive fitness (Chapter 15) of a rare mutant in a large population. Even if the altruistic allele is rare in the population, the rare mutant will interact with some neighbours that carry the same altruistic allele (le Galliard *et al.* 2003). We will now use Hamilton's method to calculate how many allele copies are passed on by a focal plant and its neighbours (a neighbourhood), taking effects of genotypes on each other into account.

With a chessboard in mind we estimate the number of neighbours N as the number of squares surrounding the focal plant. In that case N equals 8, the number of squares covered by the king in one move. To calculate the seed production of a rare altruistic mutant we have to account for the costs and benefits of altruism. Costs of altruism (see Table 17.1) are again set equal to c and the benefits are b for each altruistic neighbour. We assume there is no synergism ($s = 0$). It is easiest to think of outcrossing haploid organisms, avoiding the complications of dominance, but the same principles apply to diploids. The first altruistic mutant that arrives in the population will be on its own and can only rely on luck to produce some offspring. However, in the next step offspring of the altruistic plant will be in the vicinity and will enjoy benefits from having an altruistic neighbour. Consider a focal plant in which the altruistic allele is present. If the average coefficient of relatedness between two neighbours is r, this implies for a haploid simply that there is a probability r that a neighbour shares this allele by identical descent. Therefore on average rN neighbours of a focal plant with a rare altruistic allele will also carry the same altruistic allele (see also Box 17.1, below). For instance, with average $r = 0.25$ and $N = 8$, each focal altruistic individual has, on average, two altruistic plants in its neighbourhood from which it benefits. Because the allele is still rare in the population, the probability that, by chance, one of the six unrelated neighbours also carries the altruistic allele is negligible. Just before reproduction the focal plant carries the altruistic allele and, on average, there are two additional copies present in the neighbours. In general $1 + rN$ allele copies are present in a neighbourhood of a plant with a rare allele when the new allele is still rare in the population. We assume that the selfish type produces 100 seeds. The model assumes that the altruist has a fixed strategy and that each neighbour benefits. It would be greatly beneficial if the plant were plastic and adapted a competitive strategy towards unrelated plants and an altruistic strategy towards relatives, but this is not how Wilson (1987) set up his model. The seed production of the few selfish plants near the altruist neighbourhood is higher than 100 seeds. The fitness of the selfish type over the whole population is however still close to 100 seeds, because the neighbourhood shown in Fig. 17.2 is only a small part of a large population of mostly selfish plants. If the altruistic allele is rare each altruistic plant has c costs and benefits from rN altruistic neighbours. Its fitness is then $100 - c + rNb$ seeds. The allele for altruism invades when $100 - c + rNb > 100$ or:

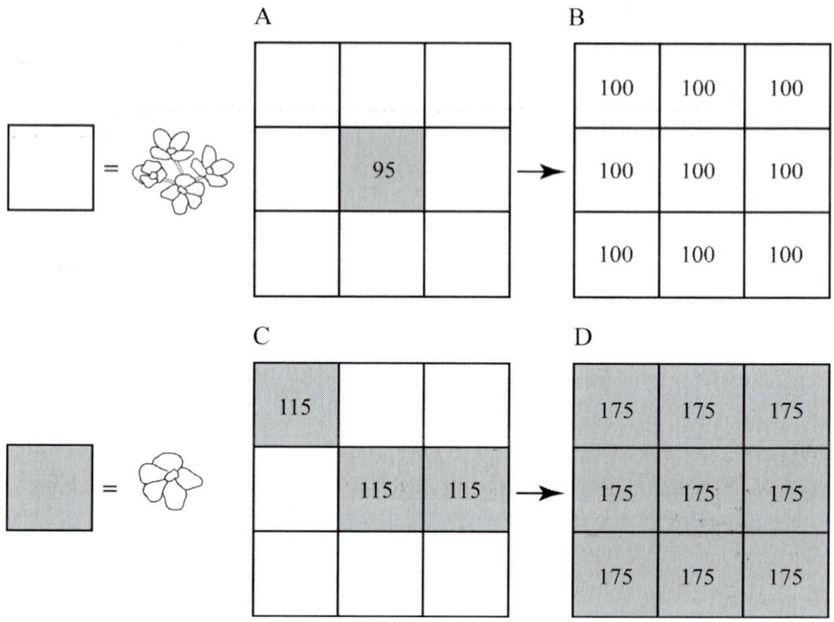

Fig. 17.2 Selfish plants (white squares) produce side branches that interfere with plants in the neighbouring square. Altruistic plants (shaded squares) have a compact growth form, which reduces seed production with 5 seeds but increases seed production of each neighbour by 10 seeds. In a monoculture of the selfish type each plant makes 100 seeds. (A) Neighbours are unrelated. An altruistic mutant only has costs and no benefits; it produces only 95 seeds and cannot invade. Seed production of the selfish plants next to the mutant is positively effected by the presence of the mutant (110 seeds), but since the mutant is rare and the figure shows only a small part of the population, the average seed production per selfish plant remains 100 seeds. (B) The altruist has disappeared. (C) Same situation as in A, but now average relatedness r between neighbours is fixed at 0.25 so that, even when the altruistic mutant is rare, it has, on average, two altruistic neighbours. Altruists can now invade; seed production per altruistic plant is $100 - 5 + 2 \times 10 = 115$. (D) The altruist has taken over. Each plant has 8 altruistic neighbours (only part of the population is shown) so that each altruistic plant produces $100 - 5 + 8 \times 10 = 175$ seeds. Productivity of the population has increased from 100 to 175.

$$\frac{b}{c} > \frac{1}{rN} \tag{17.2}$$

If a neighbourhood contains only two interacting individuals, there is only one opponent ($N = 1$) and Eq. (17.2) reduces to $b/c > 1/r$, which is a well-known inequality known as Hamilton's rule (Hamilton 1964a, b). The benefits-to-cost ratio of altruism in Eq. (17.2) has to exceed some critical value before altruistic behaviour is selected. The right-hand side of Eq. (17.2) decreases with r and with N; the more interacting neighbours there are and the stronger they are related, the sooner altruistic behaviour is favoured. Figure 17.2 shows a chess-board example ($N = 8$) with costs $c = 5$ and benefits $b = 10$ of altruistic behaviour. With these parameters altruistic behaviour is selected when $10/5 > 1/8r$ or $r > 1/16$. In the example with $r = 0.5$ this condition is easily met. The critical point is at $r = 1/16$, when, on average, an altruistic plant has 0.5 neighbours that also received the altruistic allele by identical descent. The conditions for the altruistic allele to increase in frequency are most difficult when it is rare, because the unrelated plants that receive benefits are all of the common, selfish type. When the altruistic allele is more common, complex patterns may arise and it may increase to 100% or coexist with the selfish allele. When each plant in the population is altruistic $100 - c + 8b$ seeds are produced (Fig. 17.2D). With $b = 10$ and $c = 5$ this amounts to 175 seeds. This is considerably more than the 100 seeds produced in a monoculture of the selfish type. Could this population be invaded by a free rider, a selfish mutant that reaps the benefits from the altruistic neighbours? In this scenario (Fig. 17.2C–D) it cannot, because with $r = 0.5$ a rare selfish mutant still has, on average, two selfish kin in its neighbourhood and only six altruists. Each of these rare selfish plants would then pass on 160 allele copies to the next generation, which is less than for the common altruist which passes on 175 copies. Only if the selfish plant acquired an additional mutation so that its seeds were further dispersed so that the level of relatedness between neighbours dropped to below 1/16 would it be able to invade in the altruistic population.

Rather than assigning values to relatedness as we did in Fig. 17.2, Wilson (1987) used simulations in which pollen and seeds are dispersed over different distances over the chessboard. From the simulation results Wilson suggested that the condition in Eq. (17.2) was unlikely to be met for an outcrossing annual plant. This is because the relatedness between neighbours, generated by the pollen and seed dispersal rules used, is small. For instance, if seeds disperse one square and pollen is mixed through the whole population the coefficient of relatedness r between neighbours is about 1/18 (you can check this by writing down all possibilities). Wilson considered only outcrossing annuals. For

Box 17.1 Spatial genetic structure and relatedness

Recently a number of papers have addressed fine-scale genetic structure in plant populations and how this can be used to make inferences about pollen and seed dispersal and the relatedness between neighbours (Vekemans and Hardy 2004). Geneticists typically measure if two individuals share a common allele a at the same locus, i.e. if the alleles are identical in state (IIS). If J and K are two haploid individuals and if J contains the allele a, the probability that we draw the same allele IBD from individual K is equal to the coefficient of relatedness r between J and K. Of course, K could also have received the a allele from another unrelated parent in the population, especially if the a allele is not rare but has a frequency p in the population and the alternative allele A has frequency $1-p$. The total probability that you draw from individual K the a allele is then $r+(1-r)p$. For small p this probability is r; IIS is then the same as IBD, i.e. if the allele a is rare, individuals J and K must have received this allele from a common ancestor. If we just pick a random allele from individual J, there is also a probability p that we draw the a allele. The probability Q_a that we draw the a allele from both individual J and K is then (Rousset 2002):

$$Q_a = rp + (1-r)p^2 \qquad (17.3)$$

Equation (17.3) allows an estimate of r from genetic similarity when allele frequency p in a reference population is known (Rousset 2002). However, Rousset (2002) stressed that estimation of IBD from genetic data from spatially structured populations is problematic and that it may often be unclear what constitutes the 'reference' population, necessary for estimating p. He suggested that it might be better to avoid the IBD concept altogether and focus on relatedness measures based on ratios of IIS instead. For more information, plus guidance on how to use this in the context of inclusive fitness, the interested reader is referred to the book of Rousset (2004) on this subject.

selfing or clonal plants relatedness between neighbours will be much higher and for such species the condition from Eq. (17.2) is more likely to be met, resulting in selection for altruistic, cooperative growth strategies.

17.5 Continuous versus discrete models

Game-theoretic models usually employ discrete strategies with altruistic versus selfish players. It is sometimes possible to write these models in continuous variables such as investment in an altruistic character (Day and Taylor 1998), so that we can answer the question of whether an intermediate strategy can evolve. For a continuous character like nectar production rate this is a more logical approach than the scheme in Table 17.2, with discrete characters. We will give a brief example. Suppose a focal genotype produces nectar, which is costly, but helps to attract pollinators to itself and also to the neighbours. To examine the magnitude of the neighbour effects, we could observe the number of pollinator approaches to plants and use multiple regression to examine to what extent this depends on nectar production of the focal plant itself and the nectar production of its neighbours (see also Heisler and Damuth 1987; Goodnight *et al.* 1992; Donohue 2003). Suppose a focal plant with nectar production I_1 has N_1 neighbours with the same nectar production I_1 and N_2 neighbours with nectar production I_2. Total number of neighbours (N) is: $N_1 + N_2$. The number of pollinator approaches to this focal plant could be approximated as:

$$\text{pollinator approaches} = I_1{}^{\alpha} + \beta N_1 I_1{}^{\alpha} + \beta N_2 I_2{}^{\alpha} \qquad (17.4)$$

Each plant with I_1 nectar draws $I_1{}^{\alpha}$ pollinators to itself and attracts $\beta I_1{}^{\alpha}$ pollinators extra for its neighbours. With $\beta = 0$ only the nectar of the focal plant itself matters for the number of bee approaches ($I_1{}^{\alpha}$) it receives. When $\beta > 0$ attractive neighbours induce more visits to the focal plant. This situation is similar to the one we found for flower number in *E. vulgare* (Fig. 17.1). When $\beta < 0$ attractive neighbours draw visitors away from the focal plant.

Now we calculate the inclusive fitness of a rare mutant with I_m nectar production in a population in which the majority of plants produces I nectar. We assume that the number of approaches is a good measure for fitness, for instance because seed production is strictly pollen limited, and set the basic fitness at 100 seeds. Without taking neighbour effects into account, the mutant receives $I_m{}^{\alpha}$ approaches due to its own nectar production and its absolute fitness is:

$$W_m = 100 + I_m{}^{\alpha} - cI_m \qquad (17.5)$$

Fitness is a balance between the benefits of producing nectar and the costs involved, as measured by the parameter c. Now we do take the effect of neighbours on visitation of the focal plant into account. The rare mutant has Nr neighbours of its own type and $N(1 - r)$ neighbours of the common type. The inclusive fitness of the mutant is:

$$W_m = 100 + I_m{}^{\alpha} - cI_m + \beta N r I_m{}^{\alpha} + \beta N(1 - r) I^{\alpha} \qquad (17.6)$$

The first three terms are identical to Eq. (17.5). The novel fourth and fifth terms denote the extra approaches received due to the nectar production of the neighbours. The fourth term refers to the related neighbours with nectar production I_m. The fifth term refers to unrelated neighbours which each produce I nectar when the mutant is still rare. In Eq. (17.6), W_m is the total seed output of the rare mutant with nectar I_m in a population of I-type plants. Note that Eq. 17.6 is similar to the familiar invasion fitness of the mutant (Chapter 4); the difference is only that we now take kin selection into account. As before, the candidate ESS can be calculated with Parker's method from Chapter 4, i.e. by differentiating mutant fitness (Eq. 17.6) with respect to the investment of the mutant (I_m) and setting the derivative to zero. This yields:

$$I^* = \left(\frac{c}{\alpha + \alpha\beta N r}\right)^{1/(\alpha-1)} \tag{17.7}$$

If the benefits curve is decelerating ($\alpha < 1$), the second derivative of mutant fitness is negative: I^* represents a fitness maximum, i.e. is an ESS. For instance, for $\alpha = 0.5$, Eq. (17.7) reduces to $I^* = [(1 + \beta N r)/2c]^2$. With costs $c = 0.1$ and no indirect fitness effects of neighbours ($\beta = 0$), the ESS is $I^* = 25$ (Fig. 17.3). This corresponds to individual selection. With $c = 0.1$ and $\beta N r = 0.2$, the ESS is $I^* = 36$ (Fig. 17.3). If higher nectar production affects visitation of neighbours positively and neighbours are related, this

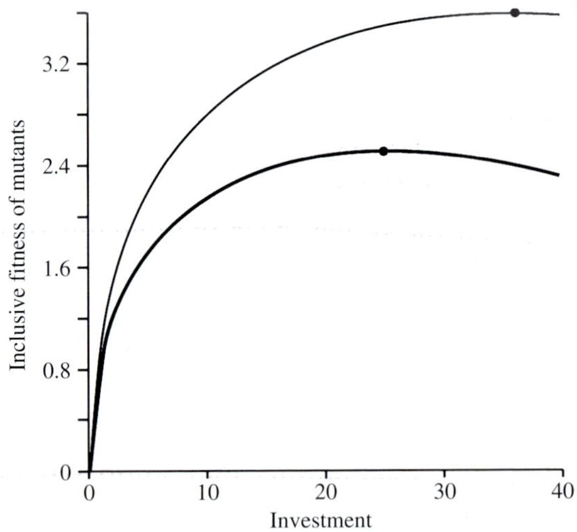

Fig. 17.3 Inclusive fitness of mutants with different nectar production rates when the common type plays the ESS (dot). The thick line indicates the situation in which neighbours are unrelated or do not affect each other (Eq. 17.5). The thin line indicates the situation with kin selection (Eq. 17.6 with $\beta N r = 0.2$; in calculations we used $c = 0.1$ and $\alpha = 0.5$). When related neighbours benefit from each other, the ESS for nectar production shifts towards higher values.

shifts the ESS for nectar production to higher values. If nectar production affects visitation of neighbours negatively ($\beta < 0$), this would select for lower ESS investment in nectar production (see also Leimar and Tuomi 1998).

The bottom line is that if neighbours do not affect each other or are unrelated, the ESS can be calculated from individual fitness (Eq. 17.5). If they do affect each other and are related, we must take the interactions into account and calculate inclusive fitness (Eq. 17.6). In this case the spatial genetic structure of the population (i.e. the relatedness between neighbouring individuals) determines what level of nectar production is selected for.

17.6 Kin selection: fact or fiction?

Griffin and West (2002) critically reviewed the rather extensive animal literature on kin selection. For plants only a handful of studies dealt with kin selection. This does not mean, however, that this process is less important in plants than in animals. Plants cannot stand up and walk away, so they interact with a small set of neighbours during their entire life. Especially for clonal plants, one would expect that the ability to avoid competition between relatives, both below and above ground, would be greatly beneficial (Holzapfel and Alpert 2003).

Two conditions must be met for kin selection to be important: reproductive success should depend not only on the phenotype of the focal plant but also on the phenotype of the neighbours and, in the natural situation in the field, these neighbours must be close relatives.

Donohue (2003) studied *Cakile edentula* (sea rocket). Seedlings of this plant grew in clumps on the beach, near the remains of the mother plant. Donohue collected seeds from different mother plants some 25 m apart. She then planted four seedlings from one sibling group, together with (1) their own sibs, (2) a sibling group from another mother or (3) a mixed group of seedling from various mother plants (Fig. 17.4). The four focal plants that grew with their own sibs produced the most fruits. Donohue (2003) suggested that plants had acquired adaptations that improved performance when plants competed with siblings, but not with non-siblings. Seedlings growing with their own sibs group were mostly of intermediate size and rarely depressed growth and reproduction of their neighbours. In the other groups a few plants dominated their neighbours. Because fruit production of the focal plants depended on the phenotype of neighbours (Fig. 17.4) and the clumps of competing seedlings were at least half sibs, conditions for kin selection were met in *C. edentula*.

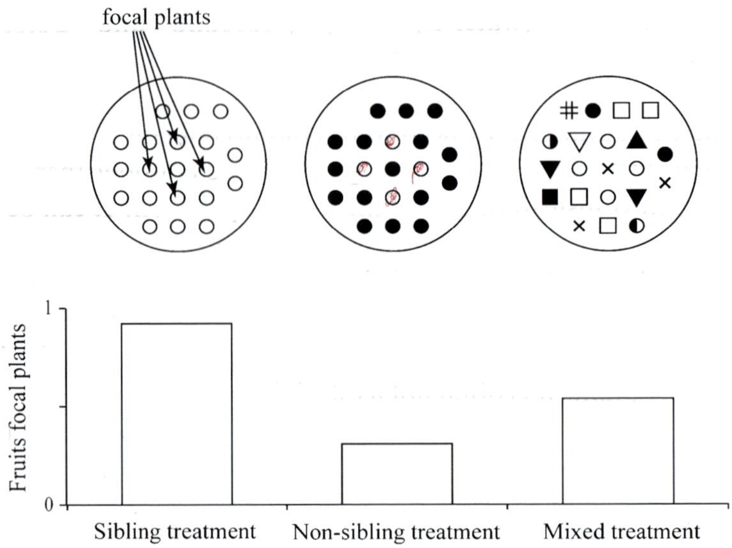

Fig. 17.4 Four sibs (focal plants) of *Cakile edentula* were grown either with a group of siblings from the same mother plant, or with a group of seeds from another plant, or with a mix of seeds from different mother plants. The four focal plants performed best when grown with their own sibs. Redrawn with permission from Donohue (2003), *American Naturalist*, University of Chicago Press.

Kelly (1996) and Stevens *et al.* (1995) examined populations of the partially selfing annual *Impatiens capensis*. They found that the height of the individual plant had a positive effect on its own reproduction but negatively affected the reproduction of the neighbours. Relatedness between neighbours was estimated with molecular methods as $r = 0.54$, about the same level as for full sibs. Conditions for kin selection were therefore met in this species. Just how much kin selection shifts average trait values, as illustrated in Fig. 17.3, is unknown in these studies on *C. edentula* and *I. capensis*.

17.7 Metapopulations

The term metapopulation has recently drawn much attention (Hanski and Gilpin 1997; Hanski 1999), but the concept is much older. For instance, van der Meijden (1979) already described the persistence of *Senecio jacobaea*, attacked by the specialist herbivore *Tyria jacobaeae*, as 'walking on ice floes'. A metapopulation is an ensemble of local populations in which there is a constant turnover, with some local populations becoming extinct while new ones become established. The local populations are loosely connected by dispersal. A good example of a plant that grows in a metapopulation is

Eichhornia paniculata (water-hyacinth) (Husband and Barrett 1998; Freckleton and Watkinson 2002). Plants grow in small transient pools near the roadside. As soon as a pool dries out the population goes locally extinct and when new pools arise they need to be colonized again.

Do optimal strategies of plants change, when we consider them in a metapopulation context instead of within the local population? It can be expected that they do if the various genotypes are affected in a different way by colonization and extinction events. It is often stated that in the metapopulation both within-group and between-group selection play a role (see also Charnov's barnacle model in Chapter 4). We adopt this terminology here but note that for plants it is often not as straightforward to distinguish groups and migration between them as it is for animals, like primates and lions, that live in social groups. A number of recent studies have addressed life histories traits in metapopulations; here we briefly discuss optimal dispersal rates, reproductive investment, optimal size for flowering and sex systems in a metapopulation context.

17.7.1 Seed dispersal

Seed dispersal is a prime example of within- versus between-group selection. Within the local population a non-dispersing type wins from a type that disperses a fraction of its seeds to outside the local population. However, dispersers are superior in establishing new local populations, i.e. they are favoured by between-group selection. Olivieri and Gouyon (1997) called these opposing selection forces the 'metapopulation effect'. The optimal strategy will be a compromise between the two selection pressures and in some limited cases one can calculate fitness of different types in the metapopulation (Metz and Gyllenberg 2001). One would expect the colonization success of dispersers to drop to zero when the metapopulation becomes more fragmented or smaller and only a single local population remains. In such a case you would expect selection against dispersal outside the local population. Shmida and Ellner (1981) argued that this is the reason why desert plants do not have seeds with adaptations for long-distance dispersal. The seeds simply have nowhere to go so within-group selection prevails.

In France, Olivieri *et al.* (1983) studied the thistles *Carduus pycnocephalus* (Italian thistle) and *C. tenuiflorus* (slender thistle). The species have two types of seeds: the plumed seed can disperse over long distances while the other seed type falls next to the parent plant. If the plumed seeds disperse to outside the local population one would expect within-population selection against dispersal. Indeed thistles growing in a late-successional habitat had a lower proportion of plumed seeds. With the degrading of natural habitats, rare species will become more and more confined to small populations in nature reserves. It would be of great interest, not only for testing the theory but also for nature management, to examine whether such trends observed by Olivieri *et al.* (1983) occur more generally in plant species for which the habitat becomes smaller and smaller.

17.7.2 Reproductive investment

A similar metapopulation effect may exist with respect to vegetative versus generative reproduction (Ronce and Olivieri 1997). In late-successional stages the success of establishment through seeds is probably small and any reproductive investment may lower the survival of the parent. Within the local population one then expects selection for vegetative growth and competitiveness. Piquot *et al.* (1988) sampled four genotypes of the aquatic *Sparganium erectum* (bur-reed) from late-successional habitats and two genotypes from an early-successional habitat. They found that typically the genotypes collected in the late, undisturbed habitats emphasized vegetative reproduction, producing a high number of shoots of which few were flowering. In another study van Kleunen *et al.* (2002) applied two generations of selection on the fraction of flowering rosettes in the clonal perennial *Ranunculus reptans* (creeping spearworth). After two years this had resulted in a 26% difference between lines selected for a low versus high fraction of flowering plants, showing that life-history characters can change rapidly under strong selection. The results of the above studies are of direct importance for nature management and conservation strategies. When you conserve only a single local population of a rare species, you may find that its life-history characters change rapidly, within a few generations. When the local population is lost through an accident, the surviving genotypes emphasizing vegetative growth may have lost their ability to reproduce successfully and thereby the ability to colonize new habitats when they arise.

17.7.3 Size for flowering

To study the optimal size for flowering de Jong *et al.* (2000) extended the single population model (Chapter 6) to a metapopulation. The species studied, *Carlina vulgaris*, is a monocarp that occurs on coastal sand dunes in the Netherlands. It only grows in a clearly defined habitat, the edges of thickets of *Salix repens* and these thickets gradually, in a period of about ten years, become unsuitable for seedling establishment so that *C. vulgaris* becomes locally extinct. In the metapopulation model, genotypes with a short generation time had an advantage because after arrival in a newly formed population, they made better use of the available safe sites (de Jong *et al.* 2000). Not surprisingly, simulations showed that the ESS optimal size for flowering was lower in the metapopulation than in the single population. A problem with these metapopulation models is that there are many uncertain parameters, like seed dispersal fractions between local populations, and it becomes more and more difficult to make exact predictions.

17.7.4 Sex systems

Androdioecy, males and cosexuals within the same population, was difficult to explain in the context of the optimization models of Chapter 8. Nevertheless a few species with this sex system exist. Pannell (2001) viewed this problem in the context of metapopulations. He studied *Mercurialis annua*, which is a dioecious annual weed in most of its range but androdioecious in certain areas of Spain and Morocco. If the colonization and extinction rates of local populations of the dioecious ancestor are high enough then female plants may often stand alone, in which case no seeds can be produced. A cosexual mutant, allocating most to female but with a few male flowers, enjoys reproductive assurance. It is a better founder of new, extremely female-biased populations and, if these colonization events happen frequently enough, it may replace the female in the metapopulation. Within a local population the females will eventually replace the cosexuals, resulting in dioecy. However, between-group selection favours cosexuals over females in Pannell's model and androdioecy may be maintained in the metapopulation.

To sum up: we have considered some very simple spatial models of plant strategies, which have hardly been tested. Kin selection may occur when related neighbours grow in each other's vicinity and affect each other. The effect of kin selection is to shift the ESS value of a trait under individual selection to higher or possibly lower values. Kin selection will be negligible in a well-mixed population of outcrossing plants, but may be important in dense stands of clonal or selfing plants and such cases deserve extensive study in the context of kin selection. While a few studies have demonstrated that the conditions for kin selection were met, it is largely unknown how important kin selection is in shaping life histories of plants and whether plants behave differently towards kin as compared to unrelated plants. In a metapopulation genotypes are favoured that score relatively better with respect to the probability of population extinction and colonization.

Questions

(1) What is kin selection?
(2) What relation would you expect between the rate of outcrossing in a plant species and the occurrence of altruistic strategies?
(3) What is meant by the metapopulation effect?
(4) How will strategies change with successional stage in a metapopulation?
(5) What is the limitation of the experiment in Fig. 17.1?

Appendix 1
Mathematical help

Simple equations can be checked using a spreadsheet like Excel. It is convenient to add a procedure like Poptools (http://www.cse.csiro.au/poptools/) to your sheet, which can help with calculating population growth and with simulating stochastic processes. For more complicated models learn a simple-as-possible programming language. Elementary rules of differentation are listed below. If you take a day to learn a program like Maple (http://www.maplesoft.com), the computer can do all the maths for you without making any mistakes.

Throughout the book the absolute fitness of the common type is denoted by W, and the absolute fitness of a rare mutant in an environment of some common type (resident) is W_{m}. Invasion fitness of a mutant, w_{m}, is defined as W_{m}/W. Further common notation includes: T = total resources available, r = relative allocation to male function, δ = inbreeding depression. The fraction of self-pollinated seeds is denoted by S; on occasion S is used for the fraction of self pollen on the stigma, which corresponds to the fraction of selfed seeds when self and outcross pollen are equally competitive.

A function f of x is denoted by $f(x)$. The derivative of $f(x)$ towards x, $\mathrm{d}f(x)/\mathrm{d}x$, is abbreviated as $f'(x)$. Commonly used rules of differentiation are listed below.

$f(x) = a + bx \qquad f'(x) = b$

$f(x) = x^n \qquad f'(x) = nx^{n-1}$

$f(x) = a^x \qquad f'(x) = (a^x)\ln a$

$f(x) = g(x) + h(x) \qquad f'(x) = g'(x) + h'(x)$

$f(x) = g(x)h(x) \qquad f'(x) = g'(x)h(x) + g(x)h'(x)$

$f(x) = g(x)/h(x) \qquad f'(x) = \dfrac{g'(x)h(x) - g(x)h'(x)}{h(x)^2}$

$h(x) = f[g(x)] \qquad h'(x) = f'[g(x)]g'(x)$

References

Ågren, J. and Schemske, D. W. (1995) Sex allocation in the monoecious herb *Begonia semiovata*. *Evolution*, **49**, 121–30.

Ainsworth, C., Parker, J. and Buchanan-Wollaston, V. (1998) Sex determination in plants. *Current Topics in Developmental Biology*, **38**, 167–223.

Aker, C. L. (1982) Regulation of flower, fruit and seed production by a monocarpic perennial, *Yucca whipplei*. *Journal of Ecology*, **70**, 357–72.

Al-Shehbaz, I. and O'Kane, S. L. Jr (2002) Taxonomy and phylogeny of *Arabidopsis* (Brassicaceae). In C. R. Somerville and E. M. Meyerowitz, eds., *The Arabidopsis Book*. Rockville, MD: American Society of Plant Biologists, pp. 1–28. doi/10.1199/tab.0009, http://www.aspb.org/publications/arabidopsis/

Alleman, M. and Doctor, J. (2000) Genomic imprinting in plants: observations and evolutionary implications. *Plant Molecular Biology*, **43**, 147–61.

Alström-Rapaport, C., Lascoux, M. and Gullberg, U. (1997) Sex determination and sex ratio in the dioecious shrub *Salix viminalis*. *Theoretical and Applied Genetics*, **94**, 493–7.

Anonymous (2003) Interview with Matthew Meselson. *BioEssays*, **25**, 1236–46.

Antonovics, J. (1980) Concepts of resource allocation and partioning in plants. In J. E. R. Staddon, ed., *Limits to Action: The Allocation of Individual Behavior*. New York: Academic Press, pp. 1–23.

Antonovics, J. and Levin, D. A. (1980) The ecological and genetic consequences of density-dependent regulation in plants. *Annual Review of Ecology and Systematics*, **11**, 411–52.

Antos, J. A. and Allen, G. A. (1999) Patterns of reproductive effort in male and female shrubs of *Oemleria cerasiformis*: a 6-year study. *Journal of Ecology*, **87**, 77–84.

Arkhipova, I. and Meselson, M. (2000) Transposable elements in sexual and ancient asexual taxa. *Proceedings of the National Academy of Sciences, USA*, **97**, 14473–7.

(2005) Deleterious transposable elements and the extinction of asexuals. *BioEssays*, **27**, 76–85.

Arroyo, J., Barrett, S. C. H., Hidalgo, R. and Cole, W. C. (2002) Evolutionary maintenance of stigma-height dimorphism in *Narcissus papyraceus* (Amaryllidaceae). *American Journal of Botany*, **89**, 1242–9.

Ashman, T. L. (2002) The role of herbivores in the evolution of separate sexes from hermaphroditism. *Ecology*, **83**, 1175–84.

Augspurger, C. K. (1980) Mass-flowering of a tropical shrub (*Hybanthis prunifolius*): influence on pollinator attraction and movement. *Evolution*, **34**, 475–88.

(1983) Seed dispersal of the tropical tree *Platypodium elegans* and the escape of its seedlings from fungal pathogens. *Journal of Ecology*, **71**, 759–71.

Baarlen, P. van, van Dijk, P. J., Hoekstra, R. F. and de Jong, J. H. (2000) Meiotic recombination in sexual diploid and apomictic triploid dandelions (*Taraxacum officinale* L.). *Genome*, **43**, 827–35.

Barrett, S. C. H. (1980) Dimorphic incompatibility and gender in *Nymphoides indica* (Menyanthaceae). *Canadian Journal of Botany*, **58**, 1938–42.

(ed.) (1992) *Evolution and Function of Heterostyly*. Berlin: Springer-Verlag.

(1993) The evolutionary biology of tristyly. *Oxford Surveys in Evolutionary Biology*, **9**, 283–6.

(2002) The evolution of plant sexual diversity. *Nature Genetics*, **3**, 274–84.

(2003) Mating strategies in flowering plants: the outcrossing–selfing paradigm and beyond. *Philosophical Transactions of the Royal Society of London B*, **358**, 991–1004.

Barrett, S. C. H. and Cruzan, M. B. (1994) Incompatibility in heterostylous plants. In E. G. Williams, R. B. Knox and A. E. Clarke, eds., *Genetic Control of Self-Incompatibility and Reproductive Development in Flowering Plants*. Dordrecht: Kluwer Academic Publishers, pp. 189–219.

Barrett, S. C. H. and Harder, L. D. (1996) Ecology and evolution of plant mating. *Trends in Ecology and Evolution*, **11**, 73–9.

Barrett, S. C. H., Lloyd, D. G. and Arroyo, J. (1996) Stylar polymorphisms and the evolution of heterostyly in *Narcissus*. In D. G. Lloyd and S. C. H Barrett, eds., *Floral Biology: Studies on Floral Evolution in Animal-Pollinated Plants*. New York: Chapman and Hall, pp. 339–76.

Barrett, S. C. H., Cole, W. W., Arroyo, M., Cruzan, B. and Lloyd, D. G. (1997) Sexual polymorphisms in *Narcissus triandrus* (Amaryllidaceae): is this species tristylous? *Heredity*, **78**, 135–45.

Bateman, A. J. (1948) Intra-sexual selection in *Drosophila*. *Heredity*, **2**, 349–68.

Bawa, K. S. (1980) Evolution of dioecy in flowering plants. *Annual Review of Ecology and Systematics*, **11**, 15–39.

Bawa, K. S. and Webb, C. J. (1984) Flower, fruit and seed abortion in tropical forest trees: implications for the evolution of paternal and maternal reproductive patterns. *American Journal of Botany*, **71**, 736–51.

Begon, M. and Parker, G. A. (1986) Should egg size and clutch size decrease with age? *Oikos*, **47**, 293–302.

Belaoussoff, S. and Shore, J. S. (1995) Floral correlates and fitness consequences of mating-system variation in *Turnera ulmifolia*. *Evolution*, **49**, 545–56.

Bell, G. (1985) On the function of flowers. *Proceedings of the Royal Society of London B*, **224**, 223–265.

Bell, G. and Koufopanou, V. (1986) The cost of reproduction. *Oxford Surveys in Evolutionary Biology*, **3**, 83–131.

Berg, H. and Redbo-Torstensson, P. (1998) Cleistogamy as a bet-hedging strategy in *Oxalis acetosella*. *Journal of Ecology*, **86**, 491–500.

Bernasconi, G. (2004) Seed paternity in flowering plants: an evolutionary perspective. *Perspectives in Plant Ecology, Evolution and Systematics*, **6**, 149–58.

Bertin, R. I. (1982) Paternity and fruit production in trumpet creeper (*Campsis radicans*). *American Naturalist*, **119**, 694–709.

(1985) Non-random fruit production in *Campsis radicans*: between-year consistency and effects of prior pollination. *American Naturalist*, **126**, 750–9.

Best, L. S. and Bierzychudek, P. (1982) Pollinator foraging on foxglove (*Digitalis purpurea*): a test of a new model. *Evolution*, **36**, 70–9.

Beukeboom, L. W., de Jong, T. J. and Pen, I. (2001) Why girls want to be boys. *BioEssays*, **23**, 477–80.

Bickel, A. M. and Freeman, D. C. (1993) Effects of pollen vector and plant geometry on floral sex-ratio in monoecious plants. *American Midland Naturalist*, **130**, 239–47.

Bierzychudek, P. (1984a) Assessing optimal life histories in a fluctuating environment: the evolution of sex-changing by jack-in-the-pulpit. *American Naturalist*, **123**, 829–40.

(1984b) Determinants of gender in jack-in-the-pulpit: the influence of plant size and reproductive history. *Oecologia*, **65**, 14–18.

(1987) Patterns in plant parthenogenesis. In S. C. Stearns, ed., *The Evolution of Sex and its Consequences*. Basel: Birkhäuser-Verlag, pp. 197–217.

Bookman, S. S. (1984) Evidence for selective fruit production in *Asclepias*. *Evolution*, **38**, 72–86.

Boudry, P., McCombie, H. and van Dijk, H. (2002) Vernalization requirement of wild beet *Beta vulgaris* ssp. *maritima*: among-population variation and its adaptive significance. *Journal of Ecology*, **90**, 693–703.

Boyce, M. (1984) Restitution of *r*- and *K*-selection as a model of density-dependent natural selection. *Annual Review of Ecology and Systematics*, **15**, 427–47.

Brommer, J. E. (2000) The evolution of fitness in life-history theory. *Biological Review*, **75**, 377–404.

Brown, J. S. and Eckert, G. C. (2005) Evolutionary increase in sexual and clonal reproductive capacity in an aquatic plant (*Butomus umbellatus*, Butomaceae). *American Journal of Botany*, **92**, 495–502.

Broyles, S. B. and Wyatt, R. (1990) Paternity analysis in a natural population of *Asclepias exaltata*: multiple paternity, functional gender, and the pollen-donation hypothesis. *Evolution*, **44**, 1454–68.

Brunet, J. (1992) Sex allocation in hermaphroditic plants. *Trends in Ecology and Evolution*, **7**, 79–84.

Budar, F., Touzet, P. and de Paepe, R. (2003) The nucleo-mitochondrial conflict in cytoplasmic male steriles revisited. *Genetica*, **117**, 3–16.

Bullock, J., Clear Hill, B. and Silvertown, J. (1994) Demography of *Cirsium vulgare* in a grazing experiment. *Journal of Ecology*, **82**, 101–11.

Bulmer, M. G. (1994) *Theoretical Evolutionary Ecology*. Sunderland, MA: Sinauer Associates.

Bulmer, M. G. and Taylor, P. D. (1980) Dispersal and the sex ratio. *Nature*, **284**, 448–9.

Burd, M. (1994) Bateman's principle and plant reproduction: the role of pollen limitation in fruit and seed set. *Botanical Review*, **60**, 83–139.

(1998) 'Excess' flower production and selective fruit abortion: a model of potential benefits. *Ecology*, **79**, 2123–32.

Burd, M. and Allen, T. F. H. (1988) Sexual allocation strategy in wind-pollinated plants. *Evolution*, **42**, 403–7.

Burd, M. and Callahan, H. S. (2000) What does the male function hypothesis claim? *Journal of Evolutionary Biology*, **13**, 735–42.

Cadet, C., Metz, J. A. J. and Klinkhamer, P. G. L. (2004) Size and the not-so-single sex: disentangling the effects of size and budget on sex allocation in hermaphrodites. *American Naturalist*, **164**, 779–92.

Campbell, C. S., Quinn, J. A., Cheplick, G. P. and Bell, T. J. (1983) Cleistogamy in grasses. *Annual Review of Ecology and Systematics*, **14**, 411–41.

Campbell, D. R. (1989) Inflorescence size: test of the male function hypothesis. *American Journal of Botany*, **76**, 730–8.

(1991) Effects of floral traits on sequential components of fitness in *Ipomopsis aggregata. American Naturalist*, **137**, 713–37.
(1998) Variation in lifetime male fitness in *Ipomopsis aggregata*: tests of sex allocation theory. *American Naturalist*, **152**, 338–53.
(2000) Experimental tests of sex-allocation theory in plants. *Trends in Ecology and Evolution*, **15**, 227–32.
Carr, D. E. (1992) The relationship between crop size and fruit removal and its implication for the evolution of dioecy. *Canadian Journal of Botany*, **70**, 1784–9.
Carter, C. and Thornburg, R. W. (2004) Is the nectar redox cycle a floral defense against microbial attack? *Trends in Plant Science*, **9**, 320–4.
Casper, B. B. (1988) Evidence for selective embryo abortion in *Cryptantha flava. American Naturalist*, **132**, 318–26.
Casper, B. B. and Charnov, E. L. (1982) Sex allocation in heterostylous plants. *Journal of Theoretical Biology*, **96**, 143–9.
Cavers, P. B. and Steel, M. G. (1984) Patterns of change in seed weights over time on individual plants. *American Naturalist*, **124**, 324–35.
Charlesworth, B. (1990) Optimization models, quantitative genetics, and mutation. *Evolution*, **44**, 520–38.
Charlesworth, B. and Charlesworth, D. (1978) A model for the evolution of dioecy and gynodioecy. *American Naturalist*, **112**, 975–97.
Charlesworth, D. (1984) Androdioecy and the evolution of dioecy. *Biological Journal of the Linnean Society*, **22**, 333–48.
(1985) The distribution of dioecy and self-incompatibility in angiosperms. In P. J. Greenwood, P. H. Harvey and M. Slatkin, eds., *Evolution: Essays in Honour of John Maynard Smith*. Cambridge: Cambridge University Press, pp. 237–68.
(2001) Evolution: an exception that proves the rule. *Current Biology*, **11**, 13–15.
Charlesworth, D. and Charlesworth, B. (1979a) The evolution and breakdown of S-allele systems. *Heredity*, **43**, 41–55.
(1979b) A model for the evolution of distyly. *American Naturalist*, **114**, 467–98.
(1981) Allocation of resources to male and female functions in hermaphrodites. *Biological Journal of the Linnean Society*, **15**, 57–74.
(1987) The effect of investment in attractive structures on allocation to male and female functions in plants. *Evolution*, **41**, 948–68.
Charlesworth, D. and Gutmann, D. S. (1999) The evolution of dioecy and plant sex chromosomes. In C. C. Ainsworth, ed., *Sex Determination in Plants*. Oxford: Biosis Scientific Publishers, pp. 25–49.
Charnov, E. L. (1979) Simultaneous hermaphroditism and sexual selection. *Proceedings of the National Academy of Sciences, USA*, **76**, 2480–4.
(1982a) *The Theory of Sex Allocation*. Princeton, NJ: Princeton University Press.
(1982b) Parent–offspring conflict over reproductive effort. *American Naturalist*, **119**, 736–7.
(1987) On sex allocation and selfing in higher pants. *Evolutionary Ecology*, **1**, 30–6.
Charnov, E. L. and Bull, J. J. (1977) When is sex environmentally determined? *Nature*, **266**, 829–30.
(1986) Sex allocation, pollinator attraction and fruit dispersal in cosexual plants. *Journal of Theoretical Biology*, **118**, 321–5.
Charnov, E. L. and Schaffer, W. M. (1973) Life-history consequences of natural selection: Cole's result revisited. *American Naturalist*, **107**, 791–2.

Charnov, E. L., Maynard Smith, J. M. and Bull, J. J. (1976) Why be an hermaphrodite? *Nature*, **263**, 125–6.

Cheplick, G. P. (1994) Life history evolution in amphicarpic plants. *Plant Species Biology*, **9**, 119–31.

(1996) Do germination patterns in cleistogamous annual grasses reduce the risk of sibling competition? *Journal of Ecology*, **84**, 247–55.

Clark, J. S., Fastie, C., Hurtt, G., *et al.* (1998) Reid's paradox of rapid plant migration. *BioScience*, **48**, 13–24.

Clausen, J. (1951) *Stages in the Evolution of Plant Species*. New York: Cornell University Press.

Clauss, M. J. and Venable, D. L. (2000) Seed germination in desert annuals: an empirical test of adaptive bet hedging. *American Naturalist*, **155**, 168–86.

Clay, K. (1982) Environmental and genetic determinants of cleistogamy in a natural population of the grass *Danthonia spicata*. *Evolution*, **36**, 335–48.

(1993) Size-dependent gender change in green dragon (*Arisaema dracontium*, Araceae). *American Journal of Botany*, **80**, 769–77.

Clay, K. and Schardl, C. L. (2002) Evolutionary origins and ecological consequences of endophyte symbiosis with grasses. *American Naturalist*, **160**, S99–S127.

Cohen, D. (1966) Optimizing reproduction in a randomly fluctuating environment. *Journal of Theoretical Biology*, **12**, 119–29.

Cole, L. C. (1954) The population consequences of life history phenomena. *Quarterly Review of Biology*, **29**, 103–37.

Collin, C. L. and Shykoff, J. A. (2003) Outcrossing rates in the gynomonoecious–gynodioecious species *Dianthus sylvestris* (Caryophyllaceae). *American Journal of Botany*, **90**, 579–85.

Cook, R. (1980) The biology of seeds in the soil. In O. T. Solbrig, ed., *Demography and Evolution in Plant Populations*, Botanical Monograph No. 15. Berkeley, CA: University of California Press, pp. 107–31.

Coomes, D. A. and Grubb, P. J. (2003) Colonization, tolerance, competition and seed-size variation within functional groups. *Trends in Ecology and Evolution*, **18**, 283–91.

Correns, C. (1928). *Bestimmung, Vererbung und Verteilung des Geschlechtes bei den Höheren Pflanzen*. Berlin: Borntraeger.

Cosmides, L. M. and Tooby, J. (1981) Cytoplasmic inheritance and intragenomic conflict. *Journal of Theoretical Biology*, **89**, 83–129.

Courtney, S. P. (1985) Apparency in coevolving relationships. *Oikos*, **44**, 91–8.

Couvet, D., Ronce, O. and Gliddon, C. (1998) The maintenance of nucleo-cytoplasmic polymorphism in a metapopulation: the case of gynodioecy. *American Naturalist*, **142**, 59–70.

Crawford, T. (1984) What is a population? In B. Shorrocks, ed., *Evolutionary Ecology*. Oxford: Blackwell, pp. 135–73.

Cresswell, J. E. (1990) How and why do nectar-foraging bumblebees initiate movements between inflorescences of wild bergamot *Monarda fistulosa* (Lamiaceae)? *Oecologia*, **82**, 450–60.

Crnokrak, P. and Barrett, S. C. H. (2002) Purging the genetic load: a review of the experimental evidence. *Evolution*, **56**, 2347–58.

Crosby, J. L. (1949) Selection of an unfavourable gene-complex. *Evolution*, **3**, 212–30.

Cruden, R. W. (1976) Pollen–ovule ratios: a conservative indicator of breeding systems in flowering plants. *Evolution*, **31**, 32–46.

Cruden, R. W. and Lyon, D. L. (1985) Patterns of biomass allocation to male and female function in plants with different mating systems. *Oecologia*, **66**, 299–306.

Cruz, A. and Moreno, J. M. (2001) No allocation trade-offs between flowering and sprouting in the lignotuberous, Mediterranean shrub *Erica australis*. *Acta Oecologica*, **22**, 121–7.

Curtis, J. and Curtis, C. F. (1985) Homostyle primroses revisited. I. Variation in time and space. *Heredity*, **54**, 227–34.

Daan, S., Dijkstra, C. and Tinbergen, J. M. (1990) Family planning in the kestrel (*Falco tinnunculus*): the ultimate control of covariance of laying date and clutch size. *Behavior*, **114**, 83–116.

Dafni, A. (1983) Pollination of *Orchis caspia*: a nectarless plant which deceives the pollinators of nectariferous species from other plant families. *Journal of Ecology*, **71**, 467–74.

(1984) Mimicry and deception in pollination. *Annual Review of Ecology and Systematics*, **15**, 259–78.

(1992) *Pollination Ecology: A Practical Approach*. Oxford: Oxford University Press.

Dalling, J. W. and Hubbell, S. P. (2002) Seed size, growth rate and gap microsite conditions as determinants of recruitment success for pioneer species. *Journal of Ecology*, **90**, 557–68.

Damgaard, C. and Abbott, R. J. (1995) Positive correlations between selfing rate and pollen–ovule ratio within plant populations. *Evolution*, **49**, 214–17.

Damme, J. M. M. van (1983) On the gynodioecy of *Plantago lanceolata*. Ph.D. thesis, State University of Groningen.

(1991) A restorer gene in gynodioecious *Plantago coronopus* subject to selection in the gametophytic and seedling stage. *Heredity*, **66**, 19–27.

Darwin, C. (1877) *The Different Forms of Flowers on Plants of the Same Species*. London: John Murray.

(1878) *The Effects of Cross- and Self-Fertilization in the Vegetable Kingdom*. London: Pickering.

David, J. L. and Pham, J. (1993) Rapid changes in pollen production in experimental outcrossing populations of wheat. *Journal of Evolutionary Biology*, **6**, 659–76.

Davidar, P. and Morton, E. S. (1986) The relationship between fruit crop size and fruit removal by birds. *Ecology*, **67**, 262–5.

Dawkins, R. (1976) *The Selfish Gene*. Oxford: Oxford University Press.

(1982) *The Extended Phenotype*. Oxford: W. H. Freeman.

Day, T. and Aarssen, L. W. (1997) A time commitment hypothesis for size-dependent gender allocation. *Evolution*, **51**, 988–93.

Day, T. and Taylor, P. D. (1998) Unifying genetic and game theoretic models of kin selection for continuous traits. *Journal of Theoretical Biology*, **194**, 391–407.

Declavijo, E. R. and Jimenez, M. J. (1993) Cleistogamy and chasmogamy in *Ceratocapnos heterocarpa* (Fumariaceae). *International Journal of Plant Sciences*, **154**, 325–33.

Delesalle, V. A. and Muenchow, G. E. (1992) Opportunities for selfing and inbreeding depression in *Sagittaria* congeners (Alismataceae) with contrasting sexual systems. *Evolutionary Trends in Plants*, **6**, 81–91.

Delph, L. F. (1990a) The evolution of gender dimorphism in New Zealand *Hebe* (Scrophulariaceae). *Evolutionary Trends in Plants*, **4**, 85–97.

(1990b) Sex-ratio variation in the gynodioecious scrub *Hebe strictissima* (Scrophulariaceae). *Evolution*, **44**, 134–42.

(1999) Sexual dimorphism in life history. In M. A. Geber, T. E. Dawson and L. F. Delph, eds., *Gender and Sexual Dimorphism in Flowering Plants*. Berlin: Springer-Verlag, pp. 217–46.

Devlin, B., Roeder, K. and Ellstrand, N. C. (1992) The effect of flower production on male reproductive success in wild radish, a hermaphrodite. *Evolution*, **46**, 1030–42.

Dijk, H. van, Boudry, P., McCombie, H. and Vernet, P. (1997) Flowering time in wild beet (*Beta vulgaris* ssp. *maritima*) along a latitudinal cline. *Acta Oecologica*, **18**, 47–60.

Donohue, K. (2003) The influence of neighbor relatedness on multilevel selection in the great lakes sea rocket. *American Naturalist*, **162**, 77–92.

Dorken, M. E., Friedman, J. and Barrett, S. C. H. (2002) The evolution and maintenance of monoecy and dioecy in *Sagittaria latifolia* (Alismataceae). *Ecology*, **56**, 31–41.

Dukas, R. and Real, L. A. (1993) Effects of recent experience on foraging decisions by bumble bees. *Oecologia*, **94**, 244–6.

Durand, B. and Durand, R. (1991) Sex determination and reproductive organ differentiation in *Mercurialis. Plant Science*, **80**, 49–65.

Eckert, C. G. (2002) The loss of sex in plants. *Evolutionary Ecology*, **15**, 501–20.

Ehrlén, J. and van Groenendael, J. (2001) Storage and the delayed costs of reproduction in the understorey perennial *Lathyrus vernus*. *Journal of Ecology*, **89**, 237–46.

Ellis, M. H., Rebetzke, G. J., Chandler, O., *et al.* (2004) The effect of height-reducing genes on the early growth of wheat. *Functional Plant Biology*, **31**, 583–9.

Ellner, S. (1985a) ESS germination strategies in randomly varying environments. I. Logistic-type models. *Theoretical Population Biology*, **28**, 50–79.

(1985b) ESS germination strategies in randomly varying environments. II. Reciprocal yield-law models. *Theoretical Population Biology*, **28**, 80–116.

(1986) Germination dimorphism and parent–offspring conflict in seed germination. *Journal of Theoretical Biology*, **123**, 173–85.

Ellstrand, N. C. and Marshall, D. L. (1986) Patterns of multiple paternity in populations of *Raphanus sativus*. *Evolution*, **40**, 837–42.

Emms, S. K. (1993) On measuring fitness gain curves in plants. *Ecology*, **74**, 1750–6.

Eppley, S. M., Stanton, L. and Grosberg, R. K. (1998) Intrapopulation sex ratio variation in the salt grass *Distichlis spicata*. *American Naturalist*, **152**, 659–70.

Falster, D. S. and Westoby, M. (2003) Plant height and evolutionary games. *Trends in Ecology and Evolution*, **18**, 337–43.

Feinsinger, P. (1983) Variable nectar secretion in a *Heliconia* species pollinated by hermit hummingbirds. *Biotropica*, **15**, 48–52.

Fenner, M. (1985) *Seed Ecology*. London: Chapman and Hall.

Fenster, C. B. (1991) Effect of male pollen donor and female seed parent on allocation of resources to developing seeds and fruit in *Chamaecrista fasciculata* (Leguminosae). *American Journal of Botany*, **78**, 13–23.

Ferdy, J. B. and Smithson, A. (2002) Geitonogamy in rewarding and unrewarding inflorescences: modelling pollen transfer on actual foraging sequences. *Evolutionary Ecology*, **16**, 155–75.

Fisher, R. A. (1930) *The Genetical Theory of Natural Selection*. Oxford: Clarondon Press.

(1941) Average excess and average effect of gene substitution. *Annals of Genetics*, **11**, 53–63.

Frank, S. A. (1989) The evolutionary dynamics of cytoplasmic male sterility. *American Naturalist*, **133**, 345–76.

(2000) Polymorphism of attack and defense. *Trends in Ecology and Evolution*, **15**, 167–71.

Frank, S. A. and Barr, C. M. (2001) Spatial dynamics of cytoplasmic male sterility. In J. Silvertown and J. Antonovics, eds., *Integrating Ecology and Evolution in a Spatial Context*. Oxford: Blackwell, pp. 219–43.

Freckleton, R. P. and Watkinson, A. R. (2002) Large-scale spatial dynamics of plants: metapopulations, regional ensembles and patchy populations. *Journal of Ecology*, **90**, 419–34.

Freeman, D. C., Harper, K. T. and Charnov, E. L. (1980) Sex change in plants: old and new observations and new hypotheses. *Oecologia*, **47**, 222–32.

Freeman, D. C., McArthur, E. D., Harper, K. T. and Blauer, A. C. (1981) Influence of environment on the floral sex-ratio of monoecious plants. *Evolution*, **35**, 194–7.

Freeman, D. C., Lovett Doust, J., El-Keblawy, A., Miglia, K. J. and McArthur, E. D. (1997) Sexual specialization and inbreeding avoidance in the evolution of dioecy. *Botanical Review*, **63**, 65–92.

Fricke, H. and Fricke, S. (1977) Monogamy and sex change by aggressive dominance in coral-reef fish. *Nature*, **266**, 830–2.

Fritz, A.-L. and Nilsson, L. A. (1996) Reproductive success and gender variation in deceit-pollinated orchids. In D. G. Lloyd and S. C. H. Barrett, eds., *Floral Biology: Studies on Floral Evolution in Animal-Pollinated Plants*. London: Chapman and Hall, pp. 319–38.

Gadgil, M. and Bossert, W. (1970) Life history consequences of natural selection. *American Naturalist*, **104**, 1–24.

Gadgil, M. and Solbrig, O. T. (1972) The evolution of *r*- and *K*-selection: evidence from wild flowers and some theoretical considerations. *American Naturalist*, **106**, 14–31.

Galen, C. and Plowright, R. C. (1985) The effects of nectar level and flower development on pollen carry-over in inflorescences of fireweed (*Epilobium angustifolium*) (Onagraceae). *Canadian Journal of Botany*, **63**, 488–91.

Gallardo, R., Dominguez, E. and Munoz, J. M. (1994) Pollen–ovule ratio, pollen size, and breeding system in *Astralagus* (Fabaceae) subgenus *Epiglottis*: a pollen and seed allocation approach. *American Journal of Botany*, **81**, 1611–19.

Galliard, J. F. le, Ferrière, R. and Dickmann, U. (2003) The adaptive dynamics in spatially heterogeneous populations. *Evolution*, **57**, 1–17.

Ganders, R. (1974) Disassortive pollination in the distylous plant *Jepsonia heterandra*. *Canadian Journal of Botany*, **52**, 2401–6.

(1979) The biology of heterostyly. *New Zealand Journal of Botany*, **17**, 607–35.

Ganeshaiah, K. N. and Shaanker, R. U. (1988) Seed abortion in wind-dispersed pods of *Dalbergia sissoo*: maternal regulation or sibling rivalry? *Oecologia*, **77**, 135–9.

(1991) Floral sex-ratios in monoecious species: why are trees more male-biased than herbs? *Current Science*, **60**, 319–21.

Geritz, S. A. H. (1995) Evolutionarily stable seed polymorphism and small-scale spatial variation in seedling density. *American Naturalist*, **146**, 685–707.

Geritz, S. A. H., Kisdi, E., Meszena, G. and Metz, J. A. J. (1998) Evolutionarily singular strategies and the adaptive growth and branching of the evolutionary tree. *Evolutionary Ecology*, **12**, 35–58.

Gersani, M., Brown, J. S., O'Brien, E. E., Maina, G. M. and Abramsky, Z. (2001) Tragedy of the commons as a result of root competition. *Journal of Ecology*, **89**, 660–9.

Ghiselin, M. T. (1969) Evolution of hermaphroditism among animals. *Quarterly Review of Biology*, **44**, 189–208.

Gibbs, P. and Sassaki, R. (1998) Reproductive biology of *Dalbergia miscolobium* Benth. (Leguminosae–Papilionoideae) in SE Brazil: the effects of pistillate sorting on fruit-set. *Annals of Botany*, **81**, 735–40.

Gigord, L. D. B., Macnair, M. R. and Smithson, A. (2001) Negative frequency-dependent selection maintains a dramatic flower color polymorphism in the rewardless orchid *Dactylorhiza sambucina* (L.) Soo. *Proceedings of the National Academy of Sciences, USA*, **98**, 6235–55.

Gigord, L. D. B., Macnair, M. R., Stritesky, M. and Smithson, A. (2002) The potential for floral mimicry in rewardless orchids: an experimental study. *Proceedings of the Royal Society of London B*, **269**, 1389–95.

Givnish, T. J. (1980) Ecological constraints on the evolution of breeding systems in seed plants: dioecy and dispersal in gymnosperms. *Evolution*, **34**, 959–72.

(1982) Outcrossing versus ecological constraints in the evolution of dioecy. *American Naturalist*, **119**, 849–65.

Godfray, H. C. J. (1995) Evolutionary theory of parent–offspring conflict. *Nature*, **376**, 133–8.

Goldberg, R. B., Barker, S. J. and Perezgrau, L. (1989) Regulation of gene-expression during plant embryogenesis. *Cell*, **56**, 149–60.

Goldman, D. A. and Willson, M. F. (1986) Sex allocation in functionally hermaphrodite plants: a review and critique. *Botanical Review*, **52**, 157–94.

Goodnight, C. J. and Stevens, L. (1997) Experimental studies of group selection: what do they tell us about group selection in nature? *American Naturalist*, **150**, 559–79.

Goodnight, C. J., Schwartz, J. M. and Stevens, L. (1992) Contextual analysis of models of group selection, soft selection and hard selection. *American Naturalist*, **140**, 743–61.

Gould, S. J. and Lewontin, R. C. (1979) The spandrels of St Marco and the Panglossian paradigm. *Proceedings of the Royal Society of London B*, **205**, 581–98.

Goulson, D. (2000) Why do pollinators visit proportionally fewer flowers in large patches? *Oikos*, **91**, 485–92.

Gouyon, P. H. and Couvet, D. (1987) A conflict between two sexes, females and hermaphrodites. In S. C. H. Stearns, ed., *The Evolution of Sex and its Consequences*. Basel: Birkhäuser-Verlag, pp. 245–61.

Gouyon, P. H., Vichot, F. and Damme, J. M. M. van (1991) Nuclear-cytoplasmic male sterility: single-point equilibria versus limit cycles. *American Naturalist*, **137**, 498–514.

Grafen, A. (1991) Modelling in behavioural ecology. In J. R. Krebs and N. B. Davies, eds., *Behavioural Ecology*. Oxford: Blackwell, pp. 5–31.

Gregorius, H. R., Ziehe, M. and Ross, M. D. (1987) Selection caused by self-fertilization. I. Four measures of self-fertilization and their effects on fitness. *Theoretical Population Biology*, **31**, 91–115.

Griffin, A. S. and West, S. A. (2002) Kin selection: fact and fiction. *Trends in Ecology and Evolution*, **17**, 15–21.

Grime, J. P. (1977) *Plant Strategies and Vegetation Processes*. Chichester: John Wiley.

Grime, J. P. and Jeffrey, D. W. (1965) Seedling establishment in vertical gradients of sunlight. *Journal of Ecology*, **53**, 621–42.

Gross, K. L. (1981) Predictions of fate from rosette size in four 'biennial' plant species: *Verbascum thapsus*, *Oenothera biennis*, *Daucus carota*, and *Tragopogon dubius*. *Oecologia*, **48**, 209–13.

Grubb, P. J. and Metcalfe, D. J. (1996) Adaptation and inertia in the Australian tropical lowland rain-forest flora: contradictory trends in intergeneric and intrageneric comparisons of seed size in relation to light demand. *Functional Ecology*, **10**, 512–20.

Gumbert, A. and Kunze, J. (2001) Colour similarity to rewarding model plants affects pollination in a food deceptive orchid, *Orchis boryi*. *Biological Journal of the Linnean Society*, **72**, 419–33.

Guth, C. J. and Weller, S. G. (1986) Pollination, fertilization and ovule abortion in *Oxalis magnifica*. *American Journal of Botany*, **73**, 246–53.

Haan, A. A. de, Hundscheid, M. P. J. and van Hinsberg, A. (1997) Effects of CMS types and restorer alleles on plant performance in *Plantago lanceolata* L.: an indication of cost of restoration. *Journal of Evolutionary Biology*, **10**, 803–20.

Hacker, J. B. (1984) Genetic variation in seed dormancy in *Digitaria milanjiana* in relation to rainfall at the collection site. *Journal of Applied Ecology*, **21**, 947–59.

Hacker, J. B. and Ratcliff, D. (1989) Seed dormancy and factors controlling dormancy breakdown in buffel grass accessions from contrasting provenances. *Journal of Applied Ecology*, **26**, 201–12.

Haig, D. (1997) Parental antagonism, relatedness asymmetries, and genomic imprinting. *Proceedings of the Royal Society of London B*, **264**, 1657–62.

(2000) The kinship theory of genomic imprinting. *Annual Review of Ecology and Systematics*, **31**, 9–32.

Haig, D. and Westoby, M. (1988) Inclusive fitness, seed resources, and maternal care. In J. L. Lovett Doust and L. Lovett Doust, eds., *Plant Reproductive Ecology*. Oxford: Oxford University Press, pp. 60–79.

(1991) Genomic imprinting in endosperm: its effect on seed development in crosses between species, and between different ploidies of the same species, and its implications for the evolution of apomixes. *Philosophical Transactions of the Royal Society of London B*, **333**, 1–13.

Hamilton, W. D. (1964a) The genetical evolution of social behaviour. I. *Journal of Theoretical Biology*, **7**, 1–16.

(1964b) The genetical evolution of social behaviour. II. *Journal of Theoretical Biology*, **7**, 17–52.

(1967) Extraordinary sex ratios. *Science*, **156**, 477–88.

Hamrick, J. L. and Godt, M. J. W. (1990) Allozyme diversity in plant species. In H. D. Brown, M. T. Clegg, A. L. Kahler and B. S. Weir, eds., *Plant Population Genetics, Breeding and Genetic Resources*. Sunderland, MA: Sinauer Associates, pp. 43–63.

Hanski, I. (1999) *Metapopulation Ecology*. Oxford: Oxford University Press.

Hanski, I. and Gilpin, M. E. (1997) *Metapopulation Biology: Ecology, Genetics and Evolution*. San Diego, CA: Academic Press.

Harder, L. D. and Thomson, J. D. (1989) Evolutionary options for maximizing pollen dispersal of animal-pollinated plants. *American Naturalist*, **133**, 323–44.

Härdling, R. and Nilsson, P. (1999) Parent–offspring and sexual conflicts in the evolution of angiosperm seeds. *Oikos*, **84**, 27–34.

Hardy, I. C. W. (ed.) (2002) *Sex Ratios: Concepts and Research Methods*. Cambridge: Cambridge University Press.

Harper, J. L. (1977) *Population Biology of Plants*. London: Academic Press.

Harper, J. L., Lovell, P. H. and Moore, K. G. (1970) The shapes and sizes of seeds. *Annual Review of Ecology and Systematics*, **1**, 327–56.

Hart, R. (1977) Why are biennials so few? *American Naturalist*, **111**, 792–9.

Hartl, D. L. and Clark, A. G. (1989) *Principles of Population Genetics*. Sunderland, MA: Sinauer Associates.

Hauert, C. (2002) Effects of space in 2 × 2 games. *International Journal of Bifurcation and Chaos*, **12**, 1531–48.

Hautekeete, N. C., Piquot, Y. and van Dijk, H. (2002) Life span in *Beta vulgaris* ssp. *maritima*: the effect of age at first reproduction and disturbance. *Journal of Ecology*, **90**, 508–16.

Heinrich, B. (1983) Do bumblebees forage optimally, and does it matter? *American Zoologist*, **23**, 273–81.

Heisler, I. L. and Damuth, J. (1987) A method for analyzing selection in hierarchically structured populations. *American Naturalist*, **139**, 582–602.

Henry, H. A. L. and Aarssen, L. W. (1999) Interpretation of stem diameter–height allometry in trees: biomechanical constraints, neighbour effects or biased regressions? *Ecology Letters*, **2**, 89–97.

Herlihy, C. R. and Eckert, G. C. (2002) Genetic cost of reproductive assurance in a self-fertilizing plant. *Nature*, **416**, 320–3.

Herrera, C. M. and Jordano, P. (1994) Recruitment of a mast-fruiting, bird-dispersed tree: bridging frugivore activity and seedling recruitment. *Ecological Monographs*, **70**, 591–615.

Herrera, C. M., Sanchez-Lafuente, A. M., Medrano, M., *et al.* (2001) Geographical variation in autonomous self-pollination levels unrelated to pollinator service in *Helleborus foetidus* (Ranunculaceae). *American Journal of Botany*, **88**, 1025–32.

Hiatt, E. N. and Dawe, R. K. (2003) The meiotic drive system on maize abnormal chromosome 10 contains few essential genes. *Genetica*, **117**, 67–76.

Hisock, S. J. and Tabah, D. A. (2003) The different mechanisms of species self-incompatibility. *Philosophical Transactions of the Royal Society of London B*, **358**, 1037–45.

Hodges, C. M. (1985) Bumble bee foraging: the threshold departure rule. *Ecology*, **66**, 179–87.

Hodgson, J. G., Grime, J. P., Hunt, R. and Thompson, K. (1995) *The Electronic Comparative Plant Ecology*. London: Chapman and Hall.

Hodkinson, D. J., Askew, A. P., Thompson, K., *et al.* (1998) Ecological correlates of seed size in the British flora. *Functional Ecology*, **12**, 762–6.

Holsinger, K. (1992) Ecological models of plant mating systems and the evolutionary stability of mixed mating systems. In R. Wyatt, ed., *Ecology and Evolution of Plant Reproduction*. London: Chapman and Hall, pp. 169–91.

Holzapfel, C. and Alpert, P. (2003) Root cooperation in a clonal plant: connected strawberries segregate roots. *Oecologia*, **134**, 72–7.

Horovitz, A. (1977) Male and female components of fitness in hermaphrodite populations. *American Naturalist*, **111**, 1210–11.

Horovitz, A. and Harding, J. (1972) Concept of male outcrossing in hermaphrodite higher plants. *Heredity*, **29**, 223–36.

Hotta, M. (1971) Study of family Araceae: general remarks. *Japanese Journal of Botany*, **20**, 269–310.

Howe, H. F. and Kerckhove, G. A. vande (1979) Fecundity and seed dispersal of a tropical tree. *Ecology*, **60**, 180–9.

Hu, X. S. and Ennos, R. A. (1997) On estimating the ratio of pollen to seed flow among plant populations. *Heredity*, **79**, 541–52.

Hurst, L. D., Atlan, A. and Bengtsson, B. O. (1996) Genetic conflicts. *Quarterly Review of Biology*, **71**, 317–64.

Husband, B. C. and Barrett, S. C. H. (1998) Spatial and temporal variation in population size of *Eichhornia paniculata* in ephemeral habitats: implications of metapopulation dynamics. *Journal of Ecology*, **86**, 1021–31.

Husband, B. C. and Schemske, D. W. (1996) Evolution of the magnitude and timing of inbreeding depression in plants. *Evolution*, **50**, 54–70.

Imbert, E. (2002) Ecological consequences and ontogeny of seed heteromorphism. *Perspectives in Plant Ecology Evolution and Systematics*, **5**, 13–36.

Iwasa, Y. (1991) Sex change evolution and cost of reproduction. *Behavioral Ecology*, **2**, 56–8.

Iwasa, Y. and Cohen, D. (1989) Optimal growth schedule of a perennial plant. *American Naturalist*, **133**, 480–505.

Iwasa, Y., de Jong, T. J. and Klinkhamer, P. G. L. (1995) Why pollinators visit only a fraction of the open flowers on a plant: the plant's point of view. *Journal of Evolutionary Biology*, **8**, 439–53.

Jacobs, M. S. and Wade, M. J. (2003) A synthetic review of the theory of gynodioecy. *American Naturalist*, **161**, 837–51.

Jain, S. K. (1982) Variation and the adaptive role of seed dormancy in some annual grassland species. *Botanical Gazette*, **143**, 101–6.

Jakobsson, A. and Eriksson, O. (2000) A comparative study of seed number, seed size, seedling size and recruitment in grassland species. *Oikos*, **88**, 494–502.

Janzen, D. H. (1976) Why bamboos wait so long to flower. *Annual Review of Ecology and Systematics*, **7**, 347–91.

(1977) Note on optimal mate selection by plants. *American Naturalist*, **111**, 365–71.

Jasieniuk, M. and Lechowicz, M. J. (1987) Spatial and temporal variation in chasmogamy and cleistogamy in *Oxalis montana* (Oxalidaceae). *American Journal of Botany*, **74**, 1672–80.

Johnson, S. D. (1994) Evidence for Batesian mimicry in a butterfly-pollinated orchid. *Biological Journal of the Linnean Society*, **53**, 91–104.

(2000) Batesian mimicry in the non-rewarding orchid *Disa pulchra*, and its consequences for pollinator behaviour. *Biological Journal of the Linnean Society*, **71**, 119–32.

Johnson, S. D. and Nilsson, L. A. (1999) Pollen carryover, geitonogamy, and the evolution of deceptive pollination systems in orchids. *Ecology*, **80**, 2607–19.

Johnson, S. D. and Steiner, K. E. (1997) Long-tongued fly pollination and evolution of floral spur length in the *Disa draconis* complex (Orchidaceae). *Evolution*, **51**, 45–53.

Johnson, S. D., Peter, C. I., Nilsson, L. A. and Ågren, J. (2003) Pollination success in a deceptive orchid is enhanced by co-occurring rewarding magnet plants. *Ecology*, **84**, 2919–27.

Jong, T. J. de (2000) From pollen dynamics to adaptive dynamics. *Plant Species Biology*, **15**, 31–41.

Jong, T. J. de and Geritz, S. A. H. (2002) The role of geitonogamy in the gradual evolution towards dioecy in cosexual plants. *Selection*, **2**, 133–46.

Jong, T. J. de and Klinkhamer, P. G. L. (1987) Selection for biennial life histories in plants. *Vegetatio*, **70**, 149–56.

(1989) Limiting factors for seed production in *Cynoglossum officinale*. *Oecologia*, **80**, 167–72.

Jong, T. J. de and Meijden, E. van der (2004) Sex ratio of some long-lived dioecious plants in a sand dune area. *Plant Biology*, **6**, 616–20.

Jong, T. J. de, Klinkhamer, P. G. L., Geritz, S. A. H. and Meijden, E. van der (1989) Why biennials delay flowering: an optimization model and field data on *Cirsium vulgare* and *Cynoglossum officinale*. *Acta Botanica Neerlandica*, **38**, 41–55.

Jong, T. J. de, Klinkhamer, P. G. L. and Staalduinen, M. J van (1992) The consequences of pollination biology for mass blooming or extended blooming. *Functional Ecology*, **6**, 605–15.

Jong, T. J. de, Klinkhamer, P. G. L. and Rademaker, M. C. J. (1999) How geitonogamous selfing affects sex allocation in hermaphrodite plants. *Journal of Evolutionary Biology*, **12**, 166–76.

Jong, T. J. de, Klinkhamer, P. G. L. and Heiden, J. L. H. de (2000) The evolution of generation time in metapopulations of monocarpic perennial plants: some theoretical considerations and the example of the rare thistle *Carlina vulgaris*. *Evolutionary Ecology*, **14**, 213–31.

Jong, T. J. de, Batenburg, F. H. D. and Dijk, J. van (2002) Seed sex ratio in dioecious plants depends on relative dispersal of pollen and seeds: an example using a chessboard simulation model. *Journal of Evolutionary Biology*, **15**, 373–9.

Jong, T. J. de, Nell, H. W. and Glawe, G. A. (2005) Heritable variation in seed sex ratio of the stinging nettle (*Urtica dioica*). *Plant Biology*, **7**, 190–4.

Joshi, J., Schmid, B., Caldeira, M. C., *et al.* (2001) Local adaptation enhances performance of common plant species. *Ecology Letters*, **4**, 536–44.

Kachi, N. and Hirose, T. (1985) Population dynamics of *Oenothera glazioviana* in a sand-dune system with special reference to the adaptive significance of size-dependent reproduction. *Journal of Ecology*, **73**, 887–901.

Kadmon, R. and Shmida, A. (1992) Departure rules used by bees foraging for nectar: a field-test. *Evolutionary Ecology*, **6**, 142–51.

Karkkainen, K., Savolainen, O. and Koski, V. (1999) Why do plants abort so many developing seeds: bad offspring or bad maternal genotypes? *Evolutionary Ecology*, **13**, 305–17.

Kaul, M. L. H. (1988) *Male Sterility in Higher Plants*. Berlin: Springer-Verlag.

Kay, Q. O. N. and Stevens, D. P. (1986) The frequency, distribution and reproductive biology of dioecious species in the native flora of Britain and Ireland. *Botanical Journal of the Linnean Society*, **92**, 39–64.

Kearns, C. A. and Inouye, D. W. (1993) *Techniques for Pollination Biologists*. Niwot, CO: Colorado University Press.

Kelly, C. K. and Purvis, A. (1993) Seed size and establishment conditions in tropical trees: on the use of taxonomic relatedness in determining ecological patterns. *Oecologia*, **94**, 356–60.

Kelly, D. (1985) Why are biennials so maligned? *American Naturalist*, **125**, 473–97.

Kelly, D. and Sork, V. L. (2002) Mast seeding in perennial plants: why, how, where? *Annual Review of Ecology and Systematics*, **33**, 427–47.

Kelly, D. and Sullivan, J. J. (1997) Quantifying the benefits of mast seeding on predator satiation and wind pollination in *Chionochloa pallens* (Poaceae). *Oikos*, **78**, 143–50.

Kelly, D., Hart, D. E. and Allen, R. B. (2001) Evaluating the wind-pollination benefits of mast seeding. *Ecology*, **82**, 117–26.

Kelly, J. K. (1996) Kin selection in the annual plant *Impatiens capensis*. *American Naturalist*, **147**, 899–918.
Kery, M., Matthies, D. and Spillmann, H. H. (2000) Reduced fecundity and offspring performance of the declining grassland plants *Primula veris* and *Gentiana lutea*. *Journal of Ecology*, **88**, 17–30.
Kleunen, M. van, Fischer, M. and Schmid, B. (2002) Experimental life-history evolution: selection on the allocation to sexual reproduction and its plasticity in a clonal plant. *Evolution*, **56**, 2168–77.
Klinkhamer, P. G. L. and Jong, T. J. de (1987). Plant size and seed production in the monocarpic *Cynoglossum officinale*. *New Phytologist*, **106**, 773–83.
(1989) A deterministic model to study the importance of density-dependence for regulation and the outcome of intra-specific competition in populations of sparse plants. *Acta Botanica Neerlandica*, **38**, 57–65.
(1993a) Phenotypic gender in plants: effects of plant size and environment on allocation to seeds and flowers in *Cynoglossum officinale*. *Oikos*, **67**, 81–6.
(1993b) Attractiveness to pollinators: a plant's dilemma. *Oikos*, **66**, 180–4.
(1997) Size-dependent allocation to male and female reproduction in hermaphroditic plants. In F. A. Bazzaz and J. Grace, eds., *Physiological Ecology*. New York: Academic Press, pp. 211–29.
Klinkhamer, P. G. L., Jong, T. J. de, Metz, J. A. J. and Val, J. (1987) Life history tactics of annual organisms: the joint effects of dispersal and delayed germination. *Theoretical Population Biology*, **32**, 127–56.
Klinkhamer, P. G. L., Jong, T. J. de and Bruyn, G. J. de (1989) Plant size and pollinator visitation in *Cynoglossum officinale*. *Oikos*, **54**, 201–4.
Klinkhamer, P. G. L., Jong, T. J. de and Meelis, E. (1990) How to test for the reproductive effort of plants. *American Naturalist*, **135**, 291–300.
Klinkhamer, P. G. L., Jong, T. J. de and Wesselingh, R. A. (1991) Implications of differences between hermaphrodite and female flowers for attractiveness to pollinators and seed production. *Netherlands Journal of Zoology*, **41**, 130–43.
Klinkhamer, P. G. L., Jong, T. J. de and Nell, H. W. (1994a) Limiting factors for seed production and phenotypic gender in the gynodioecious species *Echium vulgare* (Boraginaceae). *Oikos*, **71**, 469–78.
Klinkhamer, P. G. L., Jong, T. J. de and Metz, J. A. J. (1994b) Why plants can be too attractive: a discussion of measures to estimate male fitness. *Journal of Ecology*, **82**, 191–4.
(1997a) Sex and size in plants. *Trends in Ecology and Evolution*, **12**, 260–5.
Klinkhamer, P. G. L., Kubo, T. and Iwasa, Y. (1997b) Herbivores and the evolution of the semelparous perennial life-histories of plants. *Journal of Evolutionary Biology*, **10**, 529–50.
Klinkhamer, P. G. L., Jong, T. J. de and Linnebank, L. A. (2001) Small-scale spatial patterns determine ecological relationships: an experimental example using nectar production rates. *Ecology Letters*, **4**, 559–67.
Koelewijn, H. P. and Damme, J. M. M. van (1995) Genetics of male sterility in gynodioecious *Plantago coronopus*. I. Cytoplasmic variation. *Genetics*, **139**, 1749–58.
Koelewijn, H. P. and Hundscheid, M. P. H. (2000). Intraspecific variation in sex allocation in hermaphroditic *Plantago coronopus* (L.). *Journal of Evolutionary Biology*, **13**, 302–15.
Kohn, J. R. and Biardi, J. E. (1995) Outcrossing rates and inferred levels of inbreeding depression in gynodioecious *Cucurbita foetidissima* (Cucurbitaceae). *Heredity*, **75**, 77–83.

Kondrashov, A. S. (1984) Deleterious mutations as an evolutionary factor. I. The advantage of recombination. *Genetical Research*, **44**, 199–217.

Korbecka, G., Klinkhamer, P. G. L. and Vrieling, K. (2002) Selective embryo abortion hypothesis revisited: a molecular approach. *Plant Biology*, **4**, 298–310.

Korpelainen, H. (1991) Sex ratio variation and spatial segregation in populations of *Rumex acetosa* and *Rumex acetosella*. *Plant Systematics and Evolution*, **174**, 183–95.

(1998) Labile sex expression in plants. *Biological Review*, **73**, 157–80.

Kozlowski, J. and Stearns, S. C. (1989) Hypotheses for the production of excess zygotes: models of bet-hedging and selective abortion. *Evolution*, **43**, 1369–77.

Kozlowski, J. and Uchmanski, J. (1987) Optimal individual growth and reproduction in perennial species with indeterminate growth. *Evolutionary Ecology*, **1**, 214–30.

Kozlowski, J. and Wiegert, R. G. (1987) Optimal age and size at maturity in annuals and perennials with determinate growth. *Evolutionary Ecology*, **1**, 231–44.

Krishnamurthy, K. S., Shaanker, R. U. and Ganeshaiah, K. N. (1997) Seed abortion in an animal dispersed species, *Syzygium cuminii* (L.) Skeels (Myrtaceae): the chemical basis. *Current Science*, **73**, 869–73.

Kuiper, P. J. C. and Bos, M. (1992) Plantago: *A Multidisciplinary Study*, Ecological Studies No. 89. Berlin: Springer-Verlag.

Kunin, W. E. (1993) Sex and the single mustard: population density and pollinator behavior effects on seed set. *Ecology*, **74**, 2145–60.

(1997) Population size and density effects in pollination: pollinator foraging and plant reproductive success in experimental arrays of *Brassica kaber*. *Journal of Ecology*, **85**, 225–34.

Kunin, W. E. and Shmida, A. (1997) Plant reproductive traits as a function of local, regional, and global abundance. *Conservation Biology*, **11**, 183–92.

Kurian, V. and Richards, A. J. (1997) A new recombinant in the heteromorphy "S" supergene in *Primula*. *Heredity*, **78**, 388–90.

Lacey, E. P. (1988) Latitudinal variation in reproductive timing of a short-lived monocarp, *Daucus carota* (Apiaceae). *Ecology*, **69**, 220–32.

Larson, B. M. H. and Barrett, S. C. H. (2000) A comparative analysis of pollen limitation in flowering plants. *Biological Journal of the Linnean Society*, **69**, 503–20.

Law, R. (1979) Cost of reproduction in annual meadow grass. *American Naturalist*, **113**, 3–16.

Lecorff, J. and Horvitz, C. C. (1995) Dispersal of seeds from chasmogamous and cleistogamous flowers in an ant-dispersed neotropical herb. *Oikos*, **73**, 59–64.

Lee, T. D. (1988) Patterns of fruit and seed production. In J. Lovett Doust and L. Lovett Doust, eds., *Plant Reproductive Ecology*. Oxford: Oxford University Press, pp. 179–202.

Lee, T. D. and Bazzaz, F. A. (1982) Regulation of fruit maturation pattern in an annual legume, *Cassia fasciculata*. *Ecology*, **63**, 1374–88.

Leimar, O. and Tuomi, J. (1998) Synergistic selection in plants. *Evolutionary Ecology*, **12**, 59–71.

Leishman, M. R., Westoby, M. and Jurado, E. (1995) Correlates of seed size variation: a comparison of five temperate floras. *Journal of Ecology*, **83**, 517–30.

Lenssen, J. P. M., Herschock, C., Speek, T., During, H. J. and de Kroon, H. (2005) Experimental ramet aggregation in the clonal plant *Agrostis stolonifera* reduces its competitive ability. *Ecology*, **86**, 1358–65.

Lewis, D. (1942) On the evolution of sex in flowering plants. *Biological Review*, **17**, 46–76.

Linhart, Y. B. and Grant, M. C. (1996) Evolutionary significance of local genetic differentiation in plants. *Annual Review of Ecology and Systematics*, **27**, 237–77.

Lloyd, D. G. (1975) The maintenance of gynodioecy and androdioecy in angiosperms. *Genetica*, **45**, 325–39.

(1980) Sexual strategies in plants. I. An hypothesis of serial adjustment of maternal investment during one reproductive session. *New Phytologist*, **86**, 69–79.

(1984) Gender allocations in outcrossing cosexual plants. In R. Dirzo and J. Sarukhan, eds., *Perspectives on Plant Population Ecology*. Sunderland, MA: Sinauer Associates, pp. 277–300.

(1992) Evolutionarily stable strategies of reproduction in plants: who benefits and how? In R. Wyatt, ed., *Ecology and Evolution of Plant Reproduction*. New York: Chapman and Hall, pp. 137–68.

Lloyd, D. G. and Bawa, K. S. (1984) Modification of the gender of seed plants in varying conditions. In M. K. Hecht, B. Wallace and G. T. Prance, eds., *Evolutionary Biology*. Berlin: Springer-Verlag, pp. 255–338.

Lloyd, D. G. and Webb, C. J. (1992a) The evolution of heterostyly. In S. C. H. Barrett, ed., *Evolution and Function of Heterostyly*. Berlin: Springer-Verlag, pp. 129–50.

(1992b) The selection of heterostyly. In S. C. H. Barrett, ed., *Evolution and Function of Heterostyly*. Berlin: Springer-Verlag, pp. 151–78.

Lord, J. (1998) Effect of flowering on vegetative growth and further reproduction in *Festuca novae-zelandiae*. *New Zealand Journal of Ecology*, **22**, 25–31.

Mable, B. K. and Otto, S. P. (1998) The evolution of life cycles with haploid and diploid phases. *BioEssays*, **20**, 453–62.

MacArthur, R. and Wilson, E. O. (1967) *The Theory of Island Biogeography*. Princeton, NJ: Princeton University Press.

Manicacci, D., Atlan, A., Rosello, A. and Couvet, D. (1998) Gynodioecy and reproductive trait variation in three *Thymus* species (Lamiaceae). *International Journal of Plant Sciences*, **159**, 948–57.

Manning, A. (1956) Some aspects of the foraging behaviour of bumblebees. *Behaviour*, **9**, 164–201.

Margulis, L. and Sagan, D. (2002) *Acquiring Genomes: A Theory of the Origins of Species*. New York: Basic Books.

Marshall, D. L. (1988) Postpollination effects on seed paternity: mechanisms in addition to microgametophyte competition operate in wild radish. *Evolution*, **42**, 1256–66.

Marshall, D. L. and Ellstrand, N. C. (1988) Effective mate choice in wild radish: evidence for selective seed abortion and its mechanism. *American Naturalist*, **131**, 739–56.

Marshall, D. L. and Whittaker, K. L. (1989) Effects of pollen donor identity on offspring quality in wild radish, *Raphanus sativus*. *American Journal of Botany*, **76**, 1081–8.

Martin, F. W. (1966) Sex ratio and sex determination in *Dioscorea*. *Journal of Heredity*, **57**, 95–9.

Masuda, M., Yahara, T. and Maki, M. (2001) An ESS model for the mixed production of cleistogamous and chasmogamous flowers in a facultative cleistogamous plant. *Evolutionary Ecology Research*, **3**, 429–39.

Mattila, T. and Salonen, V. (1995) Reproduction of *Viola mirabilis* in relation to light and nutrient availability. *Canadian Journal of Botany*, **73**, 1917–24.

Maurice, S., Charlesworth, D., Desfeux, C., Couvet, D. and Gouyon, P. (1993) The evolution of gender in hermaphrodites of gynodioecious populations with nucleo-cytoplasmatic male-sterility. *Proceedings of the Royal Society of London B*, **251**, 253–61.

Maynard Smith, J. (1978) *The Evolution of Sex*. Cambridge: Cambridge University Press.

(1982) *Evolution and the Theory of Games*. Cambridge: Cambridge University Press.

(1998) *Evolutionary Genetics*, 2nd edn. Oxford: Oxford University Press.

Mazer, S. J. (1989) Ecological, taxonomic, and life-history correlates of seed mass among Indiana dune angiosperms. *Ecological Monographs*, **59**, 153–75.

Mazer, S. J. and Hultgard, U. -M. (1993) Variation and covariation among floral traits within and among four species of northern european *Primula* (Primulaceae). *American Journal of Botany*, **80**, 474–85.

Mazer, S. J., Snow, A. A. and Stanton, M. L. (1986) Fertilization dynamics and parental effects upon fruit development in *Raphanus raphanistrum*: consequences for seed size variation. *American Journal of Botany*, **73**, 500–11.

McCartney, M. A. (1997) Sex allocation and male fitness gain in a colonian, hermaphroditic marine invertebrate. *Evolution*, **51**, 127–40.

McKone, M. J. (1987) Sex allocation and outcrossing rate: a test of theoretical predictions using bromegrass (*Bromus*). *Evolution*, **41**, 591–8.

McKone, M. J., Lund, C. P. and O'Brien, J. M. (1998) Reproductive biology of two dominant prairie grasses (*Andropogon gerardii* and *Sorghastrum nutans*, Poaceae): male-biased sex allocation in wind-pollinated plants? *American Journal of Botany*, **85**, 776–83.

Meeuse, B. and Morris, S. (1984) *The Sex-Life of Flowers*. London: Faber and Faber.

Meijden, E. van der (1976) Changes in the distribution pattern of *Tyria jacobaeae* during the larval period. *Netherlands Journal of Zoology*, **26**, 136–61.

(1979) Herbivore exploitation of a fugitive plant species: local survival and extinction of the cinnabar moth and ragwort in a heterogeneous environment. *Oecologia*, **42**, 307–23.

Meijden, E. van der and Waals-Kooi, R. E. van der (1979) The population ecology of *Senecio jacobaea* in a sand dune system. I. Reproductive strategy and the biennial habit. *Journal of Ecology*, **67**, 131–53.

Melser, C. (2001) Selective seed abortion and offspring quality. Thesis, Leiden University.

Melser, C. and Klinkhamer, P. G. L. (2001) Selective seed abortion increases offspring survival in *Cynoglossum officinale* (Boraginaceae). *American Journal of Botany*, **88**, 1033–40.

Melser, C., Rademaker, M. C. J. and Klinkhamer, P. G. L. (1997) Selection on pollen donors by *Echium vulgare* (Boraginaceae). *Sexual Plant Reproduction*, **10**, 305–12.

Metcalf, J. C., Rose, K. E. and Rees, M. (2003) Evolutionary dynamics of monocarpic perennials. *Trends in Ecology and Evolution*, **18**, 471–80.

Metz, J. A. J. and Gyllenberg, J. (2001) How should we define fitness in structured metapopulation models? Including an application to the calculation of evolutionary stable dispersal strategies. *Proceedings of the Royal Society of London B*, **268**, 499–508.

Metz, J. A. J., Nisbet, R. and Geritz, S. A. H. (1992) How should we define 'fitness' for general ecological scenarios? *Trends in Ecology and Evolution*, **7**, 198–202.

Metz, J. A. J., Mylius, S. D. and Diekman, O. (1996) When does evolution optimize? On the relation between types of density dependence and evolutionarily stable life history parameters. International Institute for Applied Systems Analysis working paper WP-96–04 at http://www.iiasa.ac.at

Meyer, S. E. and Carlson, S. L. (2001) Achene mass variation in *Ericameria nauseosus* (Asteraceae) in relation to dispersal ability and seedling fitness. *Functional Ecology*, **15**, 274–81.

Michaels, H. J. and Bazzaz, F. A. (1986) Resource allocation and demography of sexual and apomictic *Antennaria parlinii*. *Ecology*, **67**, 27–36.

(1989) Individual and population responses of sexual and apomictic plants to environmental gradients. *American Naturalist*, **134**, 190–207.

Michaels, H. J., Benner, B., Hartgerink, A. P., Lee, T. D. and Rice, S. (1988) Seed size variation: magnitude, distribution and ecological correlates. *Evolutionary Ecology*, **2**, 157–66.

Miller, J. S. and Venable, D. L. (2002) The transition to gender dimorphism on an evolutionary background of self-incompatibility: an example from *Lycium* (Solanaceae). *American Journal of Botany*, **89**, 1907–15.

Mitchell, R. J. (1993) Adaptive significance of *Ipomopsis aggregata* nectar production: observation and experiment in the field. *Evolution*, **47**, 25–35.

Mitchell, R. J. and Waser, N. M. (1992) Adaptive significance of *Ipomopsis aggregata* nectar production: pollination success of single flowers. *Ecology*, **73**, 633–8.

Mitchell-Olds, T. (2001) *Arabidopsis thaliana* and its wild relatives: a model system for ecology and evolution. *Trends in Ecology and Evolution*, **16**, 693–700.

Mochizuki, A., Takeda, Y. and Iwasa, Y. (1996) The evolution of genomic imprinting. *Genetics*, **144**, 1283–95.

Mogensen, H. L. (1996) The hows and whys of cytoplasmic inheritance in seed plants. *American Journal of Botany*, **83**, 383–404.

Mogie, M. (1992) *The Evolution of Asexual Reproduction in Plants*. London: Chapman and Hall.

Montalvo, A. M. (1992) Relative success of self and outcross pollen comparing mixed-donor and single-donor pollinations in *Aquilegia caerulea*. *Evolution*, **46**, 1181–98.

Morgan, M. T. (1992) Attractive structures and the stability of hermaphroditic sex expression in flowering plants. *Evolution*, **46**, 1199–213.

(1994) Models of sexual selection in hermaphrodites, especially plants. *American Naturalist*, **144**, S100–S125.

Morgan, M. T. and Schoen, D. J. (1997) The role of theory in an emerging new plant reproductive biology. *Trends in Ecology and Evolution*, **12**, 231–4.

Morris, W. F., Price, M. V., Waser, N. M., *et al.* (1994) Systematic increase in pollen carryover and its consequences for geitonogamy in plant populations. *Oikos*, **71**, 431–40.

Müller-Schärer, H., Schaffer, U. and Steinger, T. (2004) Evolution in invasive plants: implications for biological control. *Trends in Ecology and Evolution*, **19**, 417–22.

Murayama, K., Yahara, T. and Terachi, T. (2004) Variation of female frequency and cytoplasmic male-sterility gene frequency among natural gynodioecious populations of wild radish (*Raphanus sativus* L.). *Molecular Ecology*, **13**, 2459–64.

Murray, B. G. Jr (2001) Are ecological and evolutionary theories scientific? *Biological Review*, **76**, 255–89.

Mylius, S. D. and Diekmann, O. (1995) On evolutionarily stable life-histories, optimization and the need to be specific about density dependence. *Oikos*, **74**, 218–24.

Nakamura, R. R. (1988) Seed abortion and seed size variation within fruits *Phaseolus vulgaris*: pollen donor and resource limitation effects. *American Journal of Botany*, **75**, 1003–10.

Neiland, M. R. M. and Wilcock, C. C. (1998) Fruit set, nectar reward, and rarity in the Orchidaceae. *American Journal of Botany*, **85**, 1657–71.

Nicotra, A. B. (1999) Reproductive allocation and the long-term costs of reproduction in *Siparuna grandiflora*, a dioecious neotropical shrub. *Journal of Ecology*, **87**, 138–49.

Niklas, K. J. (1994) *Plant Allometry: The Scaling of Form and Process*. Chicago, IL: University of Chicago Press.

Nilsson, L. A. (1988) The evolution of flowers with deep corolla tubes. *Nature*, **334**, 147–9.

Noordwijk, A. J. van and Jong, G. de (1986) Acquisition and allocation of resources: their influences on variation in life history tactics. *American Naturalist*, **128**, 137–42.

Norman, J. K., Sakai, A. K., Weller, S. G. and Dawson, T. E. (1995) Inbreeding depression in morphological and physiological traits of *Schiedea lydgatei* (Caryophyllaceae) in two environments. *Evolution*, **49**, 297–306.

Núñez-Farfán, J. and Schlichting, C. D. (2001) Evolution in changing environments: the 'synthetic' work of Clausen, Keck and Hiesey. *Quarterly Review of Biology*, **76**, 433–57.

Obeso, J. R. (2002) The costs of reproduction in plants. *New Phytologist*, **155**, 321–48.

O'Connell, L. M. and Eckert, C. G. (2001) Differentiation in reproductive strategy between sexual and asexual populations of *Antennaria parlinii* (Asteraceae). *Evolutionary Ecology Research*, **3**, 311–30.

Ohashi, K. and Yahara, T. (1998) Effects of flower production and predispersal seed predation on reproduction in *Cirsium purpuratum* (Asteraceae). *Canadian Journal of Botany*, **78**, 230–6.

(2001) Behavioral responses of pollinators to variation in floral display size and their influences on the evolution of floral display. In L. Chittka and J. D. Thomson, eds., *Cognitive Ecology of Pollination: Animal Behavior and Floral Evolution*. Cambridge: Cambridge University Press, pp. 274–96.

Olejniczak, P. (2001) Evolutionarily stable allocation to vegetative and sexual reproduction in plants. *Oikos*, **95**, 156–60.

Olivieri, I. and Gouyon, P. -H. (1997) Evolution of migration rate and other traits. In I. Hanski and M. E. Gilpin, eds., *Metapopulation Biology: Ecology, Genetics and Evolution*. San Diego, CA: Academic Press, pp. 293–323.

Olivieri, I., Swan, M. and Gouyon, P. -H. (1983) Reproductive system and colonizing strategies of two species of *Carduus* (Compositae). *Oecologia*, **60**, 114–17.

Ortiz-Pulido, R. and Rico-Gray, V. (2000) The effect of spatio-temporal variation in understanding the fruit crop size hypothesis. *Oikos*, **91**, 523–7.

Paige, K. N. and Whitham, T. G. (1987) Flexible life history traits: shifts by scarlet gilia in response to pollinator abundance. *Ecology*, **68**, 691–5.

Pannell, J. R. (1997) Widespread functional androdioecy in *Mercurialis annua* (Euphorbiaceae). *Biological Journal of the Linnean Society*, **61**, 95–116.

(2001) A hypothesis for the evolution of androdioecy: the joint influence of reproductive assurance and local mate competition in a metapopulation. *Evolutionary Ecology*, **14**, 195–211.

(2002) The evolution and maintenance of androdioecy. *Annual Review of Ecology and Systematics*, **33**, 397–425.

Pappers, S. M., Jong, T. J. de, Klinkhamer, P. G. L. and Meelis, E. (1999) Effects of nectar content on the number of bumblebee approaches and the length of visitation sequences in *Echium vulgare* (Boraginaceae). *Oikos*, **87**, 580–6.

Parciak, W. (2002) Environmental variation in seed number, size and dispersal of a fleshy-fruited plant. *Ecology*, **83**, 780–93.

Parker, G. A. (1985) Population consequences of evolutionarily stable strategies. In R. M. Sibly and R. H. Smith, eds., *Behavioural Ecology*. Oxford: Blackwell, pp. 33–58.

Parker, G. A. and Hammerstein, P. (1985) Game theory and animal behaviour. In P. J. Greenwood, P. H. Harvey and M. Slatkin, eds., *Essays in Honour of John Maynard Smith*. Cambridge: Cambridge University Press, pp. 73–94.

Parmesan, C. and Yohe, G. (2003) A globally coherent fingerprint of climate change impacts across natural systems. *Nature*, **421**, 37–42.

Pen, I. and Weissing, F. J. (2002) Optimal sex allocation: steps towards a mechanistic theory. In I. C. W. Hardy, ed., *Sex Ratios: Concepts and Research Methods*. Cambridge: Cambridge University Press, pp. 26–47.

Pham, J. L., Glaszmann, J. C., Sano, R., *et al.* (1990) Isozyme markers in rice: genetic analysis and linkage relationships. *Genome*, **33**, 661–8.

Philippi, T. (1993) Bet-hedging germination of desert annuals: variation among populations and maternal effects in *Lepidium lasiocarpum*. *American Naturalist*, **142**, 488–507.

Pickering, C. M. and Ash, J. E. (1993) Gender variation in hermaphrodite plants: evidence from five species of alpine *Ranunculus*. *Oikos*, **68**, 539–48.

Piper, J. P. and Charlesworth, B. (1986) The evolution of distyly in *Primula vulgaris*. *Biological Journal of the Linnean Society*, **29**, 123–37.

Piper, J. P., Charlesworth, B. and Charlesworth, D. (1984) A high rate of self-fertilization and increased seed fertility of homostyle primroses. *Nature*, **310**, 50–1.

(1986) Breeding system evolution in *Primula vulgaris* and the role of reproductive assurance. *Heredity*, **56**, 207–17.

Piquot, Y., Petit, D., Valero, M., *et al.* (1988) Variation in sexual and asexual reproduction among young and old populations of the perennial macrophyte *Sparganium erectum*. *Oikos*, **82**, 139–48.

Platenkamp, G. A. (1991) Phenotypic plasticity and population differentiation in seeds and seedlings of the grass *Anthoxanthum odoratum*. *Oecologia*, **88**, 515–29.

Pleasants, J. M. (1981) Bumblebee response to variation in nectar availability. *Ecology*, **62**, 1648–61.

Porcher, E. and Lande, R. (2005) Loss of gametophytic self-incompatibility with evolution of inbreeding depression. *Evolution*, **59**, 46–60.

Porras, R. and Muñoz, J. M. (2000) Achene heteromorphism in the cleistogamous species *Centaurea melitensis*. *Acta Oecologica*, **21**, 231–43.

Primack, R. and Stacey, E. (1998) Cost of reproduction in the pink lady's slipper orchid (*Cypripedium acaule*, Orchidaceae): an eleven-year experimental study of three populations. *American Journal of Botany*, **85**, 1672–9.

Prins, A. H., Nell, H. W. and Klinkhamer, P. G. L. (1992) Size-dependent root herbivory on *Cynoglossum officinale*. *Oikos*, **65**, 409–13.

Pyke, G. H. (1981) Optimal nectar production in a hummingbird pollinated plant. *Theoretical Population Biology*, **20**, 326–43.

(1982) Forgaging in bumblebees: rule of departure from an inflorescence. *Canadian Journal of Zoology*, **60**, 417–28.

(1991) What does it cost a plant to produce floral nectar? *Nature*, **350**, 58–9.

Queller, D. C. (1983) Sexual selection in a hermaphroditic plant. *Nature*, **305**, 706–7.

(1984) Models of kin selection on seed provisioning. *Heredity*, **53**, 151–65.

(1995) The spaniels of St Marx and the Panglossian paradox: a critique of a rhetorical programme. *Quarterly Review of Biology*, **70**, 485–9.

Rademaker, M. C. J. and Jong, T. J. de (1998) Effects of flower number on estimated pollen transfer in natural populations of three hermaphrodite species: an experiment with fluorescent dye. *Journal of Evolutionary Biology*, **11**, 623–41.

(1999) The shape of the female fitness curve for *Cynoglossum officinale*: quantifying seed dispersal and seedling survival in the field. *Plant Biology*, **1**, 351–6.

(2000) Testing sex allocation theory: flowers vs. seeds in hermaphrodite plants. *Oikos*, **91**, 115–22.

Rademaker, M. C. J. and Klinkhamer, P. G. L. (1999) Size-dependent sex allocation in *Cynoglossum officinale* for different genotypes under uniform favourable conditions. *Plant Biology*, **1**, 108–14.

Rademaker, M. C. J., Jong, T. J. de and Klinkhamer, P. G. L. (1997) Pollen dynamics in *Echium vulgare*. *Functional Ecology*, **11**, 554–63.

Rademaker, M. C. J., Jong, T. J. de and Meijden, E. van der (1999) Selfing rates in natural populations of *Echium vulgare*: a combined empirical and model approach. *Functional Ecology*, **13**, 828–37.

Raju, B. M., Shaanker, R. U. and Ganeshaiah, K. N. (1996) Intra-fruit seed abortion in a wind dispersed tree, *Dalbergia sissoo* Roxb: proximate mechanisms. *Sexual Plant Reproduction*, **9**, 273–8.

Ramsey, M. (1993) Floral morphology, biology and sex allocation in disjunct populations of Christmas bells (*Blandfordia grandiflora*, Liliaceae) with different breeding systems. *Australian Journal of Botany*, **41**, 749–62.

Reader, R. J. (1993) Control of seedling emergence by ground cover and seed predation in relation to seed size for some old-field species. *Journal of Ecology*, **81**, 169–75.

Real, L. A. and Rathcke, B. J. (1991) Individual variation in nectar production and its effect on fitness in *Kalmia latifolia*. *Ecology*, **72**, 149–55.

Reekie, E. G. (1998) An experimental field study of the cost of reproduction in *Plantago major* L. *Ecoscience*, **5**, 200–6.

Reekie, E. G. and Bazzaz, F. A. (1987a) Reproductive effort in plants. II. Does carbon reflect the allocation of other resources? *American Naturalist*, **129**, 897–906.

(1987b) Reproductive effort in plants. III. Effect of reproduction on vegetative activity. *American Naturalist*, **129**, 907–19.

(1992) Cost of reproduction in genotypes of two congeneric plant species with contrasting life histories. *Oecologia*, **90**, 21–6.

Rees, M. (1997) Evolutionary ecology of seed dormancy and seed size. In J. Silvertown, M. Franco and J. L. Harper, eds., *Plant Life Histories*. Cambridge: Cambridge University Press, pp. 121–42.

Rees, M. and Crawley, M. J. (1989) Growth reproduction and population dynamics. *Functional Ecology*, **3**, 645–53.

Rees, M. and Rose, K. E. (2002) Evolution of flowering strategies in *Oenothera glazioviana*: an integral projection model approach. *Proceedings of the Royal Society of London B*, **269**, 1509–15.

Rees, M. and Westoby, M. (1997) Game theoretical evolution of seed mass in multi-species ecological models. *Oikos*, **78**, 116–26.

Rees, M., Sheppard, A., Bries, D. and Mangel, M. (1999) Evolution of size-dependent flowering in *Onopordum illyricum*: a quantitative assessment of the role of stochastic selection pressures. *American Naturalist*, **154**, 628–51.

Rees, M., Mangel, M., Turnbull, L. A., Sheppard, A. and Briese, D. (2000) The effects of heterogeneity on dispersal and colonization in plants. In M. J. Hutchings,

E. A. John and A. J. A. Stewart, eds., *Ecological Consequences of Environmental Heterogeneity*. Oxford: Blackwell, pp. 237–65.

Reeve, H. K. and Keller, L. (1999) Levels of selection: burying the units-of-selection debate and unearthing the crucial new issues. In L. Keller, ed., *Levels of Selection in Evolution*. Princeton, NJ: Princeton University Press, pp. 3–14.

Reinartz, J. A. (1984) Life history variation of common mullein (*Verbascum thapsus*). I. Latitudinal differences in population dynamics and timing of reproduction. *Journal of Ecology*, **72**, 897–912.

Renne, I. J., Gauthreaux, S. A. and Gresham, C. A. (2003) Seed dispersal of the Chinese tallow tree (*Sapium sebiferum* (L.) Roxb.) by birds in coastal South Carolina. *American Midland Naturalist*, **144**, 202–15.

Renner, S. S. and Rickleffs, R. E. (1995) Dioecy and its correlates in the flowering plants. *American Journal of Botany*, **82**, 596–606.

Reznick, D., Bryant, M. J. and Bashey, F. (2002) *r*- and *K*-selection revisited: the role of population regulation in life-history evolution. *Ecology*, **83**, 1509–20.

Richards, A. J. (1997) *Plant Breeding Systems*. New York: Chapman and Hall.

Richman, A. D. and Kohn, J. R. (1996) Learning from rejection: the evolutionary biology of single-locus incompatibility. *Trends in Ecology and Evolution*, **11**, 497–502.

Rigney, L. P. (1995) Postfertilization causes of differential success of pollen donors in *Erythronium grandiflorum* (Liliaceae): nonrandom ovule abortion. *American Journal of Botany*, **82**, 578–84.

Roberts, H. A. and Boddrell, J. E. (1983) Seed survival and periodicity of seedling emergence in eight species of Cruciferae. *Annals of Applied Biology*, **103**, 301–9.

Robertson, A. W. (1992) The relationship between floral display size, pollen carryover and geitonogamy in *Myosotis colensoi* (Kirk) Macbride (Boraginaceae). *Biological Journal of the Linnean Society*, **46**, 333–49.

Rocha, O. J. and Stephenson, A. G. (1991) Effects of nonrandom seed abortion on progeny performance in *Phaseolus coccineus* L. *Evolution*, **45**, 1198–208.

Roldan, E. R. S. and Gomendio, M. (1999) The Y chromosome as a battle ground for sexual selection. *Trends in Ecology and Evolution*, **14**, 58–62.

Ronce, O. and Olivieri, I. (1997) Evolution of reproductive effort in a metapopulation with local extinctions and ecological succession. *American Naturalist*, **150**, 220–49.

Rose, K. E., Rees, M. and Grubb, P. J. (2002) Evolution in the real world: stochastic variation and the determinants of fitness in *Carlina vulgaris*. *Evolution*, **56**, 1416–30.

Rousset, F. (2002) Inbreeding and relatedness coefficients: what do they measure? *Heredity*, **88**, 371–80.

(2004) *Genetic Structure and Selection in Subdivided Populations*. Princeton, NJ: Princeton University Press.

Rueffler, C., Dooren, T. van and Metz, J. A. J. (2004) Adaptive walks on changing landscapes: Levins' approach extended. *Theoretical Population Biology*, **65**, 165–78.

Ruiz de Clavijo, E. and Jimenez, M. J. (1993) Cleistogamy and chasmogamy in *Ceratocapnos heterocarpa* (Fumariaceae). *International Journal of Plant Sciences*, **154**, 325–33.

Sakai, A. K., Weller, S. G., Chen, M. L., Chou, S. Y. and Tasanont, C. (1997) Evolution of gynodioecy and maintenance of females: the role of inbreeding depression, outcrossing rates, and resource allocation in *Schiedea adamantis* (Caryophyllacaea). *Evolution*, **51**, 724–36.

Salguero-Faria, J. A. and Ackerman, J. D. (1999) A nectar reward: is more better? *Biotropica*, **31**, 303–11.

Salisbury, E. J. (1942) *The Reproductive Capacity of Plants*. London: Bell.

Sallabanks, R. (1992) Fruit fate, frugivory, and fruit characteristics: a study of the hawthorn, *Crataegus monogyna* (Rosaceae). *Oecologia*, **91**, 296–304.

(2003) Hierarchical mechanisms of fruit selection by an avian frugivore. *Ecology*, **74**, 1326–36.

Samitou-Laprade, P., Cuguen, J. and Vernet, P. (1994) Cytoplasmic male sterility in plants: molecular evidence and the nucleocytoplasmic conflict. *Trends in Ecology and Evolution*, **9**, 431–5.

Samson, D. A. and Werk, K. S. (1986) Size-dependent effects in the analysis of reproductive effort in plants. *American Naturalist*, **127**, 667–80.

Satake, A. and Iwasa, Y. (2000) Pollen coupling of forest trees: forming synchronized and periodic reproduction out of chaos. *Journal of Theoretical Biology*, **203**, 63–84.

Saulnier, T. P. and Reekie, E. G. (1995) Effect of reproduction on nitrogen allocation and carbon gain in *Oenothera biennis*. *Journal of Ecology*, **83**, 23–9.

Saverimuttu, T. and Westoby, M. (1996) Seedling longevity under deep shade in relation to seed size. *Journal of Ecology*, **84**, 681–9.

Schaal, B. A. (1980) Reproductive capacity and seed size in *Lupinus texensis*. *American Journal of Botany*, **67**, 703–9.

Schaffer, W. M. (1974) Selection for optimal life-histories in plants: the effects of age structure. *Ecology*, **55**, 291–303.

Schaffer, W. M. and Schaffer, M. V. (1977) The adaptive significance of variations in reproductive habit in the Agavaceae. In B. Stonehouse, ed., *Evolutionary Ecology*. London: Macmillan, pp. 261–76.

(1979) The adaptive significance of variations in reproductive habit in the Agavaceae. II. Pollinator foraging behavior and selection for increased reproductive expenditure. *Ecology*, **60**, 1051–69.

Schmitt, J. and Gamble, S. E. (1990) The effect of distance from the parental site on offspring performance and inbreeding depression in *Impatiens capensis*: a test of the local adaptation hypothesis. *Evolution*, **44**, 2022–30.

Schnee, B. K. and Waller, D. M. (1986) Reproductive behaviour of *Amphicarpaea bracteata* (Leguminosae), an amphicarpic annual. *American Journal of Botany*, **73**, 376–86.

Schoen, D. J. (1982) Male reproductive effort and breeding system in an hermaphrodite plant. *Oecologia*, **53**, 255–7.

Schoen, D. J. and Lloyd, D. G. (1984) The selection of cleistogamy and heteromorphic diaspores. *Biological Journal of the Linnean Society*, **23**, 303–22.

Schoen, D. J. and Stewart, S. C. (1986) Variation in male reproductive investment and male reproductive success in white spruce. *Evolution*, **40**, 1109–20.

Schultz, S. T. and Ganders, F. R. (1996) Evolution of unisexuality in the Hawaiian flora: a test of microevolutionary theory. *Evolution*, **50**, 842–55.

Scott, R. J., Spielman, M., Bailey, J. and Dickinson, H. G. (1998) Parent-of-origin effects on seed development in *Arabidopsis thaliana*. *Development*, **125**, 3329–41.

Seger, J. and Eckhart, V. M. (1996) Evolution of sexual systems and sex allocation in plants when growth and reproduction overlap. *Proceedings of the Royal Society of London B*, **263**, 833–41.

Seger, J. and Stubblefield, J. W. (2002) Models of sex ratio evolution. In I. C. W. Hardy, ed., *Sex Ratios: Concepts and Research Methods*. Cambridge: Cambridge University Press, pp. 2–25.

Shaanker, R. U., Ganeshaih, K. N. and Bawa, K. S. (1988) Parent–offspring conflict, sibling rivalry, and brood size patterns in plants. *Annual Review of Ecology and Systematics*, **19**, 177–205.

Shaanker, R. U., Ravishankar, K. V., Hegde, S. G. and Ganeshaiah, K. N. (1996) Does endosperm reduce intra-fruit competition among developing seeds? *Plant Systematics and Evolution*, **201**, 263–70.

Sharma, N., Koul, A. K. and Kaul, V. (1999) Pattern of resource allocation of six *Plantago* species with different breeding systems. *Journal of Plant Research*, **112**, 1–5.

Shaukat, S. S., Siddiqui, Z. S. and Aziz, S. (1999) Seed size variation and its effects on germination, growth and seedling survival in *Acacia nilotica* subsp. *indica* (Benth.) Brenan. *Pakistan Journal of Botany*, **31**, 253–63.

Shaw, R. F. and Mohler, J. D. (1953) The selective advantage of the sex ratio. *American Naturalist*, **87**, 337–442.

Shmida, A. and Ellner, S. (1981) Why are adaptations for long-range seed dispersal rare in desert plants? *Oecologia*, **51**, 133–44.

Shykoff, J. A., Kolokotronis, S. O., Collin, C. L. and Lopez-Villavicencio, M. (2003) Effects of male sterility on reproductive traits in gynodioecious plants: a meta-analysis. *Oecologia*, **135**, 1–9.

Sigmund, K. (1993) *Games of Life*. London: Penguin Books.

Silvertown, J. and Bullock, J. M. (2003) Do seedlings in gaps interact? A field test of assumptions in ESS seed size models. *Oikos*, **101**, 499–504.

Silvertown, J. and Dodd, M. (1999) The demographic cost of reproduction and its consequences in balsam fir (*Abies balsamea*). *American Naturalist*, **29**, 321–31.

Simons, A. M. and Johnston, M. O. (2000a) Variation in seed traits of *Lobelia inflata* (Campanulaceae): sources and fitness consequences. *American Journal of Botany*, **87**, 124–32.

(2000b) Plasticity and the genetics of reproductive behaviour in the monocarpic perennial, *Lobelia inflata* (Indian tobacco). *Heredity*, **85**, 356–65.

Skellam, J. G. (1951) Random dispersal in theoretical populations. *Biometrika*, **38**, 196–218.

Smith, C. C. and Fretwell, S. D. (1974) The optimal balance between size and number of offspring. *American Naturalist*, **108**, 499–506.

Smithson, A. (2002) The consequences of rewardlessness in orchids: reward-supplementation experiments with *Anacamptis morio* (Orchidaceae). *American Journal of Botany*, **89**, 1579–87.

Smithson, A. and Gigord, L. D. B. (2001) Are there fitness advantages in being a rewardless orchid? Reward supplementation experiments with *Barlia robertiana*. *Proceedings of the Royal Society of London B*, **268**, 1435–41.

Smithson, A. and Macnair, M. R. (1997) Negative frequency-dependent selection by pollinators on artificial flowers without rewards. *Evolution*, **51**, 715–23.

Snow, A. A., Spira, T. P., Simpson, R. and Klips, R. [illegible]
geitonogamous pollination In D. G. Lloyd and [illegible]
Biology: Studies on Floral Evolution in Animal-P[illegible]
Chapman and Hall, pp. 191–216.

Sokal, R. R. and Rohlf, F. J. (1998) *Biometry: The Pr*[illegible]
in Biological Research. New York: W. H. Freem[illegible]

Soldaat, L. L., Vetter, B. and Klotz, S. (1997) Sex rat[illegible]
in relation to vegetation cover, population size a[illegible]
Vegetation Science, **8**, 697–702.

Soltis, P. S., Soltis, D. E. and Chase, M. W. (1999) Angiosperm phylogeny inferred from multiple genes as a tool for comparative biology. *Nature*, **402**, 402–4.

Stanton, M. L. (1984) Seed variation in wild radish: effects of seed size on components of seedling and adult fitness. *Ecology*, **65**, 1105–12.

Stanton, M. L., Snow, A. A. and Handel, S. N. (1986) Floral evolution: attractiveness to pollinators increases male fitness. *Science*, **232**, 1625–7.

Stanton, M. L., Snow, A. A., Handel, S. N. and Bereczky, J. (1989) The impact of a flower-color polymorphism on mating patterns in experimental populations of wild radish (*Raphanus raphanistrum* L). *Evolution*, **43**, 335–46.

Stanton, M. L., Young, H. J., Ellstrand, N. C. and Clegg, J. M. (1991) Consequences of floral variation for male and female reproduction in experimental populations of wild radish, *Raphanus sativus* L. *Evolution*, **45**, 268–80.

Stearns, S. C. (1976) Life-history tactics: review of ideas. *Quarterly Review of Biology*, **51**, 3–47.

(1992) *The Evolution of Life Histories*. Oxford: Oxford University Press.

Stephenson, A. G. and Winsor, J. A. (1986) *Lotus corniculatus* regulates offspring quality through selective fruit abortion. *Evolution*, **40**, 453–8.

Stevens, L., Goodnight, C. J. and Kalisz, S. (1995) Multilevel selection in natural populations of *Impatiens capensis*. *American Naturalist*, **145**, 513–26.

Stocklin, J. and Favre, P. (1994) Effects of plant size and morphological constraints on variation in reproductive components in two related species of *Epilobium*. *Journal of Ecology*, **82**, 735–46.

Stone, J. L. (2002) Molecular mechanisms underlying the breakdown of gametophytic self-incompatibility. *Quarterly Review of Biology*, **77**, 17–32.

Stone, J. L. and Thomson, J. D. (1994) The evolution of distyly: pollen transfer in artificial flowers. *Evolution*, **48**, 1595–606.

Sun, M. and Ganders, F. R. (1986) Female frequencies in gynodioecious populations correlated with selfing rates in hermaphrodites. *American Journal of Botany*, **73**, 1645–8.

Sutherland, S. and Delph, L. F. (1984) On the importance of male fitness in plants: patterns of fruit set. *Ecology*, **65**, 1093–104.

Taylor, D. R. (1994) The genetic basis of sex ratio in *Silene alba* (= *S. latifolia*). *Genetics*, **136**, 641–51.

(1999) Genetics of sex ratio variation among natural populations of a dioecious plant. *Evolution*, **53**, 55–62.

Taylor, D. R. and Ingvarsson, P. K. (2003) Common features of segregation distortion in plants and animals. *Genetica*, **117**, 27–35.

Taylor, D. R., Saur, M. J. and Adams, E. (1999) Variation in pollen performance and its consequences for sex ratio evolution in a dioecious plant. *Evolution*, **53**, 1028–36.

Taylor, P. D. (1984) Evolutionarily stable reproductive allocations in heterostylous plants. *Evolution*, **38**, 408–16.

(1994) Sex ratio in a stepping-stone population with sex-specific dispersal. *Theoretical Population Biology*, **45**, 203–18.

Thompson, J. D. and Dommée, B. (1993) Sequential variation in the components of reproductive success in the distylous *Jasminum fructicans* (Oleaceae). *Oecologia*, **94**, 480–7.

Thompson, J. D. and Tarayre, M. (2000) Exploring the genetic basis and proximate causes of female fertility advantage in gynodioecious *Thymus vulgaris*. *Evolution*, **54**, 1510–20.

Thompson, J. D., Piller, Y., Strasberg, D. and Manicacci, D. (1996) Tristyly in the endangered Mascarene Island endemic *Hugonia serrata* (Linaceae). *American Journal of Botany*, **83**, 1160–8.

Thompson, J. D., Manicacci, D. and Tarayre, M. (1998) Thirty-five years of thyme: a tale of two polymorphisms. *BioScience*, **48**, 805–15.

Thompson, K. and Rabinowitz, D. (1989) Do big plants have big seeds? *American Naturalist*, **133**, 722–8.

Thompson, K. and Stewart, A. J. A. (1981) The measurement and meaning of reproductive effort in plants. *American Naturalist*, **117**, 205–11.

Thomson, J. D. (1988) Effects of variation in inflorescence size and floral rewards on the visitation rates of traplining pollinators of *Aralia hispida*. *Evolutionary Ecology*, **2**, 65–76.

Thomson, J. D. and Barrett, S. C. H. (1981) Selection for outcrossing, sexual selection, and the evolution of dioecy in plants. *American Naturalist*, **118**, 443–9.

Thornburg, R. W., Carter, C., Powell, A., *et al.* (2003) A major function of the tobacco floral nectary is defense against microbial attack. *Plant Systematics and Evolution*, **238**, 211–18.

Thuijsman, F., Peleg, B., Amital, A. and Shmida, A. (1995) Automata, matching and foraging behavior of bees. *Journal of Theoretical Biology*, **175**, 305–16.

Traveset, A. (1994) Influence of type of avian frugivory on the fitness of *Pistacia terbinthus* L. *Evolutionary Ecology*, **8**, 618–27.

Trivers, R. L. (1974) Parent–offspring conflict. *American Zoologist*, **14**, 249–64.

(2002) *Natural Selection and Social Theory*. Oxford: Oxford University Press.

Turcotte, J. and Houle, G. (2001) Reproductive costs in *Salix planifolia* ssp. *planifolia* in subarctic Quebec, Canada. *Ecoscience*, **8**, 506–12.

Udovic, D. (1981) Determinants of fruit set in *Yucca whipplei*: reproductive expenditure vs. pollinator availability. *Oecologia*, **48**, 389–99.

Vamosi, J. C., Otto, S. P. and Barrett, S. C. H. (2003) Phylogenetic analysis of the ecological correlates in angiosperms. *Journal of Evolutionary Biology*, **15**, 1006–18.

Veech, J. A., Charlet, D. A. and Jenkins, S. H. (2000) Interspecific variation in seed mass and the co-existence of conifer species: a null model test. *Evolutionary Ecology Research*, **2**, 353–63.

Vekemans, X. and Hardy, O. J. (2004) New insights from fine-scale spatial genetic structure analyses in plant populations. *Molecular Ecology*, **13**, 921–35.

Venable, D. L. (1985) The evolutionary ecology of seed heteromorphism. *American Naturalist*, **126**, 577–95.

(1992) Size–number trade-offs and the variation of seed size with plant resource status. *American Naturalist*, **140**, 287–304.

Venable, D. L. and Brown, J. S. (1988) The selective interactions of dispersal, dormancy and seed size as adaptations for reducing risk in variable environments. *American Naturalist*, **131**, 360–84.

Venable, D. L. and Lawlor, L. (1980) Delayed germination and dispersal in desert annuals: escape in space and time. *Oecologia*, **46**, 272–82.

Venable, D. L., Dyreson, E. and Morales, E. (1995) Population dynamic consequences and evolution of seed traits of *Heterosperma pinnatum* (Asteraceae). *American Journal of Botany*, **82**, 410–20.

Vrieling, K., Saumitou-Laprade, P., Cuguen, J., *et al.* (1999) Direct and indirect estimates of the selfing rate in small and large individuals of the bumblebee-pollinated *Cynoglossum officinale* L. (Boraginaceae). *Ecology Letters*, **2**, 331–7.

Waller, D. (1980) Environmental determination of outcrossing in *Impatiens capensis* (Balsaminaceae). *Evolution*, **34**, 747–61.

Waser, N. M. (1988) Comparative pollen and dye transfer by pollinators of *Delphinium nelsonii*. *Functional Ecology*, **2**, 41–8.

Waser, N. M. and Price, M. V. (1981) Pollinator choice and stabilizing selection for flower color in *Delphinium nelsonii*. *Evolution*, **35**, 376–90.

Wasserthal, L. T. (1997) The pollinators of the Malagasy star orchids *Angraecum sesquipedale*, *A. sororium* and *A. compactum* and the evolution of extremely long spurs by pollinator shift. *Botanica Acta*, **110**, 343–59.

Webb, C. J. (1979) Breeding systems and the evolution of dioecy in New Zealand apioid Umbelliferae. *Evolution*, **33**, 662–72.

Weiner, J., Berntson, G. M. and Thomas, S. C. (1990) Competition and growth in a woodland annual. *Journal of Ecology*, **78**, 459–69.

Welch, D. M. and Meselson, M. (2000) Evidence for the evolution of bdelloid rotifers without sexual reproduction or genetic exchange. *Science*, **288**, 1211–15.

Werner, P. A. and Caswell, H. (1977) Population growth rates and age- versus stage-distribution models for teasel (*Dipsacus sylvestris*). *Ecology*, **58**, 1103–11.

Wesselingh, R. A. and Jong, T. J. de (1995) Bidirectional selection on threshold size for flowering in *Cynoglossum officinale* (Hound's tongue). *Heredity*, **74**, 415–24.

Wesselingh, R. A., Jong, T. J. de, Klinkhamer, P. G. L., Dijk M. J. van and Schlatmann, E. G. M. (1993) Geographical variation in threshold size for flowering in *Cynoglossum officinale*. *Acta Botanica Neerlandica*, **42**, 81–91.

Wesselingh, R. A., Klinkhamer, P. G. L., Jong, T. J. de and Boorman, L. A. (1997) Threshold size for flowering in different habitats: effects of size-dependent growth and survival. *Ecology*, **78**, 2118–32.

Westergaard, M. (1958) The mechanism of sex determination in dioecious plants. *Advances in Genetics*, **9**, 217–81.

Westoby, M. (1981) How diversified seed germination is selected. *American Naturalist*, **118**, 882–5.

Westoby, M. and Rice, B. (1982) Evolution of the seed plants and inclusive fitness of plant-tissues. *Evolution*, **36**, 713–24.

Westoby, M., Leishman, M. and Lord, J. (1997) Comparative ecology of seed size and dispersal. In J. Silvertown, M. Franco and J. L. Harper eds., *Plant Life Histories*. Cambridge: Cambridge University Press, pp. 143–62.

Wetherill, G. B. (1986) *Regression Analysis with Applications*. New York: Chapman and Hall.

Widen, B. and Widen, M. (1990) Pollen limitation and distance-dependent fecundity in females of the clonal gynodioecious herb *Glechoma hederacea* (Lamiaceae). *Oecologia*, **83**, 191–6.

Wiens, D. (1984) Ovule survivorship, brood size, life-history, breeding systems, and reproductive success in plants. *Oecologia*, **64**, 47–53.

Wiens, D., Calvin, C. L., Wilson, C. A., *et al.* (1987) Reproductive success, spontaneous embryo abortion, and genetic load in flowering plants. *Oecologia*, **71**, 501–9.

Wiklund, C. (1984) Egg-laying patterns in butterflies in relation to their phenology and the visual apparency and abundance of their host plants. *Oecologia*, **63**, 23–9.

Wilken, D. H. (1982) The balance between chasmogamy and cleistogamy in *Collomia grandiflora* (Polemoniaceae). *American Journal of Botany*, **69**, 1326–33.

Willis, K. J. and MacElwain, J. C. (2002) *The Evolution of Plants*. Oxford: Oxford University Press.

Willson, M. F. (1979) Sexual selection in plants. *American Naturalist*, **113**, 777–90.

Willson, M. F. and Price, P. W. (1977) The evolution of inflorescence size in *Asclepias* (Asclepiadaceae). *Evolution*, **31**, 495–511.

Willson, M. F. and Rathcke, B. J. (1974) Adaptive design of floral display in *Asclepias syriaca* L. *American Midland Naturalist*, **92**, 47–57.

Wilson, D. S. (2002) *Darwin's Cathedral*. Chicago, IL: University of Chicago Press.

Wilson, J. B. (1987) Group selection in plant populations. *Theoretical and Applied Genetics*, **74**, 493–502.

Wilson, P., Thomson, J. D., Stanton, M. L. and Rigney, L. P. (1994) Beyond floral Batemania: gender biases in selection for pollination success. *American Naturalist*, **143**, 283–96.

Winn, A. A. (1988) Ecological and evolutionary consequences of seed size in *Prunella vulgaris*. *Ecology*, **69**, 1537–44.

Wolfe, L. M. (1983) The effect of plant size on reproductive characteristics in *Erythronium americanum* (Liliaceae). *Canadian Journal of Botany*, **61**, 3489–93.

Young, T. P. (1990) Evolution of semelparity in Mount Kenya lobelias. *Evolutionary Ecology*, **4**, 157–71.

Zimmerman, M. (1983) Plant reproduction and optimal foraging: experimental nectar manipulations in *Delphinium nelsonii*. *Oikos*, **41**, 57–63.

Zimmerman, M. and Cook, S. (1985) Pollinator foraging, experimental nectar-robbing and plant fitness in *Impatiens capensis*. *American Midland Naturalist*, **113**, 84–91.

Index